T0275698

Reliable Maintenance Planning, Estimating, and Scheduling

Reliable Maintenance Planning, Estimating, and Scheduling

Ralph W. Peters

AMSTERDAM • BOSTON • HEIDELBERG • LONDON
NEW YORK • OXFORD • PARIS • SAN DIEGO
SAN FRANCISCO • SINGAPORE • SYDNEY • TOKYO
Gulf Professional Publishing is an imprint of Elsevier

Gulf Professional Publishing is an imprint of Elsevier
225 Wyman Street, Waltham, MA 02451, USA
The Boulevard, Langford Lane, Kidlington, Oxford, OX5 1GB, UK

Notices
Knowledge and best practice in this field are constantly changing. As new research and experience broaden our understanding, changes in research methods, professional practices, or medical treatment may become necessary.

Practitioners and researchers must always rely on their own experience and knowledge in evaluating and using any information, methods, compounds, or experiments described herein. In using such information or methods they should be mindful of their own safety and the safety of others, including parties for whom they have a professional responsibility.

To the fullest extent of the law, neither the Publisher nor the authors, contributors, or editors, assume any liability for any injury and/or damage to persons or property as a matter of products liability, negligence or otherwise, or from any use or operation of any methods, products, instructions, or ideas contained in the material herein.

ISBN: 978-0-12-397042-8

Library of Congress Cataloging-in-Publication Data
A catalog record for this book is available from the Library of Congress

British Library Cataloguing in Publication Data
A catalogue record for this book is available from the British Library

For information on all Gulf Professional Publishing
visit our website at http://store.elsevier.com

Typeset by TNQ Books and Journals
www.tnq.co.in

This book has been manufactured using Print On Demand technology.

Contents

About the Author   vii

Introduction   ix

1   Profit and Customer-Centered Benefits of Planning and Scheduling   1

2   Defining Results to Top Leaders and Operations Leaders   11

3   Leadership: Creating Maintenance Leaders, Not Just Maintenance Managers   21

4   How to Create PRIDE in Maintenance within Craft Leaders and the Technical Workforce   29

5   Define Your Physical Asset Management Strategy with The Scoreboard for Maintenance Excellence and Go Beyond ISO 55000   39

6   Planners Must Understand Productivity and How Reliable Maintenance Planning, Estimating and Scheduling (RMPES) Enhances Total Operations Excellence   67

7   What to Look for When Hiring a Reliable Planner/Scheduler   89

8   Planner Review of the Maintenance Business System—Your CMMS-EAM System   99

9   Defining Maintenance Strategies for Critical Equipment With Reliability-Centered Maintenance (RCM)   145

10   Defining Total Maintenance Requirements and Backlog   157

11   Overview of a Reliable Planning-Estimating-Scheduling-Monitoring-Controlling Process   167

12   Why the Work Order Is a Prime Source for Reliability Information   201

13   Detailed Planning with a Reliable Scope of Work and a Complete Job Package   207

14   Understanding Risk-Based Maintenance by Using Risked-Based Planning with Risk-Based Inspections   223

15   Developing Improved Repair Methods and Reliable Maintenance Planning Times with the ACE Team Process   241

16 Successful Scheduling by Keeping the Promise and Completing the
   Schedule   257

17 Maintenance, Repair, and Operations (MRO) Material Management:
   The Missing Link in Reliability   277

18 How to Measure Total Operations Success with the Reliable
   Maintenance Excellence Index   285

19 How This Book Can Apply to the Very Small Work Unit in Oil and
   Gas or to Any Type of Maintenance Operation   297

20 A Model for Success: Developing Your Next Steps for Sustainable and
   Reliable Maintenance Planning—Estimating and Scheduling   301

Appendix A   The Scoreboard for Maintenance Excellence—Version 2015   315

Appendix B   Acronyms and Glossary of Maintenance, Maintenance
             Repair Operations Stores/Inventory, and Oil and Gas Terms   369

Appendix C   Maintenance Planner/Scheduler or Maintenance
             Coordinator: Position Description, Job Evaluation Form   409

Appendix D   Charter: Format for a Leadership Driven-Self-Managed
             Team at GRIDCo Ghana   417

Appendix E   Case Study–Process Mapping for a Refinery   423

Appendix F   The CMMS Benchmarking System   439

Appendix G   The ACE Team Benchmarking Process Team Charter
             Example   453

Appendix H   Shop Load Plan, Master Schedule and Shop Schedules:
             Example Forms and Steps on How to Use   471

Appendix L   Routine Planner Training Checklist   483

Index   489

ONLINE APPENDICES (http://booksite.elsevier.com/9780123970428.)

Appendix I   Management of Change (MOC) Procedures Example   e1

Appendix J   Risk Management   e13

Appendix K   Measuring the True Value of Maintenance Activities   e23

Appendix M   Planner Viewpoints on the Question; "Is it Required to
             Have a Trades Background to be a Planner?"   e45

About the Author

Ralph "Pete" Peters is a highly recognized author, trainer and leader around the world in the areas of implementing maintenance and manufacturing best practices, developing effective productivity measurement systems and initiating long-term sustainable operational improvement processes. He has also supported both the public and private sectors during his career. His value as a consultant has been enhanced through his direct leadership and profit and loss responsibilities within large maintenance and manufacturing plant operations prior to focusing upon consulting. He is the author of two major books; *Maintenance Benchmarking and Best Practices* from McGraw-Hill and now *Reliable Maintenance Planning, Estimating and Scheduling* from Elsevier. He has written a number of e-books and five major handbook chapters plus over 200 articles and publications. And as a frequent speaker, he has delivered speeches and TrueWorkShops™ on maintenance and manufacturing excellence related topics worldwide in over 40 countries to over 5000 people.

Worldwide Maintenance Consulting and Training Services: He founded The Maintenance Excellence Institute International in 2001 and has helped such diverse operations such as British Petroleum, EcoPetrol (Columbia), Marathon Oil, Total, SIDERAR Steel (Argentina), Atomic Energy Canada Ltd, Boeing Commercial Airplane Group, Caterpillar, Campbell Soup, UNC-Chapel Hill, Ford, Honda (America), Anderson Packaging Inc., Polaroid, Lucent Technologies, Heinz, General Foods, BigLots Stores, Sheetz Inc., Sanifi Pasteur, Great River Energy, Wyeth-Ayerst (US & IR), Cooper Industries, National Gypsum, Sarasota County Government-Operations and Maintenance Division, Carolinas Medical Center, NC Department of

Transportation, NC Department of Health and Human Services, the US Department of Health and Human Health Services' Indian Healthcare Services, Air Combat Command and the US Army Corps of Engineers.

Education: He received both his BS Industrial Engineering and Masters of Industrial Engineering focused upon management information systems from North Carolina State University. He is also a graduate of the US Army Command and General Staff Course, the Engineer Officers Advanced and Basic Courses, the Military Police Officers Course and the Civil Affairs Officer Course. He is certified as a Total Quality Management Facilitator for the National Guard Bureau and the North Carolina Army National Guard.

Introduction

Operations within the oil, gas, and petro chemical sectors provide the most operational, safety, and maintenance challenges available. Also planning, estimating, and scheduling in these areas must achieve a higher level of accuracy and reliability. Thus we have used the title *Reliable Maintenance Planning, Estimating and Scheduling (RMPES)* along with estimating. However, this book has universal applications for all type operations because *if you can do it within oil, gas, and petro chemical sectors you can do it most anywhere.* This book will be much different than most others you have used, read, or researched. Of course, it will cover material and ground that has been plowed and still being cultivated before by some great author/trainers like Joel Levitt and Don Nyman in their book; *Maintenance Planning, Scheduling and Coordination* and also other great books on planning and scheduling by Tim Kister and Bruce Hawkins, Doc Palmer, and Michael Brown.

Thanks especially to Ricky Smith and Jerry Wilson who have provided the case study, Alcoa-Mt. Holly from 1997 to 2012. I have used this as the closing *Chapter 20—A Model for Success: Developing Your Next Steps for Sustainable and Reliable Maintenance Planning, Estimating, and Scheduling.* Thanks to Phillip Slater and Art Posey for *Chapter 17—MRO Material Management: The Missing Link in Reliability.*

And last but not least, thanks to a real maintenance planner, Gary Royer, who came from a giant corporation and made many things work well as a "one man band" so to speak within a small undisciplined operation. He never gave up on improving, and has made excellent progress. His story is *Chapter 19—How This Book Can Apply to the Very Small Work Unit in Oil and Gas or to Any Type of Maintenance Operation.*

This book will be about *reliable* maintenance planning, estimating, and scheduling (RMPES) within organizations having very critical and complex equipment, extreme HSSE challenges, and located in environmentally sensitive locations in most cases. For today's oil, gas, and petrochemical operations it will cover the complete scope of planning, estimating, and scheduling of work in these critical continuous process operations. It will also include in detail an added focus a planner/scheduler can have on reliability and continuous reliability improvement (CRI).

Appendix A-The Scoreboard for Maintenance Excellence Version 2015 is a major body of work within this book. The Scoreboard for Maintenance Excellence provides an extensive and prescriptive self-assessment guide to 38 best practice areas with 600 evaluation items. It was updated from the previous version to include many areas specific to oil, gas and petro chemical operations. It also goes well beyond the descriptive and often vague terms of ISO 55000 requirements with prescriptive action items. It is available from The Maintenance Excellence Institute International in easy to use

Excel format. Go to www.pride-in-maintenance.com for details to receive your own Excel copy.

Planners and schedulers (P/S) play a key role within good maintenance and reliability improvement processes. This book will further define how their roles can support CRI and how these positions when properly trained can become an important new dimension to their key position within in large and small operations.

I will refer to all leaders in this book; *Top Leaders, Maintenance Leaders, and Craft Leaders*. I will also refer infrequently to "managers of the status quo," which I have seen in many cases at all levels. Leaders at these three levels will all gain a better understanding of the professionals that perform RMPES. My underlying goal is that this book will expand the P/S role and the appreciation of this profession to another level of professional recognition.

During my career back in 1972 I was lucky enough to be on a team that interviewed, selected, personally trained, and implemented over 70 planners across the State of North Carolina for the NC Department of Transportation's Fleet Equipment Maintenance Management Division. Later I will share with you the true story of the outstanding career progressions by these mechanics that reluctantly left their tool boxes behind for a new role as planners/service managers within in a governmental agency, the NC Department of Transportation.

Our stated focus is on oil and natural gas operations and large petro chemical organizations with large complex equipment, continuous processing units, and integrated systems that present some of the greatest challenges to performing effective maintenance. Asset integrity management, process safety management, risk management, HSSE, and change management all add to a need for a fully integrated process for capital planning, day-to-day maintenance and shutdown/turnaround activities. This book will primarily cover day-to-day RMPES and how integration to shutdown, turnaround, and outage (STO) should occur.

A profit-centered maintenance strategy requires effective and reliable maintenance planning, estimating, and scheduling (RMPES) and many other best practices. RMPES is considered by me and many others as one of the most important maintenance best practices because it is a very important enabler of profit, gained value customer service, craft labor productivity, and physical asset productivity. Effective and reliable maintenance planning, estimating, and scheduling enables:

1. craft labor productivity-Improved OCE (Overall Craft Effectiveness)
2. asset productivity-Improved OEE (Overall Equipment Effectiveness)
3. reliable and safe repair methods
4. reliable planned time for knowing your TMR (Total Maintenance Requirements)
5. increased operations labor productivity
6. validated direct savings and gained value that contributes directly to the bottom line.

We will first of all define the benefits of successful RMPES to three key groups:

1. *Top leaders:* C-positions, especially the chief financial officer, VP-operations, managing directors, and also engineering and maintenance managers
2. *Maintenance leaders:* Maintenance managers, supervisors, foremen, maintenance engineers and reliability engineers

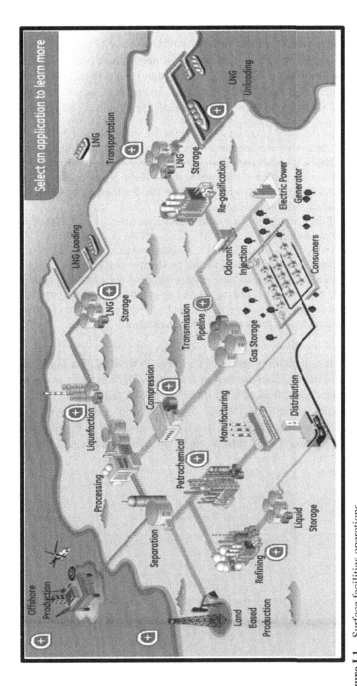

Figure I.1   Surface facilities operations.
Courtesy of General Electric.

3. *Craft leaders:* Subject matter experts, crew leaders, and other technician specialists. I like to think of a craft leader in sports term; "It's your ***Go to Guy*** when a real problem occurs."

We will define for each of these three groups the foundation on why RMPES is important. From the 500 plus plant visits and assessments, I have personally performed, less that 20% really applied RMPES effectively. Many had planners physically in the wrong place doing ineffective RMPES for the customer. That leaves 80% in both large and small operations that need help. From that real-life sample of shop level personal experience there were some obvious needs.

    a. A clear and comprehensive maintenance strategy was not defined by the maintenance leader up to their top leaders and down to craft leaders

    b. Maintenance leaders did not know "where they were in terms of best practices" and therefore did not have a clear path forward that could achieve validated and measured benefits from their existing RMPES

    c. Top leaders did not truly understand maintenance and certainly had no idea of their "total maintenance requirements."

    d. Top leaders and maintenance leaders set lower performance goals for total backlog, when in fact "total maintenance requirements" often exceeded current craft capacity.

    e. Craft leaders and craft technician were still operating in a reactive, fire fighting mode and did not really have effective planning/scheduling or did not clearly see the true benefits when planning/scheduling was in place. Figure I.1 illustrates the scope of surface facility operations within the oil, gas, and petro chemical sectors.

Our goal is that this book will benefit both large oil and gas operations as well as smaller discrete manufacturing operations. ***Ninety percent of this book will apply to all type operations, because the best practice for RMPES basic processes apply to every operation.*** Again, leaders at all three levels will gain a better understanding of the professionals that perform RMPES. My underlying goal is that this book will expand the P/S role as it is related to maintenance and reliability excellence and will increase the appreciation of this profession to an even higher level of professional recognition in the future.

Profit and Customer-Centered Benefits of Planning and Scheduling

Effective and reliable maintenance planning and scheduling that we will discuss later is an essential element for physical asset management operation. But does your view of maintenance and physical asset management process see it as a profitable in-house business within your organization? Would your current maintenance operation sustain itself as a contract maintenance organization and make a profit? These might seem to be two strange questions. But, if your current maintenance strategy, leadership philosophy, and planning and scheduling processes do not allow you to manage maintenance like a profitable internal business, you could be in trouble or heading toward serious trouble.

Top leaders that still view maintenance as a "traditional cost center" and then continuously "squeeze blood from the maintenance turnip" are on the road to major problems with physical asset management. This attitude has resulted in catastrophic failures within airlines, refineries, ships at sea, and many other operations. Maintenance requirements are everywhere and the need for effective maintenance is continuous because… **Maintenance is forever!** Maintenance of our body, soul, mind, spirit, house, car, physical assets, and infrastructures all around us will be with us forever.

Are top leaders gambling with maintenance? There can be a very high cost of gambling with maintenance, and most operations lose when they gamble with their maintenance chips. There is an extremely high cost to bad maintenance within oil and gas operations, on the plant shop floor, in combat, and everywhere the maintenance process fails in the proper care of physical assets. Look at Figure 1.1 below to gain an understanding of the P–F interval. What chances are you willing to take when you know that a failure is occurring"?

Point P above is like the saying, "You can't be a little bit pregnant." Point P can be confirmed by numerous possible means, but the P–F Interval is basically the unknown time between seeing a certain failure start and at what time the failure actually occurs. If you have not invested wisely in predictive maintenance or continuous monitoring, you may be at the point where you can manage and lead forward as a profitable internal business.

You may also be a potential takeover target for contract maintenance. Many operations have lost heavily by gambling with indiscriminate cuts to a core requirement: the resources necessary for effective physical asset management and maintenance. Quantum leaps backward will occur for the top leader that fails to view maintenance as a core business requirement. I feel strongly that indiscriminate downsizing and "dumb sizing" of maintenance is finally being recognized as a failed business practice. Where

Reliable Maintenance Planning, Estimating, and Scheduling. http://dx.doi.org/10.1016/B978-0-12-397042-8.00001-2

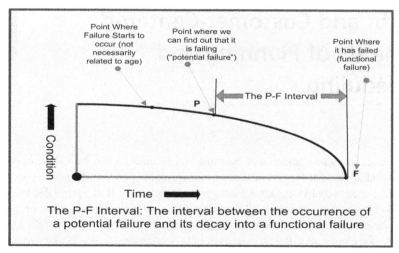

Figure 1.1 The P–F interval.

are your maintenance chips being stacked? Do not view maintenance as a cost center and not worthy of effective planning and scheduling. View it with a profit and customer-centered mentality and with an attitude that promotes initiative, customer service, profit optimization, and ownership. Invest in reliable maintenance planning, estimating, and scheduling, and the other best practices we will discuss later to ensure success.

Profit and customer-centered contract maintenance. You might say that profit and customer-centered maintenance is not possible for an in-house maintenance operation. But a profit-centered strategy does exist in the thousands of successful businesses that provide contract maintenance services everywhere we look, especially within many oil and gas operations. Maybe we invest our maintenance chips (or even real U.S. dollars) heavily in profit-centered contract maintenance providers who are truly in the maintenance for profit as business to truly serve their customers. They will expand even further if organizations continue to give up on in-house maintenance operations.

Third-party maintenance will continue to be a common practice in organizations that have continually gambled with maintenance costs and have lost. For some of the maintenance operations that I have seen as a result of hundreds of maintenance benchmarking assessments, the best answer for survival is a partnership with a contract maintenance provider. It is often a hard choice, especially when it is tempered with all the relentless pressure from unions. For some operations, quality service from qualified maintenance service providers is unfortunately the best choice available. However, we should not give up on in-house maintenance when contract maintenance could be just as bad if they operate within our current organization without effective planning and scheduling.

This is not a scare tactic that advocates third-party maintenance in total for an organization. It positively and unequivocally does not support the dumb-sizing of in-house maintenance to provide lean maintenance, which in turn fails to meet the total maintenance requirement needs. Dumb-sizing of maintenance and reengineering without true engineering has failed. Third-party maintenance in specialty areas or areas where current maintenance skills or competencies are lacking will be needed and be a growing practice. It provides real profit to the maintenance provider and savings to the customer.

Greater third-party maintenance of all types is occurring around the world, especially in the Middle East, and will continue to occur in United States' operations where maintenance is not treated as an internal business opportunity. It will obviously occur where the maintenance operations have deteriorated to the point that a third-party service is more effective and less costly than in-house maintenance staff.

Where is your chief maintenance officer? We now have more "C-positions" than we know the terms for: CEO (Chief Executive Officer), COO (Chief Operating Officer), CFO (Chief Financial Officer), CIO (Chief Information Officer), CPO (Chief Purchasing Officer). An important position that is missing is the chief maintenance officer (CMO). The evolution of the CMO position must occur to provide leadership to physical asset management within large manufacturing operations. I sincerely believe that the real maintenance leaders will begin to emerge as CMOs in the business world. This new staff addition of a CMO is desperately needed, and smart organizations will have someone near the top that is officially designated to ensure that physical assets are properly cared for. I believe that the CMO will join the ranks of the CEO, COO, CFO, and CIO in large multisite manufacturing operations to manage physical assets. This can be a real technical asset for large oil companies, for example, using the same business system and associated computerized maintenance management system such as SAP. These CMOs will manage and most importantly lead maintenance forward as a "profit center." A good CMO with "profit ability" will be in place to lead maintenance forward to profitability. A good CMO will help the CFO take the "right" fork in road as it relates to physical asset management and profitability providing consistent application of let's say SAP. Regardless of the size of the operation, every manufacturing operation needs a CMO. For smaller operations, it might be a CMO equivalent, a true maintenance leader or a maintenance leader supervisor that can really manage maintenance as a business and as an internal profit center.

Profit ability. Leadership ability is an important personal attribute, but being in the "maintenance-for-profit business" also requires an important new type of ability that we call "profit ability." To lead maintenance forward, we must learn from the leaders within the third-party maintenance business. There are many good ones out there, but one that I personally know about is a company in Oman that manages their pipeline maintenance like they own the contractor. Of course not literally, but this was a great example of true "profit ability" on both sides. This attendee who was responsible for all of Oman's pipeline maintenance was at my *"Maximizing the Value of Contracted Services"* course in Dubai. He could have taught the course. One of the key things that stood out was that the contractor was required to employ a planner/scheduler that worked closely with the pipeline company planner/scheduler. Direct measurement of contractor work, schedule compliance, plus other key metrics were at the heart of their profit-centered relationship.

Having a good CMO adds the missing link to achieving total operations success and profit = optimization. Maintenance has rapidly evolved into an internal business opportunity and can almost now stated financially correct as a true "profit center."

The change from a "run-to-failure" strategy into a proactive, planned process for asset management requires a CMO with demonstrated technical and personal leadership. Plan on becoming the CMO in your operation regardless of your organization's size or current level in your organization. Ensure that an effective planning and scheduling process is in place.

No matter how bad something is, it can always be used as an illustrative bad example. We all learn lessons either the hard way or the easy way. Therefore, bad examples are not wasted. I think we can learn important good lessons about maintenance the easy way by having an effective CMO. I think that a new breed of corporate officer will evolve. An effective CMO will be a firm requirement for organizational success. CMOs will take their place near the top with the CEO, COO, CFO, CIO, and the corporate quality gurus. I think we will start to listen closely to the maintenance messenger, our CMOs. We will not and should not shoot the maintenance messenger. Many have been seriously wounded when they have tried to state the "true state of maintenance" within an organization. Manufacturing plant managers, CFOs, COOs, and VPs of manufacturing operations who do not understand the true value of maintenance will continue to be the bad examples. The CMOs of successful organizations will have an important and unprecedented role in the success of their total operation.

The successful manufacturing company, whether discrete or continuous processing, will have true maintenance leaders, not managers of the status quo. The true maintenance leaders and CMOs of these successful companies will know the contribution to profit that their maintenance operations provide. They will view maintenance improvements and practices such as planning estimating and scheduling as value-adding investments that provide a measurable return on investment. The return on investment for RMPES plus supporting practices such as effective preventive and predictive maintenance with good parts inventory and procurement can be unbelievable. CMOs will measure the results of the maintenance process whether it is internal or outsourced maintenance. They will validate the investments they have made just as they try to validate other return on investments. The CMO will be the maintenance messenger!

The true CMO will also be the maintenance leader that understands how to operate the total maintenance process as an internal business within a business. They will be able to turn in-house maintenance into a profit center comparable to contract service providers. All true corporate top leaders must strive to understand current and future trends, take action, and proactively plan for the maintenance strategy within their total operation.

There must be a maintenance champion to manage all of a company's physical assets. The real maintenance leader readily accepts the role as champion for maintenance excellence. Likewise, integrity of purpose and the integrity of the maintenance champions must set an example for others in the organization to follow. Ralph Waldo Emerson said it very well when he remarked, "What you are thunder so loudly, I cannot hear a word you say to the contrary." Leadership by example and "walking your talk" is essential for the maintenance champion and all company leaders.

The maintenance champion as CMO understands and can communicate the true cost of deferred maintenance as well as the cost of inadequate preventive/predictive maintenance. The CMO is prepared to provide proactive leadership and support to the company's compliance to regulatory issues. The real CMO must be prepared to take bad news about the true "state of maintenance" to company leaders with courage, confidence, and most importantly with credibility. The maintenance messenger does not always bring good news to top leaders and must possess the skills and traits to be candid, honest, and credible, backed up with facts.

The effective CMO utilizes a true teaming process to bring maintenance, operations, and operators together to detect, solve, and prevent maintenance problems. This book also will show how a planner/scheduler can be another key resource for improving reliability. The effective CMO will take the lead for implementing best practices such as an effective CMMS. They will work closely with information services staff and CMMS vendors over the long term to make it work to enhance the business of maintenance.

The CMO and pride in ownership. The true CMO also encourages pride in ownership with equipment operators and maintenance as they do their part to fix and prevent maintenance problems through a cooperative team effort of operation-based maintenance. The CMO has the integrity and inspires individual integrity to the point that all employees do their jobs as if they too owned the company. Individual integrity includes pride in one's work no matter what the task. An effective CMO can help your operation go beyond the bottom line to ensure long-term total operations success of the company and the maintenance process.

Take action on this question. As a maintenance leader, you must act on this key question: "If I owned my maintenance operation, what would I do differently to make a profit?" Another question could be, "How high is your return on maintenance management?" If you begin to think like the chief maintenance officer, you will get others to think this way too. **Planner/schedulers are maintenance leaders too...so you should also think this way because you will make so much it actually happens!** You will get more people thinking "profit and customer centered." As each crafts person feels they own part of the business, you will experience a groundswell of profit ability. One key part of this answer will be to get maximum value received from your information technology tools, your CMMS, which for 95% of every Scoreboard benchmark assessment I have done over the past 40 years needed some type of improvement to gain maximum value.

Effective in-house maintenance plus high quality maintenance contractors. Profit and customer-centered in-house maintenance in combination with the wise use of high quality contract maintenance services will be the key to the final evolution that occurs. There will be a revolution within organizations that do not fully recognize maintenance as a core business requirement and establish core competencies for it. The bill will come due for those operations that have subscribed to the "pay-me-later syndrome" for deferred maintenance. There will also be a revolution within those operations that have gambled with maintenance and have lost, with no time left before profit and customer-centered contract maintenance provides the best financial option for a real solution. Whereby maintenance was once considered to be a necessary evil, it is now being viewed as a key contributor to profit in a manufacturing or service-providing operation. My goal is for this book to build your case clearly within your organization for reliable planning, estimating, and scheduling.

Where is the profit in maintenance, really? You might ask yourself after reading this far, "Where is the profit in maintenance *really* for an in-house operation trying to keep its head above water?" Later we will cover areas where planning and scheduling provides other profit and productivity improvement opportunities. For example, what if the net profit ratio of an operation is 4%? What does a 4% net profit ratio mean in terms of the amount of equivalent sales needed to generate profits? A net profit ratio of 4% requires $25 of equivalent sales for each $1 of net profit generated.

Therefore, when we view maintenance in these terms, we can readily see that a small savings in maintenance can mean a great deal to the bottom line and *equivalent pure sales revenue as shown* in Figure 1.2. Maintenance as a profit center is illustrated below showing that only a $40,000 savings is required to translate into the equivalent of $1,000,000 in sales revenue. As we will discuss in later chapters, there are many more areas such as the value of increased asset uptime, increased net capacity, and just-in-time throughput, increased product quality, and increased customer service that all contribute to the bottom line and subsequently to profit.

Maintenance as a Profit Center	
Maintenance Savings that Impact Net Profit	Equivalent Sales Required for Generating Net Profit
$1	$25
$1,000	$25,000
$10,000	$250,000
$20,000	$500,000
$30,000	$750,000
$40,000	$1,000,000
$80,000	$2,000,000
$120,000	$3,000,000
$200,000	$5,000,000

Figure 1.2 Maintenance as a profit center.

Investments in reliable maintenance, estimating and scheduling and other best practices can achieve results comparable to the following:

- 15–25% increase in critical capacity constraining equipment uptime
- 20–30% increase in maintenance productivity of the craft workforce
- 25–30% increase in planned maintenance work
- 10–25% reduction in emergency repairs

- 20–30% reduction in excess and obsolete inventory
- 10–20% reduction in maintenance repair costs

Other improvements can include:

- Improved product quality
- Improved utilization of equipment operators
- Improved equipment productivity (OEE (Overall Equipment Effectiveness)) and production throughput capacity
- Improved equipment life lower life cycle cost
- Improved productivity of the total operation and pure profit

The results listed above can be achieved by maintenance organizations who have committed to continuous maintenance improvement or other terms such as maintenance benchmarking and best practices implementation. Your organization must realize that there are no easy answers and no "quick fixes." Organizations that have invested in maintenance over the long term have realized a tangible return on that investment.

Consider what would happen if your numbers were used in the following very basic examples:

- Maintenance craft productivity increase of 20%

20% net improvement in craft productivity (craft utilization, craft performance, and craft service quality) would be 20 × 40 craftsmen × \$35,000/year = \$280,000/year.

- Increased equipment uptime of 25%

25% downtime reduced from 8% to 6%—value of increased uptime would 0.25 × baseline \$800,000 value of downtime = \$200,000/year.

- Inventory reduction in maintenance storeroom of 25%

25% reduction from \$1,000,000 to \$800,000 would be \$200,000 × 0.30 inventory carrying costs = \$60,000/year.

- Improved pricing from suppliers of 1%

1% direct price savings (not high cost of low bid buying) would be 0.01 × \$1,000,000 purchase volume/year = \$10,000/year.

- Reduction in net maintenance repair costs of 10%

10% would be 0.10 × \$750,000 annual repair cost = \$75,000/year if all required maintenance requirements were being met.

- Improved product quality of 1%

1% reduction through equipment-related scrap, rework returns, waste, and better yields would be 0.01 × \$2,000,000 value of production standard cost = \$20,000/year.

- Improved equipment life of one-half year

1/2 year longer productive asset life would be 0.5 year × $10 million capital investment × 0.10 expected capital ROI = $500,000 minus additional $200,000 additional maintenance cost = $300,000/year.

These examples all contribute to the bottom line either directly or indirectly. They illustrate briefly that tangible ROI can be significant, depending on the size of the maintenance operation and the type of organization being supported. Maintenance leaders must be able to gain support for continuous maintenance improvement by developing valid economic justifications. Take the time to evaluate the potential savings and benefits that are possible within your own organization. Gain valuable support and develop a partnership for profit with operations. Include all other key departments that will receive benefits from improved maintenance. The application of today's best maintenance practices will provide the opportunity for maintenance to contribute directly to the bottom line. However, the pursuit of maintenance excellence requires leadership.

Core requirement versus core competencies for maintenance. The core requirement for good maintenance never goes away because "maintenance is forever"! There will always be a need to maintain. Maintenance of our physical bodies, minds, souls, cars, computers, and all physical assets providing products or services will always be required. Maintenance, gravity, extinction, and change are truly forever. Yet some organizations today have neglected maintaining their core competencies in maintenance to the point that they have lost complete control. The *core requirement* for good maintenance will always remain (forever) but the *core competency* to do good maintenance can be missing. In some cases, we know that the best and often only solution is value-added outsourcing. Maintenance is a core requirement for profitable survival and total operations success. If the internal core competency for maintenance is not present, it must be regained with internal leadership of top leaders and maintenance leaders. Neglect of the past can either be overcome internally or externally. The core requirement for maintenance can be reduced, but it can never go away.

Symptoms of Ineffective Planning

Top leaders must know the true state of maintenance in their organization and understand the symptoms of ineffective planning, which one of today's most important best practices. Lack of effective planning will include the following:

1. Delays encountered by our most valuable resources, the craft workforce
   a. Gaining information about the job
   b. Obtaining permits
   c. Identifying and obtaining parts and materials
   d. Identifying blueprints, tools, and skills needed
   e. Getting all of above to the job site
   f. Waiting for required parts not in stock
2. Crafts waiting at job site for supervisor or operations to clarify work to be done
3. Delays or drop in productivity when operations request work without sufficient planning
4. Equipment is not ready, even if on a schedule

5. Number of crafts does not match scope of work
6. Coordination of support crafts; not the right skill, come too late or early and stand around watching
7. Crafts have no prior knowledge of job tasks or parts
8. Crafts leave job site for parts, go to storeroom, or wait for delivery
9. If parts to be ordered, job is left disassembled and crafts go to next job
10. Many jobs in process awaiting parts
11. Crafts cannot develop work rhythms due to start/stops and going from crisis to crisis
12. Supervisors become dispatcher for emergencies

Benefits of Effective Planning

With effective planning and scheduling, top leaders can begin to see the benefits, which include:

1. Provides central source of equipment condition, workload, and resources available to perform it
2. Improves employee safety and regulatory compliance
3. Helps achieve optimal level of maintenance in support of long- and short-term operational needs
4. Challenges work request of questionable value
5. Provides forecast of labor and material needs
6. Permits recognition of labor shortages and allows for leveling of peak workloads
7. Establishes expectations for what is to be accomplished each week, and variations from the schedule are visible
8. Improves productivity by anticipating needs and avoiding delays
9. Increases productivity of both operations and maintenance
10. Provides factual data: performance measurements, cost variations
11. Provides info to identify problems that need focused attention
12. Reduces the total cost of maintenance while improving customer service
13. Increases useful life of physical assets
14. Improves preparation, management, and control of minor and major projects
    a. Outages
    b. Turnarounds
    c. Renovations

Later we will discuss *The Scoreboard for Maintenance Excellence*, which can define the overall state of maintenance and give top leaders a road map for their overall maintenance strategy with reliable planning, estimating, and scheduling as one major cornerstone to success.

What happens when you get scared half to death.

Defining Results to Top Leaders and Operations Leaders

2

"Show me the money!" The true goals in business are results in the form of profits, along with total operation success and sustainability in the marketplace. In turn, top leaders and operations leaders must see clearly defined results. A reliable maintenance planning, estimating, and scheduling process allows the maintenance business to perform as a profit center. First, *consider* the *Harvard Business* Review's (HBR) definition of asset management, which is all about managing money. So, what about their discussions of physical asset management? There are basically no articles on physical asset management, even though physical assets provide the means to make money and profits to share dividends with stakeholders. To me, that is a definite case for improving physical asset management. However, when one searches for "physical asset management" within HBR articles, contributed by hundreds of management gurus, the results contain only articles on asset management as related to money management. That is incredibly unbelievable to me.

However, things are changing. Remember that change and gravity are constant. New circumstances now require maximum uptime, throughput, and quality. With many low-cost producers competing around the world, the previously dominant and status quo organizations are fighting for survival. During these challenging and economically difficult times, all operations are looking for improvement opportunities.

Top leaders must recognize that an important area for improvement is the maintaining of facilities and other physical assets. In the past, the maintenance department may have been viewed as a "necessary evil"—costly "wrench turners," often viewed unjustly as those who sat in the shop and waited for equipment to fail. That outlook is now history for smart companies that are finding ways to detect and prevent catastrophic failures before they occur. Predictive maintenance technologies such as vibration analysis, infrared imaging, acoustic testing, and preventive maintenance help companies to maximize profits by minimizing downtime. For example, the information from the P–F interval, as illustrated in Chapter 1, can be analyzed when deciding whether to shut down a piece of equipment for planned maintenance. Planned maintenance is up to three times more productive than purely reactive, "firefighting" maintenance strategies.

We must show top leaders accurate results so that they will realize the potential increases in profits resulting from the best practice of reliable maintenance planning, estimating, and scheduling. This is especially necessary in large and small oil and gas surface maintenance facilities. In addition to identifying potential failures, we must also focus our resources on correcting potential problems before a catastrophic failure occurs. With decreasing work forces and increasing responsibilities, this can easily become a second priority. However, as less work is completed, more failures occur and our time is spent repairing failures, not on preventing failures from happening.

Reliable Maintenance Planning, Estimating, and Scheduling. http://dx.doi.org/10.1016/B978-0-12-397042-8.00002-4

It is very important to understand the differences between productivity and efficiency. We can be very efficient at firefighting and performing reactive maintenance repair, but are we truly being productive? In other words, like the government at times, we can be very efficient at doing the wrong things. Later, in Chapter 6, we will discuss productivity in greater detail, reviewing plant labor productivity (standard labor variances), physical asset productivity, and craft labor productivity, which The Maintenance Excellence Institute International (TMEII) has coined as *overall craft effectiveness* (OCE) .The way to break this cycle is to approach maintenance planning, estimating, and scheduling as a part of a new external profit center. To do this, we must develop a new, disciplined approach to identifying, prioritizing, and completing properly planned maintenance work. The total operation must understand that an effective maintenance planning and scheduling program will result in greater productivity of the craft workforce by increasing uptime, minimizing costs, and increasing overall production throughput and capacity. Survival is a keyword, and this best equates to the highest possible profit margins for all for-profit operations.

Here again, the increased productivity of people and physical assets is a direct result of planning and scheduling. It is basically a disciplined approach for utilizing existing maintenance resources to increase craft productivity and asset productivity by increasing uptime and minimizing the overall production costs. This is accomplished in many ways, such as the following:

- Prioritization of work
- Developing the physical steps to complete the job
- Procuring the necessary tools and materials
- Scheduling the work to be done
- Completing the work with reasonable performance levels
- Identifying any additional work to be completed on the equipment
- Filing written documentation for equipment history

From my past experience as a plant manager of two large manufacturing plants, I see the primary and true goal as being "total operations success." Maintenance must be very customer centered, with operations and production being the internal customers. When we view maintenance as an internal business, the strategy and philosophy become more apparent to all. For top leaders, maintenance leaders, and craft leaders, what are the underlying results from maintenance and reliability excellence? Your obvious answer might be to improve the long-term viability of your facility due to the reduced costs associated with maintenance and equipment downtime. That translates into total operations success. But for some craftspeople and technicians in the trenches, that can often mean very little. Later chapters address how change can be viewed in a more positive light. Nonetheless, how many times have you been called out in the middle of the night to deal with an equipment failure? Or had a vacation cancelled due to issues with critical equipment? Or had a technician cover work when he or she had an important event to attend? Better planning and scheduling practices will reduce the frequency of these catastrophic failures. Later in this book, I will explore the philosophy of pride in maintenance for maintenance technicians and pride in ownership for operators.

One can very easily sit back and play the victim by claiming there are too few staff members to keep the area running or complete the total backlog of work. You can manage the status quo for just so long, then you must decide to lead. If you do not lead and have a disciplined system in place, then you do not know how many people it truly takes to maintain your equipment. Planners are there to help you determine one very important number: the total maintenance requirements or your total backlog of work, which includes completing 100% of your preventative maintenance (PM) tasks, deferred maintenance, and maybe minor projects. With reliable information from the planner, the maintenance leader can then take facts to the top leaders. Reliable estimating methods add greatly to the accuracy of your backlog photos. All of this may or may not lead to more staffing, but facts on total maintenance requirements can support the various best practices necessary to improve the productivity of existing resources; people assets, and physical assets.

During my 40+ years in manufacturing and maintenance, I have seen in detail more than 400 different types of plants. Many were in the oil-gas-petrochemical sector or were discrete manufacturing facilities, including pharmaceuticals, universities, hospitals, and fleet operations. One was Boeing Commercial Airplanes, which at that time was improving its maintenance processes significantly during the 1990's. Their approach then was to have what they called "the plan of the day." It was a relatively simple system that worked. They were also moving rapidly into predictive maintenance and condition monitoring—an important first step in establishing an effective planning and scheduling program.

First, we must identify the current planning and scheduling practices. From my experience in performing on-site scoreboards for maintenance excellence audits, I have seen very few sites with true world-class planning, estimating, and scheduling. Improving your process requires getting out in the operating areas and speaking with supervisors, planners, and maintenance and hourly employees. Process mapping of work initiation through work completion and work order closeout are excellent tools to visually see the entire current process. From this graphical representation, you can quickly see areas for improvement, especially by applying key principles from this book.

It is very likely that you will run into a lot of frustrations around what is currently in place. In this case, a third-party audit can be extremely beneficial and cost-effective. Just like a third-party maintenance consultant, you will get an opportunity to hear a multitude of excuses for why it cannot work. You must keep digging for information to help build a successful system. Here is a very important recommendation for newly hired maintenance leader: have a third-party conduct a due-diligence audit to determine the current state of maintenance that you will be inheriting. This can be related to the new ISO 5001 or it can be an audit based upon the 2015 version of The Scoreboard for Maintenance Excellence, included as Appendix A in this book. PAS 55: 2008, which recently led to the ISO 55000 standards, is a very descriptive narrative that is often unclear. The Scoreboard for Maintenance Excellence has X best practices areas rating of the y evaluation criteria and is very prescriptive as to what is required for world-class maintenance. For planning, estimating, and scheduling, here are some things are some things to look for:

- Is your backlog of maintenance work extremely large?
- Are reliable estimates and estimating techniques being used?
- Do you seem to accomplish very little with the people you have?

- Do you have repetitive failures on critical equipment?
- Are all jobs estimated and scheduled based upon the Noah's Ark method (e.g., 2 × 2 with two employees, 2 h; 2 × 4 with two employees, 4 h)?
- Are parts on the job before it is started?
- Do employees understand what is expected of them?

These are just some of the symptoms of a planning and scheduling system that is not functioning well. For example, I have seen planners who went to the storeroom and retrieved parts for jobs, while others labored over data entry and struggled with scope of work, poor storeroom support, and gaining cooperation with production to schedule PMs and other work.

When we look at planning, estimating, and scheduling closely, we see that it is not rocket science. Other best practices complement planning and scheduling, such as effective PM and predictive maintenance (PDM), plus a responsive and customer-driven maintenance storeroom. For planning and scheduling, the goal is to get the most critical jobs completed in the most productive way. If you have attempted a planning and scheduling project and it failed or did not achieve its full potential, then learn from those mistakes and drive down a different path next time.

Problems with well-defined roles, responsibilities, and written guidelines often have been encountered in past attempts to fully implement planning, estimating, and scheduling. Someone once said, "As soon as the man with the briefcase left, we quickly reverted to the comfort of how it was done in the past." You must make sure the newest variation of a planning and scheduling program becomes a way of natural operation, where you get serious about creating clear roles and responsibilities for everyone in the organization. This includes everyone from the mechanics to craft leaders, maintenance leaders, and most importantly, top leaders.

To lead maintenance forward as a business, a key challenge is to make sure that the new roles and responsibilities of your planners/schedulers (Chapter 7) become a part of the way you do business. You must take a realistic approach to what can be sustained and implemented for long-term benefits. I love the term of being "rigidly flexible," which personally I am. Your system also has to be flexible enough to accommodate the skill set and resources of each area. Your planners will have a key role with very well-defined responsibilities. As stated in the introduction, planners can be a viable resource to support maintenance and reliability improvement, but they cannot neglect their primary roles and responsibilities; rather, they must integrate the areas related to reliability improvement into their daily routines. Much of the work they should be doing relates to reliability.

It has been said that, in general, planning consumes approximately 10% of project time versus implementation, which consumes 90% of the time. Creating a planning and scheduling process on paper is easy using this book and many other books like it. All are excellent guidelines for planning based upon your unique site requirements and your computerized maintenance management system. However, how many times does a program of the month fail when it runs into the "not this again" mentality of the workforce? We have "cried wolf" so many times when implementing new types of systems or programs that the majority of the workforce has quit listening. Before you roll the system out, make sure you have documented lessons learned (as to why you failed in the past, if that is the case) and what you will do differently to succeed now. Again, I recommend that you conduct a thorough assessment of the total physical asset management process. Do not form a committee but rather use a leadership-driven, self-managed team that is chartered for a specific reason, such as how planning and scheduling should be conducted at your site. This is a cross-functional team from across the entire operation, including technicians and craft leaders. Also, this team should have a clear charter sponsored by top leaders. A sample charter for a leadership-driven, self-managed team is included in Appendix D. You can began with just some informal polling, shop floor observations, and even develop a process map of the vision of your future system. Selected sections of an actual process map for the British Petroleum (BP) refinery in Texas City, Texas are included in Appendix E. Find out what the perception is of the current or previous planning and scheduling systems. Make notes of previous weaknesses and address them specifically in your implementation plan. Also, please do not think your operation is the only one having problems with this best practice. Approximately 75% of the plant sites that I have audited with the Scoreboard for Maintenance Excellence have had serious challenges with planning and scheduling.

I have seen and facilitated a number of cross-functional teams made up of middle managers, front-line supervisors, and even some hourly employees and current planners, plus key support staff, for CMMS. Remember that an existing planner should be classified as a super-user of your system. The employees selected should be some of the most influential and respected members of their respective work areas. The team should be sponsored by a top leader to develop the system and a define timeline for rollout. The rollout should include defined team member roles and responsibilities (especially the planner/scheduler), as defined in Chapter 7.

You must decide on appropriate metrics, meeting frequencies and agendas, and sustainability plans. Off-site time of a week or so can useful for team training and solution discussions with a third-party facilitator to help train and work through the entire system. This can be a very good investment, especially if there are other best practices that must be implemented to complement the planning and scheduling process. In this case, a detailed process map of the current system should be developed as a team deliverable, providing a graphic representation of the document flow and actions required by all involved. A process map of the recommended solution also serves as an important project deliverable.

There may be some very heated discussion on what will or will not work, but in the end the goal is to develop a plan that can be implemented and sustained over time as a normal business practice. It is recommended that this same team be part of the implementation solution, as either resource persons or active participants in the operation of the planning and scheduling system. Normally, a consensus can be achieved on the system without creating unsolvable issues within operations or within the maintenance department. Also, a strong team leader should be named by the team sponsor or nominated by the actual team members. During the team's activities, every effort should be taken to define and quantify benefits from planning and scheduling, as well as other best practices that may be needed to complement planning and scheduling. These values may be in direct savings, cost avoidances, or gained value as shown as increases in craft labor productivity (i.e., OCE). A dollar range of benefits can be used. For example, one client manufacturing a wide range of cosmetics had an annual potential savings/gained value of $3,000,000. If our estimate of benefits was only 50% accurate, then savings and gained value would be $1,500,000, which is not a bad return on investment for a short amount of consulting and training time. Also, the team can look back and say, "Wow, we did it ourselves!"

The logical next step is to present the team's findings and plans to the team sponsor for final approval. This is a very crucial step because without support of top leaders and operational/production leaders, the system will never be sustainable. Sustainability is one important evaluation category that was included our 2015 Scoreboard for Maintenance Excellence. It is also highly recommended that the proposal be signed off by key top leaders, including the vice president (VP) of operations. When the management team signs the cover page to indicate their approval, support, and commitment, this covers new ground by doing something may never have been done before for maintenance. This signoff can help set the expectations that the system will be implemented across the total operation this time by committed top leaders. Remember, there is a big difference between involvement and commitment: It is like an egg and bacon breakfast—the chicken was involved, but the pig was committed.

Now the real work of implementation begins, which I stated earlier was 90% of the work. At this point, we have an agreed-upon document for how the operation will plan and schedule maintenance. Dedicated resources may be needed to successfully roll out the system. This should be defined in the report to top leaders. The developmental team becomes responsible for the individual area rollouts. This was one of the key reasons that we selected some of the most influential and well-respected members from each operational area. Because they helped to develop the system, they have built-in ownership, so they then become the drivers at the lowest level in their periods. Communication to the shop floor is a very important element for success, up and down and all across the organization.

In some cases, area team members can hold short meetings with their areas to review the system, including a line-by-line review of the system, allowing for a question-and-answer session. It is recommended that assigned planners and the third-party facilitator also attend these meetings. Questions can be brought back to the team as a whole to get a consensus decision, but most should have been having addressed in the document as it stands. In the end, a complete and comprehensive description of the system is a very effective way to communicate details as well as to maintain the integrity of the system into operations.

Sustainability is important to our nation's infrastructure as well as an ongoing business. So our new scoreboard includes a new evaluation category Number 34 for Sustainability. Once the system is rolled out, the main focus must be on sustainability. The first 6 months is the crucial timeframe for establishing a foothold in the organizational culture. Each work area must feel a constant presence to ensure that using the new or improved planning and scheduling system becomes second nature. Customers must also see value-added results. The implementation team members should be encouraged to attend area meetings at least weekly. As planning progresses into scheduling, with weekly and daily schedules, the team can also attend some of these meetings, along with the planner and employees who are having work schedules for their respective areas. The team should also conduct monthly meetings to allow team members to discuss issues and advancement of the process. By openly and regularly sharing, you will be able to make decisions on the progress and status of the system. In turn, you can keep top leaders well informed of the shop-level results. This also helps the team members to stay aware of how the system is progressing, and maybe even inspiring some competition between areas to try to be the best.

To ensure sustainability, you may use *The Scoreboard for Maintenance Excellence*™ as a system audits tool. This tool can very easily be completed as a team-based self-assessment performed at, for example, 6-month intervals to show progress. This is another way to document results to top leaders as well at the grassroots level and to your most important maintenance asset—your craftspeople. Basically, our scoreboard is a checklist of 38 benchmark/best practice areas. Each best practice has number of evaluation criteria that are rated on a scale of 5–10, with 5 = poor, 6 = below average, 7 = average, 8 = good, 9 = very good, and 10 = so-called world class. The scoreboard is very easy to use, with each criterion defined in clear prescriptive terms.

Ratings are not absolute values, but the scoreboard results can be a total overall score of 38 best practices or just scores for the two work management and control categories and the two planning and scheduling categories. The score for each best practice category is a category progress score, and it provides valuable benefits information to all involved. There can also be specific checklists for areas, such as conducting scheduling meetings with operations, completeness of work scope of major job packages, or a checklist to help identify well-planned jobs in the backlog. This becomes a real tool when you see an area start to backslide a little. One good practice is for the planner to get periodic feedback from supervisors and craft workforce surveys related to his or her work.

I was personally amazed one time when a VP of operations attended a kickoff meeting for a Scoreboard audit for the world's largest contract packaging operation. It made the group snap to attention when the VP came into the audit kickoff meeting just to see how things were going. A few well-placed questions about some part of the system that is lacking always goes a long way to show the team that the expectations are real and will be followed up on. When I returned 8 months later to train their prospective planners for just 3 days, he was plant manager/president of this single plant operation. I heard him tell all of the second-shift employees about their new crafts skill training program and the great opportunities ahead to move from packaging into maintenance.

Later, I will discuss how to measure the effectiveness of your planning and scheduling process, as well as your crafts skills training program. It will also be an important part of the overall measurement tool—the reliable maintenance excellence index (RMEI), which is detailed in Chapter 25. Here, we develop metrics to measure the impact of the planning and scheduling system, plus a range of 10–15 other key metrics, including financial, physical assets, maintenance storeroom, quality, and craft productivity items. During our own scoreboard audits, the RMEI is developed and reviewed with the client. In general, the current client selects from metrics that we propose and determines what levels of achievement that they are driving toward for each metric. Using a one-page Microsoft Excel document, the client reports and monitors on a monthly basis. The implementation team's planner representative, as a super-user of the CMMS, would normally maintain the RMEI. Based on the agreed-upon metrics to measure planning and scheduling system utilization, the RMEI also integrates metrics from operations and financial areas. This goes a long way in helping everyone to understand the total scope of planning and scheduling, as well as the scope of other improvement activities that may be ongoing. Just like the broad-based scoreboard audits, the RMEI goes down to both shop and plant levels. It is a way to keep everyone actively monitoring the overall goals within the RMEI.

Here I note that there has been much discussion about the balanced scorecard. I have both books by the originators of the concept, and neither book mentions the words "maintenance" or "physical asset management" anywhere. Also, as I discussed before, HBR does not include articles related to physical asset management. If physical asset management is not part of a "balanced scorecard," there a great imbalance for sure.

There is a good saying within the industrial engineering community (as well as others) that states: "What gets measured gets done" Once your system has been rolled out and people understand that it is not going away, you will see amazing things start to happen. People adapt to a system that helps them find ways to make it easier to use in their areas. You will see a crossover stage by operations, with their roles and responsibilities being accepted, and team members will enhance the system to make their jobs easier.

For example, planners and schedulers provide a vital service to support a plant shutdown, turnaround, or outage, especially in oil and gas operations that have both planners and schedulers. As planners scope out work for shutdowns for the equipment, they could very easily tag each job during the scoping process. It could also identify the mechanic assigned, the trade crews involved, and even contractor support personnel. This tag also may include where the parts are located, such as in a job box, a controlled laydown area, staged in the storeroom, or (best of all) delivered to the jobsite. This can help reduce the "morning rush" of a shutdown when supervisors are running to get mechanics on their respective jobs and to answer any questions when contractors' are clocking into the refinery (or any site for that matter). With the use of radiofrequency identification (RFID) tags that contain chips for significant data, the sky is almost the limit as to what you can do with RFID tags on a wireless basis. Your hourly mechanics should give very positive feedback on the system, and they may even leave more notes than usual on what they found and the necessary follow-up work. Nonetheless, all work completed during a shutdown turnaround or outage should be documented on a formal work order and captured in the equipment history file. Examples like this can often be seen across the total operation once each area takes ownership of the system and it becomes part of the everyday culture and way of supporting the business of maintenance.

Once you start to see people taking the system to the next level, you will have succeeded with your initial implementation. Now, it is time to take another look in the mirror for continuous improvement, find your weaknesses, and focus on ways to continue to improve them. Those who do not continue to improve will be left behind. Also, do not forget to celebrate the successes that you have achieved along the way. These will be readily denoted on your RMEI on a monthly basis, where direct savings and gained value can be calculated on improvements from your baseline timepoint.

Conclusion

Planning and scheduling can be exciting when applied properly. The concepts of an effective planning and scheduling system have been around for years. Your charter today may be to successfully develop and implement a new system or to revive a system that has to achieve its maximum value. You may very well become a part of the way your site does business for years to come. If you are currently a planner/ scheduler, do not just sit back and watch what happens—be motivated to make good things happen.

In a world where only the strong survive, it is imperative to focus on the costs that are within your control. We all must change the perception that maintenance is a business and that planning and scheduling are essential to success. This book includes a number of chapters that appear to be outside the scope of reliable maintenance planning and scheduling, but there is a purpose for including them; they will cover some areas that other books have neglected to cover. As stated in the introduction, by the inclusion of "reliable" in the book's title, I wanted to emphasize my belief that a planner/scheduler is a next step for a greater leadership role within a maintenance department. Therefore, these additional topics support professional development of a planner or scheduler (or a combined planner/scheduler position).

Five key things to for you to consider and remember are as follows:

1. Always know your total maintenance requirements in terms of job types, job status, man hours required, and specifically the high cost of deferred maintenance throughout the organization.
2. Use a reliable method of estimating that is based on improving repair methods and providing estimated job times for scheduling by using a system such as the ACE Team Process developed by TMEII (discussed in detail in Chapter 17).
3. Remember, "What gets measured get done."
4. P.T. Barnum of circus fame said, "Without publicity, a terrible thing happens—nothing." Therefore, publicize and communicate your success across and up and down your organization.
5. It has been said that, "Nothing happens until someone sells something." This is a true statement for those managing the status quo, never selling the top leaders on the benefits of planning and scheduling.

Leadership: Creating Maintenance Leaders, Not Just Maintenance Managers

Maintenance planners must be leaders or develop into true leadership for their future success. We still need managers, but more maintenance leaders are needed within the physical asset management arena. Figure 3.1 summarizes some of the basic differences between maintenance managers and maintenance leaders.

Leadership is rarely an innate quality. Rather, it is a combination of hard work, conviction, and instinctive strategy, which needs to be developed and nurtured. When you see someone who is naturally charismatic and inspiring, you are likely disregarding an immense amount of work that has occurred behind the scenes. This is precisely the reason why we are witnessing an increasing demand for cultivating this talent at the earliest possible age. Whether in sports, business, or entrepreneurship, today's youth are striving to sow the seeds of leadership in hopes of future success.

In my case, I served 26 years within the US Army Corps of Engineers, as a trained civil affairs officer and last as a military police officer. I can say without question that my formal education did nothing to expand my leadership capabilities. The US Army offered some of the best opportunities for teamwork and to build leadership skills. Below are some key points for leadership in the business of maintenance:

1. It all starts with a well-defined and measurable vision

> *People buy into the leader before they buy into the vision.*
>
> *John Maxwell*

The true essence of leadership begins with envisioning a set of personal or professional goals. Do not just have a vague image in your mind; rather, define the target with focused clarity. Think through the final result over and over to make sure you will be committed until the end. As discussed in Chapter 2, a team process brings together a synergy of cross-functional members of an organization.

However, stating objectives is not enough. Enforcing the purpose and mission are equally important. Provide a clear and realistic path forward to your team. Believe in yourself and be persistent when things look difficult. Without John F. Kennedy's ambitious vision, Neil Armstrong would not be the first man on the moon. No dream is too big until you have realized it.

Reliable Maintenance Planning, Estimating, and Scheduling. http://dx.doi.org/10.1016/B978-0-12-397042-8.00003-6

Maintenance Leader	versus	Maintenance Manager
• Proactive		• Reactive
• Makes Things Happen		• Wonders What Happened
• Has Vision		• May Wear Blinders
• Promotes Growth		• Maintains Control
• Promotes Continuous Improvement		• Maintains Status Quo

Figure 3.1 Basic differences between maintenance leaders and maintenance managers.

2. Communicate often and clearly. Publicize your goals when it is appropriate. Report periodic reviews of your personal goals and the positive status of ongoing improvement goals

> *Great leaders are almost always great simplifiers, who can cut through argument, debate, and doubt, to offer a solution everybody can understand.*
>
> *General Colin Powell*

Communication is the fundamental link between vision and reality. Deliver the message concisely and with conviction so that it permeates through all levels of the organization. Your people need to understand why they are working on a task, what they should be doing, and where it will lead them. This entails having good presentation skills, being a good listener, and facilitating problem-solving on the leadership-driven self-managed teams established to improve the planning and scheduling process. Effective communication skills make a standout leader.

3. Do not underestimate the power of optimism. Become a member of an Optimist Club International and do not be nominated to join a local pessimist club.

> *If opportunity doesn't knock, build a door.*
>
> *Milton Berle*

A great story in this regard is Walt Disney's remarkable story. It was 1928 in New York when Walt learned that his distributor hired most of Disney's animators to start a new studio. He practically lost everything, including his staff, the contract, his income, and the hit character Oswald the Rabbit. He immediately sent a telegram to his brother Roy saying, "Don't worry. Everything okay. Will give details when I arrive." On his 3-day journey back to Hollywood, Walt took out his sketchbook and created the character of Mickey Mouse. Within a year, Mickey was the most popular cartoon in the world.

Optimism helps channel the negative energy of fear and uncertainty towards driving innovation. As a leader, you will be surrounded by skeptics. Reject pessimism and turn the volume up on positivity.

4. Motivate and empower

If your actions inspire others to dream more, learn more, do more and become more, you are a leader.

John Quincy Adams

Without the right kind of stimulus, people produce mediocre work and drain out quickly. Some get inspired by power, some by incentives, some by appreciation, and some by interesting work. It is your responsibility to identify specific motivation factors in your employees and empower them. There are three main types of motivation: fear motivation, incentive motivation, and attitude motivation. All three at times may play a part within our professional lives and careers. As I have said many times before, a successful planner is on a career path to becoming a supervisor or a maintenance leader at a higher level. As a leader, your efforts to nourish and empower your team will also indicate that you care for them, which in turn is a great fuel to boost productivity and loyalty.

5. Accept feedback generously.

Leadership and learning are indispensable to each other.

John F. Kennedy

One of the best ways to grow and improve is by graciously accepting constructive feedback. Planners may receive feedback from their services through surveys conducted or direct/indirect communication with supervisors and operations staff. Each planner must be able to take constructive criticism and not let it be a negative factor. Many managers, especially chief executive officers (CEOs), by way of their power, find it demeaning to be advised by their juniors. However, your people hold the key to invaluable information that can make you more successful. Therefore, leave your ego behind, and ask what you can do better. You may choose to do that in an informal setting or through a defined 360-degree feedback model.

6. Lead by example.

You don't lead by pointing and telling people some place to go. You lead by going to that place and making a case.

Ken Kasey

Planners and schedulers in essence are leading their organization on a path forward to productive and cost-effective execution of the maintenance mission. Teaching by force and directive orders is passé. This is the generation of producing future leaders by walking the talk. Do not waste hours trying to convince people. Instead, demonstrate the benefits of a particular decision by your own actions. You cannot expect others to do what you would not do. Besides garnering respect and trust, you will be able to set higher standards and achieve better results. The easiest way to begin is by thinking of your role model. Who would you want to emulate? What kind of traits does that person have?

7. Take responsibility and own up!

> *A good leader is a person who takes a little more than his share of the blame and a little less than his share of the credit.*
>
> *John Maxwell.*

As the top maintenance leader, you may have to come clean on the declining state of maintenance, if it does exist. While serving as a major within the 30th Infantry brigade, I worked for a brigadier general with a full staff of lieutenant colonels commanding armor, infantry, artillery, and direct support maintenance battalions. My unit was a larger-than-normal company responsible for combat engineering support to the brigade, which included ribbon bridge construction. During the final assault, which crossed our ribbon bridge to the objective, one commander of an armor battalion failed to make the final river assault crossing due to bad maintenance. At the general's staff meeting for an after-action review, this commander was relieved and fired right on the spot.

When things go wrong with critical equipment and the plan was not completed as scheduled due to many factors, there will be finger-pointing. In your role as planner, you must realize things will go wrong, and it is not for you to assess who is to blame. Say no to passing blame onto others—that is the most diminishing quality any leader can possess. Being at the top implies taking ownership of your vision and your team's actions. In spite of having a robust set of internal controls, any organization will have its share of slip-ups and errors. You will need a whole lot of courage to apologize for mistakes and take measures to improve upon them. I have also said this many times: Maintenance leaders must have courage. You must have courage to be the messenger to top leaders about total maintenance requirements not been accomplished and a declining state of maintenance (if it is present).

8. Use your power to drive change

> *Anyone can hold the helm when the sea is calm.*
>
> *Publilius Syrus*

A planner can drive change and in essence create a sense of so-called power within the maintenance organization. By having a finger on all the factors related to a specific job, the planner can answer many questions. Having the right answers as to when repair jobs can be completed builds a base of knowledge. As the craftspeople complete schedules regularly on time, the credibility of maintenance plus planning, scheduling, and execution of work builds the so-called power to provide greater customer service.

In the book *Onward: How Starbucks Fought for Its Life Without Losing its Soul*, Starbucks CEO Howard Schultz shared his remarkable story, giving us many leadership lessons. Eight years after stepping down from the daily oversight of Starbucks,

Schultz returned as CEO in 2008. His aim was to bring back the core values that Starbucks was originally known for. He made some drastic decisions, including closing 900 stores and shutting the remaining 11,000 US stores for a day to retrain 115,000 people. The media questioned the relevance of these changes, but Schultz explained, "It was honest, it was authentic and it was necessary."

Maintenance leaders and planners are often faced with challenges that require bold and unconventional decisions. Trust your instincts and use your authority to your advantage, but always bring the facts related to the situation. Change is necessary to establish an environment for continuous growth. And, just like gravity, change is constant.

9. Cultivate patience

> *Patience and perseverance have a magical effect before which difficulties disappear and obstacles vanish.*
> *John Quincy Adams*

Have you ever heard the saying to have the patience of Job? During Job's return to a righteous way of life, his three friends came to give him their thoughts and advice as to why he had lost everything in life. They sat with Job for seven days and seven nights without saying a word. Maybe this standoff led to the slogan, "He has the patience of Job."

Successful leaders are proactive yet patient. They understand that a lifespan consists of periods of sprint followed by periods of recovery time. Many of us are prone to making snap decisions under deadlines and pressure. Be careful when you are influenced by excitement and wishes to see quick results. This especially holds true for small businesses and start-up phases of projects, where patience can make or kill. The Dutch often say that a handful of patience is worth a bushel of brains.

10. There is no one leadership style.

> *In matters of style, swim with the current; in matters of principle, stand like a rock.*
> *Thomas Jefferson*

When there are no two people in this world exactly alike, how can there be a single way to lead? Daniel Goleman studied around 3000 mid-level managers, uncovering six different leadership styles:

a. Commanding
b. Visionary
c. Affiliative
d. Democratic
e. Pacesetting
f. Coaching

Leadership style depends on underlying emotional intelligence competencies, when the style works best, and the overall impact on climate or culture.

Emotional intelligence is the driver for each of these techniques, and it can have a deep impact on your organizational climate and culture. Effective leaders have all of these cards up their sleeve and address the demands of the particular situation. They are flexible and keep switching from one style to the other.

11. Opportunity never knocks

Opportunity never calls, and it never stays the night. If you want a seat at the table, you have to hunt down every opportunity yourself. You are not entitled to anything. For maintenance planners, it is very important that they possess the initiative and the desire to create their own opportunities for job improvement and service to their customers. Maintenance staff who are nominated and recruited to become planner schedulers are beginning a key new role for their future within the maintenance organization.

12. It is better to be best than to be first

If you watch the Hollywood version of success, you can easily get duped into thinking you have to be the first to hit the market in order to win. That is true if you are a reporter and want credit for breaking a story. Otherwise, keep in mind that Myspace predated Facebook.

The first person out the gate may get the competitive advantage and land early. However, when the honeymoon phase is over, people want quality. As long as you focus on creating quality, you will always have something to offer. Maintenance must be focused on quality repairs and not create unrealistic planning times, in which case craftspeople are forced to take shortcuts that result in callbacks to a job that was initially done incorrectly.

13. Leaders believe in serving others over themselves

Maintenance is a vital service for total operation success. The planner's service to both maintenance and operations puts this position in a very important service role. In turn, planners must believe that their service is making a difference within the organization and is creating value that can be measured.

14. Quality is important

One of the three elements of overall craft effectiveness is what I call craft service quality (CSQ), which will provide a good measure of the quality of repairs within maintenance. A truly successful person is not successful because of their position in life. It does not matter if you are a janitor or a CEO—success is defined by how content you are with where you are. Kevin O'Leary (a self-made millionaire and TV personality) will tell you that success means being rich, while Gandhi successfully led a revolution and freed both India and Pakistan while living poor.

Quality of life is not defined by what you own or how high up the ladder you have climbed. It is defined by your satisfaction with what you have. No matter where you are in your life, strive to create quality experiences for those around you. Planners create expectations with the development and publishing of a weekly or daily schedule. Craftspeople who truly exhibit a pride in maintenance attitude normally can meet expectations.

15. Execution trumps ideas

Everyone has great ideas. Some executed idea have brought great success. For example, someone thought there should be a Website where people can socialize online, so Facebook was invented. Someone said it would be cool if you could shop online; now Amazon exists for fast service from online shopping. John Lennon's and Paul McCartney's best songs used only a handful of basic chords. Those names rose to prominence because they thought of something no one else did. They took action and accomplished something no one else did, and most of them continue doing so to this day.

Ideas are important, but anyone can come up with ideas. Backing those thoughts with action is how you create success. The idea of planning and scheduling along with good estimating is not rocket science and is a very good idea, but execution and implementation are the key.

16. Respect is something you earn

Planners may or may not begin their work with respect from all within maintenance or operations. The advice I have heard the most in my life—at home, in school, in the military, in corporate America—is that respect is something that is earned. You are not entitled to respect. You are entitled to common courtesy and politeness, but you have to prove yourself worthy of peoples' respect. It does not come from a title; it comes from your daily actions and attitude. Respect everyone, act ethically, and always follow through. People will respect such traits in a planner or scheduler.

17. Leaders believe in their place in history

A successful person knows his or her place within the work environment and is comfortable with it. Whether or not you have made a blip in the history books, you have your own history, a work history, family history, and a history in your community. With time comes memory. People remember your actions in the past, and they judge you in the present based on them. If you understand your place in your organization, you will be prepared for successful results.

18. Quitting is the only failure

Some people talk about treading water or keeping your head above water, but that is only enough to remain in the same spot. To actually reach your goal, you cannot tread water—you have to keep swimming.

You can only see by looking back and forward, but that gives the confidence to keep swimming.

19. Success is about more than money

Money does have its uses. While it may be the root of all evil, it is also a resource that can be used to enact good change. If you define your worth by how much money you have, you have a long way to go before you are as valuable as anyone on the Forbes billionaire list. You will also likely never reach that billionaire list, because it takes a belief in your own value to reach that level. This brings me to the final point.

20. Believe in yourself

Successful people think that they are successful—it is what makes them successful. Perspective is everything in life, and the only way to reach success is to move with a successful perspective. You become what you think. If you do not believe in yourself, no one else will.

How to Create PRIDE in Maintenance within Craft Leaders and the Technical Workforce

This chapter strives to understand the importance of planners readily soliciting and accepting ideas from the crafts workforce. Planners who have moved from being a good craftsperson into one as a planner at the next level will have an important advantage compared with a brand-new mechanical engineer being assigned as a planner. This chapter also examines why planners not only support reliability improvement but help to create People Really Interested In Developing Excellence (PRIDE) in maintenance within craft leaders and the technical workforce.

I begin this chapter with something I wrote while serving as plant manager of manufacturing operations for the Crescent Xcelite tool plant in Sumter, South Carolina. PRIDE in maintenance is something that every maintenance leader must practice and believe in. Planners are in the category of maintenance leaders as well. This introduction talks about PRIDE in excellence, which parallels all we will talk about for achieving PRIDE in maintenance in this chapter. Late in 1983 when this was written to our plant employees, little did we know that more than 100 of the 500 employees from the total plant direct labor workforce would face layoffs (in late 1984 and 1985) due to the dumping of foreign hand tools into US markets. Because South Carolina is a right-to-work state, the layoff was based upon employee seniority. The good news was that we were able to call back almost everyone by early 1986. We began the layoff process with positive expectations that the business would return, which it did. So, what about PRIDE in excellence?

PRIDE in Excellence: The very famous German sports car manufacturer, Porsche, had a slogan that said, "Excellence is expected." To me, this is what customers, distributors, and group sales/marketing people expect. Excellence is expected from the world-famous Crescent and Xcelite trade name brands.

In our employee communication meetings, the topic of foreign competition and the real threat that it presented to the American hand tool industry at that time was an issue that all employees had to keep in mind. Crescent/Xcelite was continually addressing both foreign and domestic competition through an aggressive capital investment effort to upgrade old equipment, secure new equipment, and improve manufacturing methods, implementing a closed-loop MRP (Materials Requirements Planning) system (pilot plant in Cooper Industries) and improving processes by whatever means available. We were also scheduling a vibration analysis service across this large plant with a large portfolio of rotating equipment. At that time, we did not have planners; the two supervisors did a good job reporting at weekly operations meetings on upcoming PM's (preventive maintenance) and critical jobs.

Reliable Maintenance Planning, Estimating, and Scheduling. http://dx.doi.org/10.1016/B978-0-12-397042-8.00004-8

However, although money buys the new equipment and raw materials, people are the most important assets for developing continued excellence in a company's products. Personalized human skills, ranging from die making to final inspection, were the assets that determined the outcome of the "big game" with foreign and domestic competitors. To win the "big game," each employee at Crescent/Xcelite was encouraged to consider taking "PRIDE in excellence" as a renewed effort for the 1984–1986 period, which was very challenging.

Below, I listed a few of the thoughts that I used with the production and maintenance employees at Crescent/Xcelite in 1984. They will still apply to every job at your plant as well as to your personal family life. Add one or two to your list of resolutions.

1. **Do every job as if you owned the plant, the department, or the piece of equipment you operate or maintain**. Every employee in the plant is, in a real sense, a manager of a small business, regardless of the operation. Crescent/Xcelite as a plant is made up of many small plants or teams. Be a proud competitor on your team.
2. **Develop a commitment to excellence in everything you do**. Have fun and seek justice against poor quality where it is due. If work is thought of as a hobby, such as golf or fishing, imagine how much fun it would be to meet that 7:00 a.m. starting time at the plant.
3. **Develop PRIDE in your work, regardless of the task**. Give 105% to your performance whenever possible to make up for the times you were at only 95%.
4. **Maintain a sincere belief in your capabilities, as well as the potential of those you meet each day**. Practice positive reinforcement on yourself and others.
5. **Practice the golden rule**. If it does not work the first time, then Practice!, Practice!, Practice!
6. **Practice good maintenance in all areas: physical, spiritual, family, equipment, mental, financial, etc**. Plan to wear out rather than rust out. I have a saying that "maintenance is forever."
7. Particularly for planner/schedulers, it is your job to "make things happen" rather than "watch things happen." Try not to be in the group of people that wonders what happened. Reduce work-in-process and stamp out rework-in-process.
8. **Develop PRIDE in yourself, your company, and your country**.
9. **Establish written and specific written personal goals in all areas of your life**.
10. If none of the above items works for you, just keep smiling. People will really wonder what you are up to!

PRIDE in excellence is a very personal matter that we as individual Americans must address. The late Congressman Larry McDonald (who died in a Korean Air Lines accident when the plane was shot down by Soviet interceptors in 1983) made a closing statement in the documentary film *No Place to Hide* with these words: "Freedom is not free." Excellence, quality, and success likewise are not free; all take hard work and commitment from people.

PRIDE, on the other hand, is free when the work is done. When the game is over, you can look back and say, "I did the very best I could do." A departmental team can say, "We did the best we could do!" Our goal should be to look back a year later and be able to honestly say, "We did our best!" If a planner can do this, he or she can and will be successful.

Think about leading your maintenance operation with the above as part of your personal philosophy for achieving PRIDE in maintenance. Planning for excellence will help create positive expectations for every planner/scheduler (P/S) serving their customers all around the world.

PRIDE in Maintenance: The term "PRIDE in Maintenance" was first used in 1981 while I was on the Cooper Tools Group staff as group manager of industrial engineering, striving to improve maintenance processes across seven plant sites. The first step, along with creating the original Scoreboard for Maintenance Excellence, was to sponsor a groupwide training session on maintenance best practices. This was a weeklong session held in Greenville, Mississippi, on the Mississippi River, way down south at two of the Nicholson saw plants. We had plant managers attend for one day and then engineering managers, maintenance supervisors/foreman, planners and storeroom managers for the rest of the week. We also had two predictive maintenance equipment vendors come in to discuss this new technology and to demonstrate the benefits of vibration analysis and other technologies. To our good fortune during an in-plant demo, one vendor actually found a bad bearing on a major piece of equipment. This was in the mill for rolling specialty steel for hacksaw and band saw blades. The training event was a success at creating a better understanding about the importance of maintenance for top leaders and maintenance leaders. It also helped to avoid a catastrophic bearing failure at the Nicholson steel mill.

However, we neglected to include any craftspeople in this weeklong event. The consultant that I used for the training was Mr. George Smith from Orange, New Jersey, who had just recently purchased Marshall Institute from the founders. George discussed many topics related to the crafts workforce throughout his presentations. Therefore, I said to George, "We forgot a very important group. We did not get your message down to the crafts level. I will talk to the plant manager here at the Greenville plant. We will see if we can get the crafts together for at least a 1 h session and you can talk to them too before you leave." The plant manager agreed, and George talked to 20 or more craft people on the topic that I called "PRIDE in Maintenance." George was a Navy pilot in World War II, and he had some real maintenance stories to tell about why PRIDE in maintenance was important .For example, he always told the aircraft mechanic working on his plane to be ready for a test to "see how she's running." We videotaped George's presentation, and his words apply today just as much as they did back in the 1980s.

Planners must not forget the crafts people either. The craft workforce must not be forgotten in today's world of fancy new terms and technological advances. Your crafts workforce will be really interested in developing excellence in maintenance when they feel that they are a true member of the team. Our domain name, www.Pride-in-Maintenance.com, reflects our belief about the important of maintenance, the value of maintenance people and the work they do, and how we must change attitudes about the profession of maintenance.

I now would like to reinforce some key points made in previous chapters that have been leading toward our main subjects: reliable maintenance planning, estimating, and scheduling. Your improvement strategy must include all maintenance resources, equipment, and facility assets, as well as the craftspeople and equipment operators. It must also include MRO parts and material assets, maintenance informational assets, and the added value resource of synergistic team-based processes. Maintenance leaders and top leaders must support their most important maintenance resource of all, the crafts workforce, which should be considered as your most valuable people. Figure 4.1 illustrates the support role, which must start with top leaders realizing that the foundation for maintenance excellence begins with PRIDE in maintenance.

Our vision is to help achieve **PRIDE in Maintenance** from within the craft workforce and their maintenance leaders. In addition, top leaders should realize the true value of their total maintenance operation (Chapter 1) and then take positive actions to support the maintenance leaders, craft leaders, and the craft workforce. The Maintenance Excellence Institute International provides a wide range of consulting with our maintenance excellence services, temporary operational services and customized and public training for maintenance excellence offerings we call TrueWorkShops™. We support all types of maintenance operations. However, the bottom line is that PRIDE in maintenance within your craft workforce is the foundation for your success for building long-term maintenance and reliability excellence and sustainability. Your craft workforce can be a valuable source of new ideas and positive reinforcement during your journey toward maintenance and reliability excellence.

During each Scoreboard for Maintenance Excellence assessment, we always interview a number of craftspeople. We get very candid input and many comments that support the improvement needs of the overall maintenance operation. However, we lacked a focused method that could bring out more ideas and concerns from the craft workforce. Following an assessment, one client requested that we conduct a session with the craft workforce and support staff to get ideas directly from this group. They did not want any supervisors or managers present, which might hinder open discussion

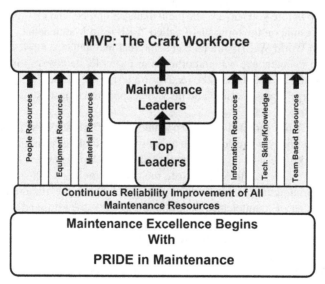

Figure 4.1 Build your foundation upon PRIDE in maintenance. Most valuable player (MVP).

of ideas and concerns. As a result, the first **PRIDE in Maintenance** sessions for the craft workforce and support staff were developed. This client wanted to make sure that their craft and support employees (approximately 350 in total) had the opportunity to express both their concerns and ideas. It is important to:

1. Understand the client's goals for maintenance improvement per assessment results, which should include a new P/S, for example.
2. Provide ideas that the crafts people thought were important and needed by the operation.

From the results of this first session, more than 300 good ideas were generated. Therefore, we feel that it is well worth the time to bring crafts and support staff together for a session devoted to sharing ideas and concerns. We firmly believe that maintenance excellence begins with PRIDE in maintenance. It is important to have people at all levels with PRIDE in maintenance—**P**eople **R**eally **I**nterested in **D**eveloping **E**xcellence **in Maintenance**. These sessions can help you gain people with greater PRIDE in maintenance. They will be your own craftspeople who can add greater value to your maintenance operation by sharing their ideas and being a vital part of helping you implement today's proven best practices for maintenance excellence.

A Positive and Proven Approach: PRIDE in Maintenance sessions with your craft workforce should begin only after the consultant has a clear understanding of the client's current improvement goals, current challenges, and past successes. To do this, a Scoreboard for Maintenance Excellence assessment is recommended. The following key steps (illustrated in Figure 4.2) should be taken to help your craft workforce become a valuable source of new ideas and attitudes.

1. Develop your results from the Scoreboard for Maintenance Excellence Assessment.
2. Develop your PRIDE in Maintenance session materials. Gain your approval of materials and your commitment to begin.
3. Conduct PRIDE in Maintenance sessions for the craft workforce and support staff to share goals and key challenges (1 h).
4. Conduct PRIDE in Maintenance team exercises with the craft workforce focused on your key challenges (1 h).
5. Have the teams present their recommendations. Presentations from each team should be videotaped and a summary of all team recommendations prepared.
6. Have the client review the assessment results and crafts team recommendations, then determine their strategic, tactical, operation, and "do it now" commitments.
7. Provide implementation support from The Maintenance Excellence Institute as needed.
8. Continue with continuous reliability improvement and chartered cross-functional teams as a possible next step.
9. Measure and validate results.

Gaining Support from the Craft Workforce: Without support from the craft workforce, achieving maintenance excellence can be extremely difficult. Planners especially must have the craft workforce on their side and gain respect from them. In Chapter 13, we will define some of the traits to look for when hiring a P/S. PRIDE in Maintenance sessions developed specifically for the crafts workforce and other maintenance support staff can add much to your maintenance improvement efforts. They can serve as one way to introduce the concepts of your P/S. They can serve as a means

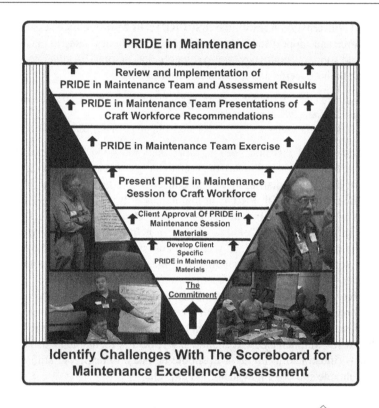

Figure 4.2 Your craft workforce is a valuable source of ideas.

to gain craft level support for the P/S, to achieve better understanding and greater cooperation for current and future maintenance improvements. It is important that maintenance leaders provide very positive reinforcement to the crafts workers that their job is important and that their ideas are needed and are welcomed. Because they perform such a mission-essential role to success, the need for their positive input, ideas, and active participation is critical.

PRIDE in Maintenance sessions also help instill a philosophy of profit-centered maintenance into the thinking and attitudes of each participant. For public service as well as oil and gas operations, it is about maximizing customer service. PRIDE in Maintenance sessions help to support internal teamwork as well as to eliminate the fear of changing the status quo. It is important for your craftspeople to understand their contribution to greater profit and service and how they fit into your P/S process. You must challenge and support them to do each job as if they owned the company. PRIDE in Maintenance materials should be customized to the operation, whether the goal is to maximize profit, service, or both. The client should be able to review and approve all client-specific materials that are developed for the presentation. Session materials, including participant handouts, case studies, and additional maintenance excellence references, are provided to each attendee. Each session can include up to a maximum of 25 participants, allowing

Figure 4.3 Maintenance leadership team at UNC-Chapel Hill.

for three teams of eight craftspeople across typical craft functions. We like to videotape all team presentations and give each client reproduction rights for future use of their PRIDE in Maintenance video and all custom materials prepared for their session. Determining the true value of training often seems difficult. However, when you can incorporate ideas from your crafts workforce into various plans of action, the value of this type of training can be readily demonstrated. Scarce craft resources are a terrible thing to waste. So, take action now to consider investing in your most valuable maintenance resources to get their ideas.

The following is a case study example from the Facilities Services Division for the University of North Carolina (UNC) at Chapel Hill. It shows a section from their newsletter, which is one of the communications tools they used to spread the word about their workforce's involvement in maintenance improvement. Each of the more than 300 recommendations from their initial PRIDE in Maintenance sessions were reviewed by the Maintenance Leadership Team, who are shown in Figure 4.3. This team looked at each recommendation, and their goal while reviewing each employee's idea was to answer one of these three questions:

1. What can I (we) do now to put this idea into action?
2. How can this idea be put into action based upon *additional internal* review and help?
3. How can this idea be put into action with support and cooperation from departments outside the UNC-Chapel Hill (UNC-CH)'s Facilities Services Division and *additional external* review?

Based on question 3, some recommendations needed support and cooperation from other departments. The following is an excerpt from UNC-CH newsletter, describing how they created a cross-functional team to address a major improvement opportunity. This newsletter column started out as PRIDE in Maintenance, but was quickly changed to PRIDE in Maintenance and Construction because minor construction and renovation work was an important part of the division's direct responsibility.

PRIDE in Maintenance and Construction

Excerpt from the Resource Newsletter of UNC-CH Facilities Services Division

PRIDE in Maintenance and Construction has continued with follow up on each of the over 300 recommendations received from employees during the initial PRIDE in Maintenance sessions. The Team headed by Steve Copeland with Stanley Young, Bob Woods, Bob Humphreys, Mark McIntyre and Joe Emory completed the review of *each and every recommendation received* back in November 2003.

Systems Performance Team Chartered: The review of all employee recommendations led to a number of recommendations where support and cooperation from departments outside the Facilities Services Division was necessary. Therefore, the Systems Performance Team was chartered. Its job was to provide additional external review and recommendations for a top priority issue that was common across all PRIDE in Maintenance sessions.

The objective of this team is to recommend methods to improve overall building systems performance. The scope includes design and construction phases and the commissioning process of new facilities. The team started in January 2004 and has been meeting biweekly. It has looking at ways that Building Services can be more closely involved with design and construction processes and how to eliminate additional work that typically must be performed on new facilities constructed by contractors. This team addressed specific quality control improvement opportunities during the design/construction phases to improve contractor accountability. It considered how to better support the construction administration process and how to support the final user's requirement for the facility during commissioning.

Team leader Joe Wall from Construction was supported by Ricky Robinson—Maintenance/Plumbing, Rodney Davis—Life Safety and Access, Robbie Everhart—HVAC, Terry Bowers—Housing Support, Eddie Short—Construction Administrator, Joel Carlin—Engineering Information Services, Ed Willis, PE—Manager of Construction Management, Cindy Shea—Sustainability Coordinator, Julie Thurston, PE—A/E Services, Pete Peters—The Maintenance Excellence Institute, and Carole Questa and Keith Snead—Facilities Planning/Design.

The Systems Performance Team prepared its final recommendations and a presentation by team members was planned during April 2004. Their work has had very positive actions on important employee recommendations from our PRIDE in Maintenance session held last year.

Results from the Systems Performance Team helped to ensure that maintenance was involved during the design, preconstruction, construction, final inspection, and commissioning phases of new facilities on the UNC-CH campus. Facilities maintenance staff from each craft area became an important part of the total team. Significant costs were tried to be eliminated by ensuring contractors did the job right the first time and by not having to call maintenance in to correct contractor mistakes after commissioning. Tracking of warranted items was also improved with vendors and contractors being held more accountable.

Summary: Achieving PRIDE in Maintenance requires many things within an organization. It requires a top leader who understands the value of maintenance and its challenges. It requires maintenance managers, supervisors, and foreman that are true maintenance leaders within the important profession of maintenance. It requires crafts leaders and a craft workforce trained and dedicated to profit and customer-centered service. Effective storeroom and support staff all combine to perform the business of maintenance. Across all these people resources there must be dedication to the maintenance profession and PRIDE in Maintenance that comes from teamwork, personal motivation, good leadership, and good maintenance practices. Your maintenance operation will continue to improve its progress on the journey toward maintenance excellence. Maintenance excellence is truly not a final destination but rather a continuous journey.

Define Your Physical Asset Management Strategy with The Scoreboard for Maintenance Excellence and Go Beyond ISO 55000

5

Previous chapters have provided a prelude of important topics that the planner and others in maintenance should understand especially the practice of true leadership true leadership. The planner should also understand how his or her current operation compares with others in regards to best practices that support the planning scheduling process, such as the maintenance storeroom, their parts supply chain, preventive and predictive maintenance (PdM), reliability process improvements, and many others we will detail in Appendix A—The Scoreboard for Maintenance Excellence version 2015. Here are many new best-practice categories specific to oil, gas, and petrochemical operations and any complex operation with extreme health, safety, security, and environmental (HSSE) challenges. It is also important for top leaders to understand their HSSE challenges as well as to know the scope of the Scoreboard practices as well as the return on investment that can be achieved with an effective planning and scheduling process. Before beginning or renewing our material specific to developing or improving a planning, estimating, and scheduling process, we must get down to the detailed level of "determining where we are" with actually applying today's best practices.

You may have developed a viable physical asset management strategy defined in PASS 55: 2008 Part 1 and 2, now morphed into the ISO 55000 standards. However, you will need to take these open-ended descriptive requirements down to the shop level with prescriptive actions to be taken and measured. Your planners, storeroom, preventive maintenance (PM)/PdM, and all programs for reliability improvement must be fully integrated for gaining maximum value from your planning/scheduling process. This chapter introduces the 2015 version of **The Scoreboard for Maintenance Excellence**, focused on oil, gas, and petrochemical plant maintenance as your **global benchmark or guideline** for maintenance best practices. The 2015 version has 38 best-practice categories and 600 specific evaluation items. It has evolved since 1981 (then with 10 benchmark categories and 100 benchmark criteria) into today's most complete benchmarking tool to define where you are on your journey toward maintenance and reliability excellence. Combined with the other three benchmarking tools of The Maintenance Excellence Institute International (TMEII), we feel sure that this book provides today's best process, starting from global, external benchmarking

Reliable Maintenance Planning, Estimating, and Scheduling. http://dx.doi.org/10.1016/B978-0-12-397042-8.00005-X

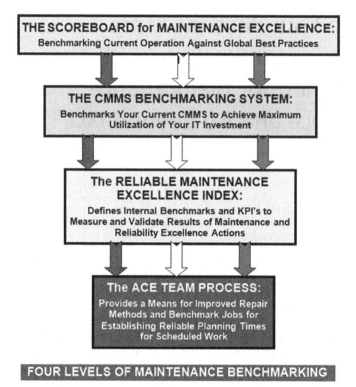

Figure 5.1 Maintenance benchmarking from four important levels.

and going down to the craft workforce level. We will discuss in detail how reliable benchmarks for craft repair times are with the ACE (a consensus of experts) Team Benchmarking Process. This is included later in Chapter 15, "Developing Improved Repair Methods and Reliable Maintenance Planning Times with the ACE Team Process." The four levels of maintenance benchmarking are illustrated in Figure 5.1.

Understanding the Types of Benchmarking

A **benchmark** is a point of reference for a measurement. The term presumably originates from the practice of making dimensional height measurements of an object on a workbench using a graduated scale or similar tool and using the surface of the workbench as the origin for the measurements. In surveying, **benchmarks** are landmarks of reliable, precisely known altitude and location and are often manmade objects, such as features of permanent structures that are unlikely to change or special-purpose "monuments," which are typically small concrete obelisks set permanently into the earth. Does your current operation have a current benchmark or assessment as to "where you are" with your physical asset management process? Also, how do you compare to implementing a physical asset management strategy? If you desire to be committed to continuous reliability improvement and progressing toward a profit- and customer-centered operation, here is where we begin to help you get started, especially with the key elements required for successful planning, estimating, and scheduling.

As a planner is beginning their new position, they can then see "where their organization stands" and what may be missing from within their own operation. During my visits to many, many plants, I have seen serious gaps existing that caused the maintenance planning and scheduling process not to achieve its expected results. Therefore, I encourage every planner or maintenance leader to use this document or review any type of other audits, such as one resulting from an ISO 55000 audit, and like Nike, Inc., says, "Just do it." Now that is obviously easier said in a book than doing it in the often hectic maintenance environment no matter how well planned. We will now review parts of the 2015 version of **The Scoreboard for Maintenance Excellence** contained within Appendix A and enhanced for use within the oil, gas, and petrochemical sectors of business. This new Scoreboard for Maintenance Excellence addresses 38 major benchmark categories with 600 total benchmarking evaluation criteria. The benchmark categories are recognized maintenance best practices from around the world. Some are easier than others and basically can be "do it now" items. Many are strategic, tactical, or even operational tasks and improvement opportunities that may require added internal and external resources to implement. **The Scoreboard for Maintenance Excellence** provides the first of four benchmarking tools introduced in this book that the planner can use. First is the global one that "benchmarks where you are with applying external best practices that other successful operations recognize and use or regulatory agencies require. Results from a Scoreboard assessment will identify the gaps and allow development of complete strategic, tactical, operational, or "do it now" plans for physical asset management.

Benchmarking is a very versatile tool that can be applied in various ways to meet a range of requirements for improvement. The following is a summary of different terms used to distinguish the various ways of applying benchmarking and how benchmarking tools from TMEII fit these summary definitions.

1. **Strategic benchmarking**: Improving an organization's overall performance by examining the long-term strategies and general approaches that have enabled high-performers to succeed. It involves high-level aspects such as core competencies, developing new products and services, changing the balance of activities, and improving capabilities for dealing with changes in the culture of the organizational environment. This type of benchmarking may be difficult to implement, and the benefits are likely to take a long time to see results. I consider an ISO 55000 audit to be in this category as well as best practices from the total scope of the Scoreboard.

The Scoreboard for Maintenance Excellence is *strategic benchmarking* and what we term *global benchmarking* because it applies to the total maintenance process and overall best practices for the "business of maintenance." Results from using **The Scoreboard for Maintenance Excellence** will include a strategic, tactical, and operational level of difficulty in improvement opportunities. There will almost always be "**do it now**" opportunities that can be immediately implemented.

2. **Performance benchmarking or competitive benchmarking**: A total organization considers their positions in relation to the performance characteristics of key products and services compared with benchmarking partners (or competitors such as contract maintenance

providers) from the same sector. In our scope, this is physical asset management. In the commercial world, companies tend to undertake this type of benchmarking through trade associations or third parties to protect confidentiality. Very seldom will you see an in-house maintenance operation openly benchmark with a contract maintenance provider. Conversely, contract maintenance providers base their entire business case upon comparing/benchmarking *what they can do for your organization* compared *with what you are now doing or not doing.* The planner should also strive to plan and schedule contractors' work and especially make use of the estimated repair times and performance measurement just as we will talk about measuring our own maintenance workforce. As stated several places in this book, you may always takeover a target for contract maintenance providers in one form or another. With the Scoreboard, it can serve performance/competitive benchmarking as it did within the Boeing Commercial Airplane Group, in which all manufacturing units all across the United States were assessed using the same Boeing Scoreboard for Maintenance we developed with specific Boeing criteria added to the 1993 Scoreboard. And because this audit had a 5% impact on maintenance in the top leader's annual performance reviews, this was very competitive benchmarking.

3. **Process benchmarking**: Focuses on improving specific critical processes and operations. Benchmarking partners are sought from best-practice organizations that perform similar work or deliver similar services. Process benchmarking invariably involves producing **process maps** to facilitate comparison and analysis. This type of benchmarking can result in short- and long-term benefits. Solomon Associates for refinery studies provide standardized peer groupings for performance comparisons within major geographic areas, operating regions, size/complexity groups, and those of similar configuration within the operating region. Customized peer group selection lets you request more narrowly defined peer groups for each of their refineries. Therefore, as in the military when you know that a command maintenance management inspection (CMMI) is coming or that a Solomon Associates audit is planned, this always brings maintenance to attention as well as operational concerns for each Solomon audit.

We feel very, very strongly that process benchmarking must go beyond analysis paralysis and lead directly to successful implementation and measured results. This gets down to the focused implementation of tactical and operational plans of actions that are based on priorities defined from **The Scoreboard for Maintenance Excellence** benchmarking or assessment results. It may involve vendors of predictive technology equipment, integrated suppliers, or external consultant resources to help recruit, train, and install a planning function; to help install a CMMS; or to help modernize an MRO (maintenance repair operations) storeroom. Process benchmarking is the complete business case analysis. Results of process benchmarking plus implementation must be nailed down with a valid measurement process that your financial folks can readily understand and agree upon. Here is where **The Reliable Maintenance Excellence Index** (RMEI) comes into play as a key benchmarking tool at the shop level as the example shown in Figure 5.2. How to develop your own RMEI will be in discussed in Chapter 25, "How to Measure Total Operations Success with The Reliable Maintenance Excellence Index."

4. **Functional benchmarking or generic benchmarking**: This is benchmarking with partners drawn from different business sectors, public/private operations, military organizations, and different maintenance environments to find ways of improving similar functions or work processes. This sort of benchmarking related to planning and scheduling can lead to innovation and dramatic improvements. Although this book focuses on the oil, gas, and petrochemical sectors, it can be easily applied universally across all types of operations. Because planning and scheduling within these areas is so critical, there are special best practices necessary for total operations success.

The Reliable Maintenance Excellence Index													Example: 13 Performance Measures
A. Performance Measures	1. Actual Maintenance Cost Per Unit of Production	2. % Major Work Completed within 5% of Cost Estimate	3. % Overall Maintenance Budget Compliance	4. Overall Schedule Compliance	5. % Overall PM Compliance	6. % Planned Work	7. % Craft Time For Customer Charge Back	8. % Work Orders With Reliable Planned Time	9. % Critical Asset Availability	10. % Wrench Time (Craft Utilization)	11. % Craft Performance	12. % Inventory Accuracy	13. Number of Stock Outs of Inventoried Stock Items	F. Performance Level Scores
B. Current Month	1.30	90	94	90	94	68	75	50	90	34	90	90	15	Perf Level
C. Performance Goal	1.00	95	98	95	100	80	85	60	95	50	95	98	10	10
	1.05	94	96	94	98	78	83	58	94	48	94	97	11	9
	1.10	93	94	93	96	76	81	56	93	46	93	96	12	8
	1.15	92	92	92	94	74	79	54	92	44	92	95	13	7
	1.20	91	90	91	92	72	77	52	91	42	91	94	14	6
D. Baseline Performance Levels	1.25	90	88	90	90	70	75	50	90	40	90	93	15	5
	1.30	89	86	89	88	68	73	48	89	38	89	92	16	4
	1.35	88	84	88	86	66	71	46	88	36	88	91	17	3
	1.40	87	82	87	84	64	69	44	87	34	87	90	18	2
	1.45	86	80	86	82	62	67	42	86	32	86	89	19	1
	1.50	85	78	85	80	60	65	40	85	30	85	88	20	0
E. Performance Level Score	4	5	8	5	7	4	5	5	5	2	5	2	5	I. Total
G. Weighted Value of Metric	10	6	6	7	11	7	6	7	13	8	8	6	5	RMEI Score
H. Performance Level Score (E) x Weight (G)	40	30	48	35	77	28	30	35	65	16	40	12	25	481
J. Total MEI Value Over Time	Date	7/08	8/08	9/08	10/08	11/08	12/08	1/09	2/09	3/09	4/09	5/09	6/09	
	Score	481												

Figure 5.2 The RMEI example.

The various Scoreboard versions available from TMEII include The Scoreboard for Maintenance Excellence (manufacturing plants), The Scoreboard for Facilities Management Excellence (pure facilities maintenance), The Healthcare Facilities Scoreboard for Maintenance Excellence (hospitals and other healthcare facilities), and The Scoreboard for Fleet Management Excellence (equipment fleet operations). These four represent unique maintenance processes requiring the art and science of maintenance but applications within different work environments. The sharing of practices and innovations across these maintenance practice areas often can have dramatic improvements or can enlighten one practitioner because exposure to other types of maintenance environments is made via benchmarking.

5. **Internal benchmarking**: This involves seeking partners from within the same organization (e.g., from business units located in different parts of the world). The main advantages of internal benchmarking are (1) access to sensitive data and information and standardized data are often readily available, (2) less time and fewer resources are usually needed, and (3) the same CMMS such as SAP may be in use. There may be fewer barriers to implementation because practices may be relatively easy to transfer across the same organization. However, real innovation may be lacking, and best-in-class performance is more likely to be found through external benchmarking.

The Scoreboard for Maintenance Excellence is ideal for this internal benchmarking by making comparisons between maintenance operations within a larger organization. *A Strategy for Developing a Corporate-Wide Scoreboard* in this chapter defines how multiple site operations can provide internal benchmarking with TMEII tools. A Scoreboard for Maintenance Excellence assessment will define "where you are" and helps you to then define a physical asset management strategy to cover the gaps in your current strategy.

We also consider the **RMEI**, as was shown in Figure 5.2, as an important internal benchmarking process, but this tool is at the grassroots level—the shop floor. It should also focus not just on maintenance but rather the success of the total operations, whether a plant, a pure facilities complex, a fleet operations, or healthcare facility, as well as the equipment maintenance process. It is here from the RMEI that we will measure the success of our planning and scheduling process.

6. External benchmarking: Outside organizations that are known to be best in class are sought out to provide opportunities of learning from those who are at the leading edge. For example, the North American Maintenance Excellence award is given to nominees each year, and in 1999, the Marathon Ashland Petroleum refinery in Robinson, IL, received this prestigious award. During most of 1998, I was fortunate to work with Marathon to evaluate and select a CMMS system, which resulted in the selection of Meridian's solution, which included robust reliability-related data analysis. The Robinson refinery also had a world-class planning, estimating, and scheduling process at that time.

External benchmarking keeps in mind that not every best-practice solution can be transferred to others. This type of benchmarking may take up more time and resources to ensure the comparability of data and information, the credibility of the findings, and the development of sound recommendations. External learning is also often slower because of the "not invented here" syndrome.

The Scoreboard for Maintenance Excellence is also an external benchmarking tool, defining "where you are" with overall maintenance and physical asset management best practices at a single site or across multiple sites within the same organization. In fact, at Marathon we conducted Scoreboard assessments at each of their five refineries to help determine functional requirements for a future CMMS and other best-practice needs.

7. International benchmarking: This is used when partners are sought from other countries because best practitioners are located elsewhere in the world and/or there are too few benchmarking partners within the same country to produce valid results. The Toyota production system is a prime example. Globalization and advances in information technology (IT) are increasing opportunities for international projects. However, these can take more time and resources to set up and implement, and the results may require careful analysis because of national differences.

However, our work with many international maintenance operations has shown that we can learn or have our so-called best practices reconfirmed when we see maintenance in other countries. For example, my experience with Boeing, SIDERA Argentina, Ford Argentina, Coke Argentina, Avon Brazil, and AC Smith Mexico reconfirms that a very simple best practice—**cleanliness in shops and plant areas**—is an important contributor to maintenance excellence and pride in maintenance. I truly believe this and have seen it; therefore, planners get your maintenance leader to periodically have you schedule some shop clean-up time as needed.

Our Scoreboards all began originally as *The Scoreboard for Excellence in Maintenance and Tooling* services in 1981 while I was the industrial engineering manager for the Cooper Tool Group, the hand tool division of Cooper Industries. One of my many roles on a very small corporate staff was to improve maintenance processes at our seven plants: *Crescent-Xcelite, Weller, Plumb, Nicholson File, Nicholson Saw, Wiss, and Lufkin*. Ironically, after learning the manufacturing operations in each of these high-quality hand tool plants, I got the chance to be plant manager of the one making the "Knuckle Buster" Crescent Wrench. I was hired at the Cooper Tool Group corporate level, not only for my industrial engineering background but also for my previous maintenance experience installing a very extensive fleet maintenance management across 100 county shops from 1973 to 1975 for the North Carolina Division of Highway's Equipment Unit in the North Carolina Department of Transportation.

We recruited, selected, trained, and installed over 66 planners across the 14 divisions within the North Carolina Department of Transportation.

How Do You Get "There" If You Do Not Know Where There Is When You Start?

That question was my first concern as I tried to spread the maintenance gospel. A map is completely useless if you do not know your current location—"where you are right now on the ground." For the seven Cooper Tool Group plants, my first task was to determine "where we are" with regards to current maintenance practices. Therefore, research began to put together an assessment tool to benchmark where we were with best practices. Tooling services were included in this first Scoreboard because all plants had extensive tool rooms. For example, Nicholson File in Cullman, AL, made 80% of their repair parts for their custom-built file-making machines. It was a parts manufacturing business and in the business of maintenance. We began with a self-assessment, which helped to get each plant to understand the basic best practices that were to come as we later got down to supporting each plant. The evolution of The Scoreboard for Maintenance Excellence is summarized in Figure 5.3.

The 1993 version of The Scoreboard for Maintenance Excellence, shown graphically in Figure 5.4, evolved and was used until 2003, when the next enhancement was made in 2010 using Excel and allowing up to five evaluators to be analyzed for level of consensus.

Over 30 Years of Application and Evolution: You can see how this external, global benchmarking process has evolved from over 30 years of successful application to many different types of public and private organizations. Its counterpart, the Scoreboard for Facilities Management Excellence™, was also developed for pure facilities maintenance operations.

We will define now a recommended approach to develop your overall maintenance strategy and your own unique *Scoreboard for Maintenance Excellence* for continuously evaluating progress on your journey toward physical asset management excellence. Of course, the very first step is for the top leaders, maintenance leader, and the overall plant to have a firm commitment to improving the total maintenance operation. This is something we discussed in Chapter 1 and is a main ingredient for ISO 55000 success. Supply and maintenance are high priorities across all five of the U.S. military services. Without a C-position (Chief Maintenance Officer) as a champion for maintenance, the total operation may not view and accept as core business requirement.

However, it is very important to avoid taking a piecemeal approach. A consistent piecemeal approach I and others can also confirm is that a new CMMS is not "the solution." Others within the organization must also understand this and share the commitment for best practices that are enhanced by a good CMMS. One key element of success is having a commitment from top-level leaders across the organization and craft leaders. Establishing a formal maintenance excellence strategy team is highly recommended, including having a current planner. This high- to

Evolution of the Scoreboard for Maintenance Excellence™				
Date	Scoreboard Version	Benchmark Categories	Benchmark Criteria	Focus
1981	Scoreboard for Excellence	10	100	Plant Maintenance and Tooling Services
1993	Scoreboard for Maintenance Excellence	18	200	Plant Maintenance
2003	Scoreboard for Maintenance Excellence	27	300	Plant Maintenance
2003	Scoreboard for Facilities Management Excellence	27	300	Facilities Maintenance
2003	Scoreboard for Healthcare Maintenance Excellence	27	300	Healthcare Facilities Maintenance
2004	Scoreboard for Fleet Maintenance	27	300	Fleet Maintenance
2010	Scoreboard for Maintenance Excellence	27	300	Plant Maintenance-Updated Excel Design
2015	Scoreboard for Maintenance Excellence	38	600	Updated for Improved Oil & Gas Use and Many Categories Expanded

Figure 5.3 Evolution of the Scoreboard for Maintenance Excellence.

mid-level, leadership-driven, cross-functional team can be made up from maintenance leaders, key operations leaders, shop-level maintenance staff, IT, engineering, procurement, operations/customer, and financial and human resources staff. Some may be designated as resources only when they are needed. At least one craft employee is always a recommended member of the **maintenance excellence strategy team**. Figure 5.5 illustrates continuous reliability improvement steps to improve the total operation.

The Maintenance Excellence Strategy Team: The mission of this team is to develop, lead, and facilitate the overall maintenance improvement process and to ensure measurement of results. Top leader support and required resources must be provided. This team can also sponsor other teams within the organization (e.g., to support implementation operational plans for an individual site's planning and scheduling process). One of the very first things that this team should do is to sponsor a comprehensive evaluation of the total physical asset management and maintenance operation. Again, this is the first step to help determine "where you are." A sample charter for a maintenance excellence strategy team is included in Appendix D. This example was used for a multisite operation that was also undergoing the installation of a new CMMS, storeroom modernizations, and maintenance planner selection and

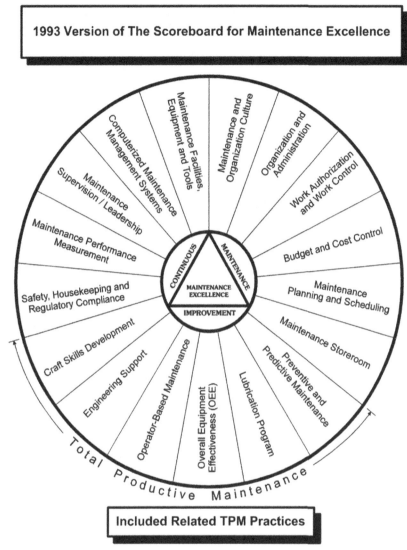

Figure 5.4 The 1993 Scoreboard for Maintenance Excellence.

training. Another case study charter for a new CMMS for Ghana's national power transmission company is also included in Appendix D.

How to Determine "Where You Are": The real goal is improvement of your total maintenance operation to better support profit- and customer-service of the total operation. Maintenance leaders must clearly understand and **remember—your maintenance operation is not the tail wagging the dog, and you are working for your production customers as a total maintenance operation**. Within plant maintenance operations, this is maintenance and repair of all production and facility

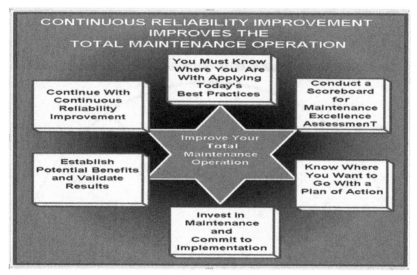

Figure 5.5 Continuous reliability improvement steps to improve the total operation.

assets; supporting infrastructure; overhaul and renovation activities; engineering support processes; and all material management and procurement of typical repair parts, supplies, and contracted services. You should benchmark your current operation against today's best practices for PM and PdM, especially planning and scheduling, effective spare parts control, work orders, work management, and the effective use of computerized systems for maintenance business management. The 2015 Scoreboard also includes 13 new categories, most of which apply to the oil and gas sectors as well as large, complex, continuous processing operations. This step is important because it gives you a baseline as to your starting point for making improvements and for validating results. It will help to ensure that you are taking the right steps for taking care of your mission-essential physical assets. They are included in Figures 5.6 and 5.7.

In most cases, an independent evaluation helps to reinforce the local maintenance leader's desire to take positive action in the first place and to do something to improve the overall maintenance process. For multiple site operations, this provides a great opportunity for developing standard best practices that can be used across the corporation and for all sites.

The Scoreboard for Maintenance Excellence: You can also develop your own *Scoreboard for Maintenance Excellence* as Boeing did using the TMEII 1993 Scoreboard as the baseline. You then begin with a self-assessment, but we normally recommend getting help from a third-party consulting resource and beginning with at least a pilot plant site. For example, for Boeing, we tested the Boeing Scoreboard at the Portland plant before it was used on all five manufacturing regions (at that time with central leadership at corporate level). Therefore, if you plan a corporate-wide assessment/audit, define your own Scoreboard and test it on what you see as your best plant and especially the best planning and scheduling process. Personally, I will wager a significant sum of money that the best plant will have the best planning and

THE SCOREBOARD FOR MAINTENANCE EXCELLENCE™ 2015 Part 1

MAINTENANCE BENCHMARK EVALUATION SUMMARY

Category Number	Benchmark Category Descriptions & Rating Values for Each Evaluation Criteria 5=Poor, 6=Below Average, 7=Average, 8=Good, 9=Very Good & 10= Excellent — New Benchmark or Expanded Category	Number of Criteria	Total Assessment Points per Category
1.	Top Management Support to Maintenance and Physical Asset Management	10	100
2.	Maintenance Strategy, Policy and Total Cost of Ownership	30	300
3.	The Organizational Climate and Culture	9	90
4.	Maintenance Organization, Administration and Human Resources	18	180
5.	Craft Skills Development and Technical Skills	12	120
6.	Operator Based Maintenance and PRIDE in Ownership	10	80
7	Maintenance Leadership, Management and Supervision	12	120
8	Maintenance Business Operations, Budget and Cost Control	15	150
9	Work Management and Control: Shop Level Maintenance Repair (M/R)	12	120
10.	Work Management and Control: Shutdowns, Turnarounds and Outages (STO)	26	260
11.	Shop Level Reliable Planning, Estimating and Scheduling M/R	30	300
12.	STO and Major Planning /Scheduling and Project Management	16	160
13.	Contractor Management	31	310
14.	Manufacturing Facilities Planning and Site Property Management	9	90
15.	Production Asset and Facilities Condition Evaluation Program	6	60
16.	Storeroom Operations and Internal MRO Customer Service	12	120
17.	MRO Materials Management and Procurement	12	120
18.	Preventive Maintenance and Lubrication	18	180
19.	Predictive Maintenance and Condition Monitoring Technology Applications	15	150
20	Reliability Centered Maintenance (RCM)	34	340

Figure 5.6 Part 1: Scoreboard for Maintenance Excellence.

THE SCOREBOARD FOR MAINTENANCE EXCELLENCE™ 2015 Part 2			
Category	Benchmark Category Descriptions &Rating Values for Each Evaluation Criteria 5=Poor, 6=Below Average, 7=Average, 8=Good, 9=Very Good & 10= Excellent New Benchmark Category or Expanded Category	Number of Criteria	Total Assessment Points Per Category
21.	Reliability Analysis Tools: Root Cause Analysis (RCA), Root Cause Corrective Action (RCCA Failure Modes Effects Analysis (FMEA), Root Cause Failure Analysis (RCFA) and Failure Reporting Analysis and Corrective Action System (FRACAS)	17	170
22.	Risk Based Maintenance (RBM)	24	240
23.	Process Control and Instrumentation Systems Technology	9	90
24.	Energy Management and Control	12	120
25.	Maintenance Engineering and Reliability Engineering Support	9	90
26.	Health, Safety, Security and Environmental (HSSE) Compliance	15	150
27.	Maintenance and Quality Control	9	90
28.	Maintenance Performance Measurement	12	120
29.	Computerized Maintenance Management System (CMMS) as a Business System	18	180
30.	Shop Facilities, Equipment, and Tools	9	90
31.	Continuous Reliability Improvement	15	150
32.	Critical Asset Facilitation and Overall Equipment Effectiveness (OEE)	15	150
33.	Overall Craft Effectiveness (OCE)	6	60
34.	Sustainability	11	110
35.	Traceability	19	190
36.	Process Safety Management (PSM) and Management of Change (MOC)	26	260
37.	Risk Based Inspections (RBI) and Risk Mitigation	29	290
38.	PRIDE in Maintenance	8	80
	Total Evaluation Items:	600	
	Scoreboard Total Possible Points:		6000
	Actual Total Benchmark Value Score of All Ratings:		
	Assessment Performed By:		Date:

Figure 5.7 Part 2: Scoreboard for Maintenance Excellence.

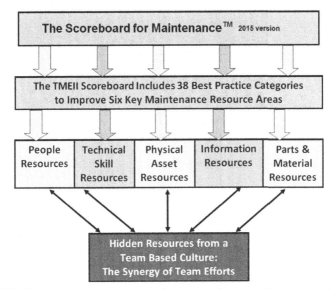

Figure 5.8 Six maintenance resources to improve maintenance in total, not piecemeal.

scheduling process. Figure 5.8 illustrates how The Scoreboard for Maintenance Excellence includes these six key maintenance resources and how hidden resources can evolve from the synergy of team efforts.

The 2015 version of The Scoreboard for Maintenance Excellence was updated by adding or expanding these 15 important best-practice categories shown in Figure 5.9.

The Scoreboard for Maintenance Excellence provides a means to evaluate how we are managing our six key maintenance resources: people, technical skills, physical assets, information, parts and materials, and our hidden resources—the synergy of team efforts. The following shows how the 38 evaluation categories are broken down across the six key maintenance resource areas:

1. People Resources

- Top management support to maintenance and physical asset management
- The maintenance organizational climate and culture
- Pride in maintenance
- Maintenance organization, administration, and human resources
- Contractor management
- Operator-based maintenance and pride in ownership
- HSSE compliance
- Shop facilities, equipment, and tools
- Risk-based inspections and risk mitigation
- Process safety management (and management of change)
- Risk-based maintenance

SCOREBOARD FOR MAINTENANCE EXCELLENCE VERSION 2015

Scoreboard Category Number	\New or Expanded Benchmark Categories for Scoreboard Version 2015	Number of Criteria	Total Assessment Points per Category
1.	Top Management Support to Maintenance and Physical Asset Management	10	**100**
2.	Maintenance Strategy, Policy and Total Cost of Ownership	30	**300**
10.	Work Management and Control: Shutdowns, Turnarounds and Outages (STO)	26	260
11.	Shop Level Reliable Planning, Estimating and Scheduling M/R	30	300
12.	STO and Major Planning /Scheduling and Project Management	16	160
26.	Health, Safety, Security and Environmental (HSSE) Compliance	15	150
13.	Contractor Management	31	310
20.	Reliability-Centered Maintenance (RCM)	34	340
21.	Reliability Analysis Tools: Root Cause Analysis (RCA), Root Cause Corrective Action (RCCA Failure Modes Effects Analysis (FMEA), Root Cause Failure Analysis (RCFA) and Failure Reporting Analysis and Corrective Action System (FRACAS)	17	170
22.	Risk Based Maintenance (RBM)	24	240
34.	Sustainability	11	110
35.	Traceability	19	190
36.	Process Safety Management (PSM) and Management of Change (MOC)	26	260
37.	Risk Based Inspections (RBI) and Risk Mitigation	29	290
38.	PRIDE in Maintenance	8	80

Figure 5.9 Fifteen new or expanded Scoreboard for Maintenance Excellence benchmark categories in version 2015.

2. Technical Skill Resources

- Craft skills development and technical skills
- Maintenance engineering and reliability engineering support
- Reliability centered maintenance (RCM)
- Overall craft effectiveness

3. Physical Assets and Equipment Resources

- Production asset and facility condition evaluation program
- PM and lubrication
- Predictive maintenance and condition monitoring technology applications
- Process control and instrumentation system technology
- Energy management and control
- Maintenance and quality control
- Sustainability
- Critical asset facilitation and overall equipment effectiveness

4. Information Resources

- Maintenance strategy, policy, and total cost of ownership
- Maintenance business operations, budget, and cost control
- Work management and control: maintenance and repair
- Work management and control: shutdowns, turnarounds, and outages (STOs)
- Shop-level maintenance planning and scheduling
- STO and major planning/scheduling and project management
- Manufacturing facilities planning and property management
- Maintenance performance measurement
- Traceability
- CMMS and business system

5. Parts and Material Resources

- Storeroom operations and internal MRO customer service
- MRO materials management and procurement

6. Hidden Resources—The Synergy of Team Efforts

- Maintenance leadership, management, and supervision
- Reliability analysis tools: root cause analysis, root cause corrective action, failure modes effects analysis, root cause failure analysis, and failure reporting analysis and corrective action system
- Continuous reliability improvement

Your Global Benchmark: An assessment using The Scoreboard for Maintenance Excellence provides a "global, external" benchmark against today's best maintenance practices. It is very important to note that the overall total score for the assessment is not an absolute value. However, the assessment results and the overall total score do represent an important benchmark and a baseline value. Results will identify **relative strengths** within an operation and **opportunities for improvement** from among the 38 best-practice areas and 600 prescriptive evaluation items. The assignment of baseline values to each evaluation item is not an exact science. It should be based on an objective assessment of the best practice item when observed at a point in time.

Benchmarking Results: Overall assessment results fall into five possible overall rating levels as shown in Figure 4.6: excellent, very good, good, average, and below average. Each of the five overall ratings levels represents at least a nine percentage point spread.

A summary of baseline value assignment for each rating level is as follows (Figure 5.10):

Baseline Value 10: Excellent: The practices and principles for this evaluation item are clearly in place, and this rating provides an example of world-class application of the practice. There is evidence that this practice has received high-priority support in the past to achieve its current level.

For an overall baseline rating of excellent, the range of total baseline points is 5400–6000 points or 90–100% of the total possible baseline points of 6000.

TOTAL POINT RANGE	SCOREBOARD FOR MAINTENANCE EXCELLENCE™: BENCHMARK RATING SUMMARY
90% to 100% of Total Points	EXCELLENT: 5400 to 6000 Points
80% to 89% of Total Points	VERY GOOD: 4800 to 5399 Points
70% to 79% of Total Points	GOOD: 4200 to 4899 Points
60% to 69% of Total Points	AVERAGE: 3600 to 4199 Points
50% 59% of Total Points	BELOW AVERAGE: 3000 to 3599 Points
49% and Less%	POOR: Less than 2999

Figure 5.10 Benchmark rating summaries by total point range.

Baseline Value 9: Very Good: This rating denotes that current practices are approaching world-class level as a result of high-priority focus on continuous improvement for this evaluation item. This practice continues to be a high priority for improvement within the organization.

For an overall very good baseline rating, the range of total baseline points is 4800–5399 points or 80–89% of the total possible baseline points of 6000.

Baseline Value 8: Good: This rating denotes that the practice is clearly above average and above what is typically seen in similar maintenance operations. There is a need for additional emphasis, to reassess priorities, and to reconfirm commitments to improvement for this evaluation item.

For an overall good baseline rating, the range of total baseline points is 4200–4899 points or 70–79% of the total possible baseline points of 6000.

Baseline Value 7: Average: This rating represents the typical application of the practice as seen within a maintenance operation that has had very little emphasis toward improvement. It reflects an operation that typically is just maintaining the status quo. The organization should conduct a complete assessment of maintenance operations and review this practice in detail for improvement or for implementation if it is not currently in place.

For an overall average baseline rating, the range of total baseline points is 3600–4199 points or 60–69% of the total possible baseline points of 6000.

Baseline Value 6: Below Average: This rating (a score of 6) denotes that the application of this practice is below what is typically seen in similar maintenance operations. This practice may not be currently in place and should be considered as part of the organization's continuous improvement process. Immediate attention may be needed in some areas to correct conditions having an adverse effect on customer service, safety, regulatory compliance, or maintenance costs.

For an overall baseline rating of below average, the total baseline points are from 3000 to 3599 points of the total possible baseline points of 6000.

Baseline Value 5: Poor: This rating denotes that the application of this practice is below what is typically seen in similar maintenance operations. Similar to a below average rating, this practice may not be currently in place and should be considered as part of the organization's continuous improvement process. Immediate attention is to correct conditions having an adverse effect on customer service, safety, regulatory compliance, or maintenance costs.

For an overall baseline rating of poor, the total baseline points are less than 2999, or 49% or less of the total possible baseline points of 6000.

Figure 5.11 provides key comments for each of the Scoreboard for Maintenance Excellence benchmark rating levels.

Appendix A includes a complete Scoreboard for Maintenance Excellence formatted as shown in the following example in Figure 5.12.

Benchmark Criteria Number 1: This example also illustrates the very first benchmark evaluation criteria on the Scoreboard, which is related *to whether or not maintenance is a considered a priority in your operation*. Does your organization include maintenance within its plant goals as a key item?

TOTAL POINT RANGE	SCOREBOARD FOR MAINTENANCE EXCELLENCE™: RATING SUMMARY COMMENTS
90% to 100% of Total Points	**5400 to 6000 points Excellent:** Practices and principles in place. for achieving maintenance excellence and World Class performance based on actual results. Reconfirm overall maintenance performance measures. Maintain strategy of continuous improvement. Set higher standards for maintenance excellence and measure results.
80% to 89% of Total Points	**4800 to 5399 Points -Very Good:** Fine-tune existing operation and current practices. Reassess progress on planned or ongoing improvement activities. Redefine priorities and renew commitment to continuous improvement. Ensure Top Leaders see results and reinforce performance measurement process.
70% to 79% of Total Points	**4200 to 4799 Points-Good:** Reassess priorities and reconfirm commitments at all levels of improvement. Evaluate maintenance practices, develop, and implement plans for priority improvements. Ensure that measures to evaluate performance and results are in place... Initiate strategy of continuous reliability improvement.
60% to 69% of Total Points	**3600 to 4199 Points-Average:** Conduct a complete assessment of maintenance operations and current practices. Determine total costs/benefits of potential improvements. Develop and initiate strategy of continuous reliability improvement. Define clearly to Top Leaders where deferred maintenance is increasing current costs and asset life cycle costs. Gain commitment from Top Leaders to go beyond maintenance of the status quo.
50% to 59% of Total Points	**3000 to 3599 Points-Below Average:** Same as for Average, plus, depending on the level of the rating and major area that is Below Average, immediate attention may be needed to correct conditions having an adverse effect on life, health, safety, and regulatory compliance. Priority to key issues, major building systems and equipment where increasing costs and deferred maintenance are having a direct impact on the immediate survival of the business or the major physical asset. The capabilities for critical assets to perform intended function are being severely limited by current "state of maintenance". Consider possible leadership as required for business survival and for achieving the core requirements for maintenance services. Determine if the necessary investments for internal maintenance improvements is going to be made
49% or Less	**2999 or Less Points-Poor:** Same as for Below Average, plus, depending on the level of the rating and major area that is Poor, immediate attention may be needed to correct conditions having an adverse effect on life, health, safety, and regulatory compliance. An overall Total Score indicates that contract maintenance should be considered Priority to key issues, major building systems and equipment where increasing costs and deferred maintenance are having a direct impact on the immediate survival of the business or the major physical asset. The capabilities for critical assets to perform intended function are being severely limited by current "state of maintenance". Consider immediate contract services as required for business survival and for achieving the core requirements for maintenance services. Determine if the necessary investments for internal maintenance improvements are going to be made. **An overall poor rating requires a close look at the effectiveness of leadership at all levels.**

Figure 5.11 Rating summary comments for results from a Scoreboard for Maintenance Excellence assessment.

1. TOP LEADER'S SUPPORT TO MAINTENANCE & PHYSICAL ASSET MANAGEMENT		
ITEM #	BENCHMARK ITEM Rating: Excellent – 10, Very Good – 9, Good – 8, Average – 7, Below Average – 6, Poor – 5	RATING VALUE
1.	The organization has a strategic plan that is the starting point for development of the asset management and maintenance strategy, policy, objectives and plans	9

Figure 5.12 Example format of benchmark evaluation criteria.

Lesson Learned: As a plant manager of a high-quality hand tool manufacturing plant in the mid-1980s, I was fortunate to be a part of an operation that did have maintenance as a key written part of our plant goals, as listed below:

Plant Operating Goals: Crescent-Xcelite Plant (A Division of Cooper Industries)

1. Aggressive investigation and follow-through of potential new products and markets. Maintain a close working relationship with sales and marketing functions to enable Crescent-Xcelite to react to customer and marketplace changes.
2. New manufacturing technology will be integrated into our manufacturing methods and process. This will allow Crescent-Xcelite to be competitive in the basic areas of quality, price, and service.
3. Maintain technical competence or workforce (internal and external training).
4. Clean, orderly, safe workplace.
5. Create an atmosphere of pride in workmanship (quality awareness and quality improvement programs).
6. Balanced line product flow through manufacturing phases using hard automation or robotics when economically feasible.
7. Control all inventory levels using automated reporting and Materials Requirement Planning (MRP) system and capacity planning.
8. PM to monitor and adjust equipment to eliminate production disruptions.
9. Simplified quick-change tooling to accommodate short runs and reduce Materials Requirement Planning (MRP).
10. Equipment and tooling monitored to achieve 100% first-run quality capability.
11. Create a no-crisis atmosphere for all employees.

In our case, effective PM and PdM was one of our key operating goals for the plant and was one of the factors contributing our quality and service level during a period of intense competition from foreign hand tool manufacturers. Our PM included not only heavy machining equipment, heat treat, plating, molding, and automated forging equipment, but it also included all of the tooling fixtures, cutters, and dies used in a high-quality machining operation. I vividly remember that our contracted vibration analysis service saved us from having a major catastrophic failure with a new automatic side polishing machine

Never, never, never give up: If you do not have a top leader that appreciates and values maintenance as a key contributor to profit and customer service, then do not give up. As Jim

Valvano (the late coach of the 1983 North Carolina State National Champions) and Great Britain's World War II leader Winston Churchill both said, "Never, never, never give up!" Continue the good fight, get the facts for true return on maintenance investments, and continue to educate top leaders that pride in maintenance is a core requirement for total operations success. As a new or experienced planner, you play a key role in leading maintenance forward with proactive maintenance and greater productivity. Know what best practices you need or just the often very plain adjustments needed to make your planning, estimating, and scheduling a reliable process in your operation.

We strongly believe in basic maintenance best practices as the foundation for maintenance excellence. Our improvement process includes all maintenance resources, equipment, and facility assets as well as the craftspeople and equipment operators. It also includes MRO materials management assets, maintenance informational assets, and the added value resource of synergistic team-based processes. Continuous reliability improvement improves the total maintenance operation.

The Scoreboard for Maintenance Excellence concept and the various Scoreboard versions have been used to perform over 200 maintenance assessments/audits, and over 5000 organizations have requested and received copies for their internal use. For example, The Scoreboard for Maintenance Excellence was used by plant maintenance operations at Honda of America after making slight modifications and then using it extensively as a self-assessment to help direct their maintenance strategy. It was then translated into Japanese for presentation to key Japanese executives visiting Honda plants in the United States. Another excellent example as discussed previously is where the Boeing Commercial Airplane group combined elements from this same Scoreboard with their company-wide maintenance goals to develop *The Boeing Scoreboard for Maintenance Excellence*. At Boeing, over 60 manufacturing maintenance work units at region, group, and team levels were then evaluated with structured on-site visits. The use of this comprehensive best-practice guideline, specifically tailored to the maintenance of aircraft manufacturing equipment (and the associated manufacturing and test facilities complexes) across the United States, is still one of the largest internal benchmarking efforts ever undertaken.

A Complete Scoreboard Self-Assessment is Recommended, So "Just do it!": For example, MRO materials management, storeroom operation, and procurement may be an area needing special attention. Shop-level planning and scheduling is often a typical need and can be a primary focus area. Regardless of the different areas creating the obvious concerns and "organizational pain," a short-term, piecemeal approach to an evaluation is not recommended. A complete benchmark evaluation of the total maintenance operation is highly recommended. There are 600 specific evaluation items that are evaluated through direct shop floor interviews, close observations, and review of information or procedures. Each one is important; some provide more value more than others. However, each of the 600 items on *The Scoreboard for Maintenance Excellence* is part of establishing a solid foundation for profit- and customer-centered maintenance. Long-term continuous reliability improvement is also a very important connecting link that ensures that we consider all maintenance resources in our improvement process. The journey is not **to maintenance excellence** but rather a journey that **continuously moves toward maintenance excellence**. Another key point is all about what this book encourages you to do with all topics in this book. That key point borrowed from the Nike commercial says to "Just do it."

Several organizations such as TMEII stand ready to support your mission-essential plant maintenance operation with an assessment performed by well-qualified staff. Although a self-assessment has many benefits, we believe an assessment conducted by an outside resource provides a greater sense of the "big picture" in terms of objectivity and completeness. Regardless of your situation, it is important that you do something to "determine where you are." Should you want to begin with an internal self-assessment of maintenance, here are some guidelines to consider when using *The Scoreboard for Maintenance Excellence*. You can do a self-assessment at any time by using Appendix A in this book. See Figures 5.13 and 5.14 for key steps for your self-assessment.

Obtain Leadership Buy-In

1. Establish a firm commitment from the organization's top leadership for conducting a total maintenance operation assessment. Figure 5.2 illustrates this important key to your success. Not only must the commitment be from top leadership, it must also begin with the maintenance leader.
2. Establish a firm commitment from the organization's top leadership to take action based on your current needs. Every organization is different and may have a need to focus on one or more specific best-practice categories. For example, in this case, the self-assessment can

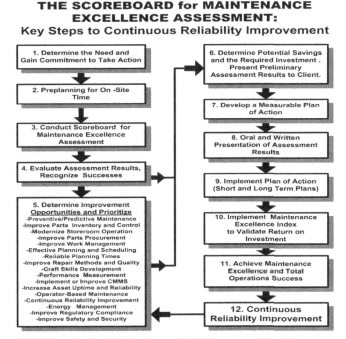

Figure 5.13 Key steps for the Scoreboard for Maintenance Excellence assessment.

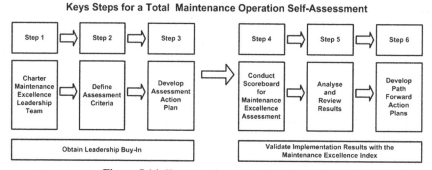

Figure 5.14 Key steps for your self-assessment.

focus on areas such as improving storeroom operations or planning and scheduling processes. However, the assessment must take a complete look at all Scoreboard categories and get a complete "where we are now" picture of the total maintenance process.

3. Maintenance leaders must be brave enough and prepared to share good news and bad news on the basis of results of the evaluation to top leaders. When we perform an assessment, we strive to make it a very positive process for all involved. There will always be successes to highlight and really no bad news; rather, we focus on opportunities.

Step 1: Charter Maintenance Excellence Strategy Team

(a) Establish a maintenance excellence strategy team to guide and promote improved maintenance practices within your organization whether at single or multiple sites

(b) Use a team-based approach with a cross-functional evaluation team specifically chartered for conducting and preparing the results of your evaluation

(c) Have at least one team member (or designated resource person) with a solid background and knowledge in each of the 38 evaluation categories

(d) Consider bringing on board third-party outside support to be a part of your team and the assessment process.

Step 2: Understand the Evaluation Categories and Evaluation Criteria

(a) Gain complete understanding of each of the 600 evaluation items

(b) Modify existing evaluation criteria as required for your organization

(c) Define the importance and weighted value of evaluation categories

(d) Add or delete evaluation criteria as required for your unique operation

(e) Ensure that all team members understand the scoring process for each evaluation item

(f) Ensure that consistency in scoring each evaluation category is applied using standard guidelines

Step 3: Develop Assessment Action Plan

(a) Determine baseline information requirements, persons to interview, and observations needed before the start of the evaluation. A pre-assessment checklist follows:

Pre Assessment Checklist for Baseline Information

The following checklist is not all inclusive of the information required for an assessment. It does represent very important areas that we try to get before performing an assessment. All of the data/information listed in the following checklist may not be readily available. As much information as possible should be assembled and reviewed before the start date of the assessment. When we perform an assessment, this step allows us to gain a better understanding of the client's operation before the on-site visit and helps save a lot of time when on site.

Organization Charts/Job Descriptions

- Mission statement/quality statement for your overall organization
- Mission statement for your maintenance operation
- Staff directory of personnel and contact information/e-mail, etc
- Organization chart for your plant/facility
- Maintenance organization charts by each craft area
- Head counts in each craft area
- Position descriptions of key maintenance leaders—managers, supervisors, foremen, maintenance engineers, planners, team leaders, storeroom supervisor, etc.
- Craft job descriptions—sample from each craft area

Craft Skills Development

- Information on craft training completed in past several years (i.e., in-house programs, vendor)
- Seminars or ongoing apprenticeship programs, etc.
- Results of past craft skills assessments or employee surveys on training needs, etc.

Craft Labor Rates/Overtime History

- Average hourly rate by craft area
- Average fringe benefit percentage factor for your organization
- Overtime rates
- Past 3 to 5-year history of overtime
- Overtime hours (by craft area, if available)
- Overtime payroll costs (by craft area, if available)

Maintenance Budget and Cost Accounting

- Total maintenance budget (3 years)
- Total craft labor cost (3–5 years)
- Total parts/materials (3–5 years)
- Contract maintenance costs (3–5 years) (by type of service if available)
- Copy of current maintenance budget
- Copy of previous year's maintenance budget
- Procedures for charging craft labor and parts/materials to equipment history or to maintenance budget accounts
- Procedures for monitoring equipment/process uptime

PM, PdM, RCM, and Total Productive Maintenance (TPM) Processes

- Sample lubrication services checklists or charts
- Sample PM checklist/instructions

- List of equipment having PdM services for vibration analysis, oil analysis, infrared or other technologies
- Any summary results of major success with PM/PdM and reliability improvement at the plant/facility site
- Summary of experiences with RCM-type processes
- Summary of experiences with TPM-type processes

Maintenance Storeroom and MRO Purchasing Operations

- Total dollar value of current maintenance-related inventory (ABC classification if available)
- Inventory dollar value of critical spares/insurance-type items
- Inventory dollar value of all other items
- Estimated value of items not on inventory system
- Total number of stocked items (stock-keeping units)
- Total number of critical spares/insurance items
- Total stock items—all other parts not classified as critical spares
- Copy of storeroom procedures for
  - Purchase requisitions/purchase orders
  - Additions to stock/establishing stock levels
  - Issuing/receiving
  - Receiving requirements for incoming quality validation
  - Obsolete parts
  - Parts/material inventory classification
- Results of most recent physical inventory or cycle count results
- Accuracy level, write-offs, or adjustments
- Copy of storeroom catalog (if online will review on-site)
- Information on vendor stocking plans and vendor partnerships currently in place

Work Orders and Work Control

- Copy of work order currently being used and priority system description
- Copy of work order/work control procedures
- Current backlog by craft area (if available)
- Planning and Scheduling Procedures
- Work management procedures
- Time-keeping methods (operations personnel and maintenance personnel)

CMMS

- System name and date initially installed
- CMMS system administrator's name and contact information
- Names of other key staff with responsibilities for data integrity:
  - Parts information
  - Equipment/asset information
  - PM/PdM procedures
  - Maintenance budgeting and service charge backs
  - Shop-level planning and scheduling
  - Project-type planning and scheduling
- Your primary CMMS vendor support person and contact information

Maintenance Performance Reports

- Copy of any reports or current information that is being used to evaluate maintenance performance
- Summary of key performance indicators that you feel are needed
- Operations performance report
- Copy of any reports or information used to evaluate operations performance that uses maintenance-related data (i.e., equipment uptime [availability])
    (b) Develop schedule and implementation plan for the assessment
    (c) Develop and implement a communication plan within the organization to inform all about the process

Step 4: Conduct Assessment of Total Maintenance Operation
    (a) Assign team members to specific evaluation categories (ideally, in two-person teams for each category)
    (b) Conduct kickoff meeting, firm up specific interview and observation schedules, etc.
    (c) Conduct the assessment, record observations, and assign scores to each evaluation item
    (d) Ensure CMMS is an effective business management tool for maintenance

Step 5: Analyze, Review, and Present Results
    (a) Review all scoring for consistency
    (b) Develop final results of the assessment and document in a written report
    (c) Determine strengths/weaknesses and priorities for action
    (d) Define potential benefits, either direct or indirect savings, or gained value from existing resources
    (e) Gain internal team consensus on methodology for determining benefits and the value and type of savings
    (f) Present results to top leaders with specific benefits and improvement potential clearly defined
    (g) Refine results based on feedback from top leaders
    (h) Gain commitment from top leaders for investments to implement recommendations

Step 6: Develop Path Forward for Maintenance Excellence
    (a) Develop a strategic plan of action for implementing best practices
    (b) Define tactical plans and operational plan of actions
    (c) Define key performance measures, especially those that will validate projected benefits
    (d) Implement methodology to measure performance and results
    (e) Measure benefits and validate Return on Investment (ROI)
    (f) Maintain a continuous reliability improvement process (i.e., repeat assessment process)
        – Follow-up initial use of Scoreboard for Maintenance Excellence with periodic assessments every 6–9 months
        – Follow-up initial use of CMMS benchmarking system
        – Continuously validate results with Maintenance Excellence Index

Invest in External Resources: It is extremely important to know where your organization stands on physical asset management and maintenance issues and challenges so that it can quickly identify areas for improvement. Every delay along the way delays receiving the potential benefits and added value. Self-assessments are recommended and very good starting points when nothing else is available for using external support. However, a more comprehensive, objective assessment performed by external consulting resources (or possibly qualified corporate-level staff with decades of maintenance-focused expertise) is highly recommended. In the long run, external resources will provide additional justification and measurable results. Therefore, take a look at using an external resource to support this essential first step after your organization makes the initial commitment.

Typical Project Plan of Action: The recommended path forward offers an excellent opportunity for immediate results at the pilot site plus the time to learn from this assessment before conducting future assessments. TMEII highly recommends having a maintenance excellence strategy team in place to provide overall leadership, support, and direction. The measurement of results ensures that initial projections of benefits are achieved and that the ROI for this pilot effort exceeds expectations. Included in Figure 5.15 is a typical project schedule in which the self-assessment includes two sites.

Recommended Next Steps after the Scoreboard for Maintenance Excellence Assessment

Document Assessment Results: After *The Scoreboard for Maintenance Excellence* assessment has been completed, a written and oral report to top leaders will document the results with a presentation of recommendations and a plan of action. Key areas of the report presentation will help you to

- Determine strengths/weaknesses and priorities for action
- Benchmark your CMMS installation
- Maximize benefits of CMMS
- Develop maintenance as a profit center
- Define potential savings
- Develop recommended plan of action (and implement)

Path Forward Action Items	Weeks after Project Initiation					
	Pre-Work	Week 1	Week 2	Week 3	Week 4	Week 5
Step 1: Define Impact on Quality, Capacity and Customer Service	■					
Step 2: Establish Client Scoreboard	■					
Step 3: Conduct Pilot Assessments						
• Site One		■				
• Site Two			■			
Step 4: Define Specific Opportunities		■	■			
Step 5: Define Potential Benefits				■	■	■
Step 6: Plan for Pilot Implementation					■	■
Step 7: Establish Measure Pilot Results				■		
Step 8: Refine Client Scoreboard					■	■
Step 9: Plan Additional Assessments						■
Step 10: Present Pilot Assessment Results					■	■
Step 11: Begin Other Assessments					After Week 5	

Figure 5.15 Typical project schedule in which the assessment includes two sites.

- Develop method to measure and validate results
- Initiate a maintenance excellence index

Determine Strengths/Weaknesses and Priorities for Action: After an objective assessment is completed, it is very easy to identify strengths and weaknesses, which then leads to defining the priorities for action. In some operations it is very often back to the basics, such as

- PM has been neglected; no time to do it
- The understanding of predictive technologies is limited
- The application of continuous reliability improvement was never initiated
- The parts storeroom was never given the proper attention it needed
- The accountability for craft time is not being done
- The charge back to the customer was not done or is incomplete
- A reactive, fire-fighting repair strategy is in place
- Valuable craft time is wasted (e.g., chasing parts/materials, waiting, unplanned work)
- There is never time to do the job right the first time
- The asset uptime is uncertain and the manufacturing operation is not reliable
- Quality is inconsistent because of maintenance processes
- There is never time for craft training
- The CMMS was purchased as "the solution" and not "the tool."
- The existing CMMS functionality is not being fully used

Very often, the CMMS takes the hit as the cause of all of the weaknesses. CMMS is blamed for not being able to do this and that, causing all types of problems and extra work. This attitude will generally always be the case when the CMMS was purchased as "the solution" and not "the tool". The bottom line here is that most systems are underutilized, and when fully used with all of their intended functionality they will serve their primary IT purpose.

Therefore, just as we can benchmark a total maintenance operation and its best-practice application with the 38-category Scoreboard for Maintenance Excellence, we can also benchmark the CMMS that is in place. We really need to evaluate the CMMS and its current application as to its effectiveness in supporting all best practices. Is your CMMS enhancing current and future best practices or not? Are we getting maximum value from this IT investment? Is your CMMS truly a maintenance business management system? How can we improve the current use of the system? In Chapter 8, "Planner Review of a Critical Planning Tool Evaluating Your CMMS Systems Effectiveness," we will answer and take action on these key questions for reliable maintenance planning, estimating, and scheduling.

Planners Must Understand Productivity and How Reliable Maintenance Planning, Estimating and Scheduling (RMPES) Enhances Total Operations Excellence

6

This chapter is extremely important for the planner to understand how they impact productivity improvement within the craft workforce as well as across the total operation. The profit and customer-centered maintenance operation must clearly understand productivity and be committed to measuring craft productivity. All the best practices we discuss in this book contribute either directly or indirectly to improving craft productivity and physical asset productivity. I know for sure that it applies to the profit and customer-centered maintenance contractor striving to meet contract terms and conditions. Defining overall productivity from the total maintenance process of the total operations can be a slippery task. Attempting to increase productivity can be even more difficult if managers mistake efficiency for productivity.

We can be very efficient doing the wrong things such as firefighting and reacting to emergency repairs. A focus on increasing work order completion rates may sacrifice quality of repair work. Cutting costs by specifying cheap, prone-to-fail equipment or parts yields the high cost of low-bid buying and maybe achieving short-term budget requirements. **Doctors and lawyers practice medicine and law; maintenance and good engineering need to get it right the first time!**

Today's economic climate requires that we define and then improve maintenance and engineering productivity. Every employee must take ownership in the organizational mission and the maintenance mission. One manager for the decision support for operations and maintenance program at the *Pacific National Laboratory* in Richland, Washington, categorizes productivity into four levels: (1) **Survival**: In a low-productivity environment, the goal is survival. Chronic operations and maintenance problems will plague such an organization, and low reliability or outright equipment failures characterize this category. (2) **Adequacy**: Uncertainty might characterize this category, but the department is keeping equipment running. Low efficiency and low productivity are classic traits of this category. (3) **Accuracy**: Departments in this segment are secure in their knowledge of operating and maintaining their facilities but would like better performance measurements and want to know how their operations and maintenance affect facility processes. (4) **Optimized**: These departments know the slope of their performance curve. They search for ways to optimize the state of an already effective maintenance process.

What is productivity? One definition of productivity is that it is a measure relating to quantity or quality of output compared to the inputs required to produce it. When

Reliable Maintenance Planning, Estimating, and Scheduling. http://dx.doi.org/10.1016/B978-0-12-397042-8.00006-1

we look at just three elements of total plant productivity, production labor, production equipment productivity, and craft labor productivity all have the same three key factors in common. The same applies to craft employees performing repair and preventive maintenance (PM) to support production operation customer. My definition includes three key factors for productivity:

1. The Effectiveness Factor—*Doing the right things.*
2. The Efficiency or Performance Factor—*Doing the right things and giving the best personal performance as possible.*
3. The Quality Factor—*Doing the right things giving the best personal performance as possible with high quality results.*

These three factors apply to production labor productivity, to asset productivity overall equipment effectiveness (OEE), and to craft labor productivity as shown in Figure 6.1.

Supporting the improvement of craft productivity can easily provide a return on investment of six equivalent craft employees to one qualified planner. In addition, craft labor productivity can be measured for all three elements. Surveys consistently show that wrench time (craft utilization (CU)), one element of craft productivity within a reactive, firefighting maintenance environment, is in the range of 30–40%. A study using work sampling of a major corporation's maintenance operation revealed 26%. My surveys from speech and seminar attendees over the past 15 years shows some people that think their true wrench time is 15%, 20%, or 25% and much lower than 30%. Therefore, if your baseline pure wrench time baseline is conservatively at 40%, this means that for a 10-h day, there are only 4 h of actual hands-on, wrench time. You can do the math on other levels, 10%, 20%, or 30%. **Craft workforce should not**

PRODUCTIVITY: THREE KEY FACTORS			
Productivity Factors	Production Employees	Production Equipment	Craft Employees
Effectiveness	% Time on Direct Labor Adding Value	% Availability to Add Value	% True Wrench Time or Craft Utilization
Efficiency	% Performance Against a Standard Time	% Performance Against Design Speed	% Craft Performance Against a Standard Planning Time
Quality	% Good Compared to Total Produced	% Good Compared to Total Produced	% Good Repairs Compared to Total Repairs
	Standard Labor Cost Variance	*Overall Equipment Effectiveness (OEE)*	*Overall Craft Effectiveness (OCE)*

Figure 6.1 Three key productivity factors compared for three elements of total plant productivity. Note: Total plant productivity would also include material variances, actual throughput compared to capacity, and net profit.

take all the blame: Typically, most of the lost wrench time is not the fault of the craft workforce. Lost wrench time can be attributed to the following reasons:

1. Running from emergency to emergency; a reactive, firefighting operation.
2. Waiting on parts and finding parts or part information.
3. Waiting on other information, drawings, instructions, manuals, etc.
4. Waiting for the equipment to be shut down.
5. Waiting on rental equipment to arrive.
6. Waiting on other crafts to finish their part of the job.
7. Travel to/from job site.
8. Lack of effective planning and scheduling.
9. Make ready, put away, clean up, meetings.
10. Troubleshooting, however, is another story: (1) With a "hit-and-miss approach" it can be wasted time. (2) But it can also be scheduled by the planner, performed by a qualified person to really determine the scope of work and parts needed so a true planned job can occur.

There is a question by some maintenance gurus about whether we can or should measure craft productivity directly. My answer to this is a resounding yes! A reliable planner will be able to establish reliable repair times for many reasons other than measuring craft labor performance. We will describe in detail a method that is easy to use and acceptable to the craft workforce that we call the ACE (a consensus of experts) Team Benchmarking Process™ to be discussed in detailed in Chapter 15, *Developing Improved Repair Methods plus Reliable Maintenance Planning Times with the ACE Team Process.* Craft resources are becoming more difficult to find in many areas and there is a true crisis and shortage of craft labor. One question we want to answer for you is, "How can we get maximum value from craft labor resources and achieve higher craft productivity?"

First, we must be able to measure it: Maintenance operations that continue to operate in a reactive, run-to-failure, firefighting mode and disregard implementation of today's best practices will continue to waste their most valuable asset and very costly resource—craft time. Often true wrench time is within the 20–30% range for operations covering a wide geographic service area. An example is PM and maintenance of a countrywide pipeline like in Oman. With best practices such as effective maintenance planning/scheduling, preventive/predictive maintenance, more effective storerooms and parts support, delivery to job site, crafts skill training, and the best tools, we can contribute to proactive, planned maintenance, and more productive hands-on "wrench time." The sticky question of craft productivity measurement and improvements in union plants often comes up. Consistent craft productivity measurement and improvements in union manufacturing in U.S. plants must occur.

Union plants cannot increase hourly rates higher and higher with little or no productivity. This is not mathematically possible forever. I have said before that we cannot perform maintenance off shore, except for maybe troubleshooting via online monitoring/operation of process control systems and worst case, moving the plant. Maybe we can send items offshore for rebuilding, etc. What are our options when current productivity of both production operators and crafts people do not increase as union labor rates continuously go higher and higher? *At some point, higher wages without productivity gains will lead to plant closings due to profits and offshore relocation for cheaper*

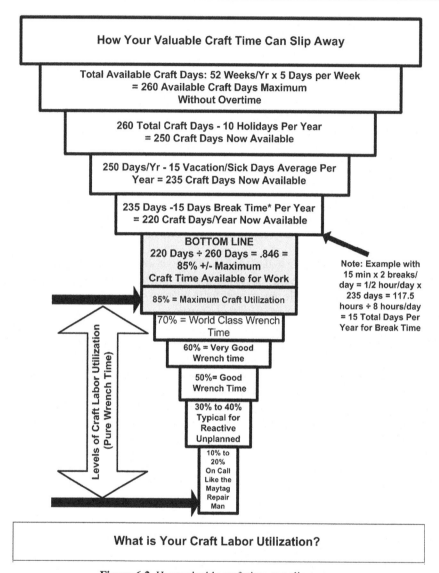

Figure 6.2 How valuable craft time can slip away.

labor. We might replace union maintenance with contractor maintenance or even move plants to right-to-work states. After seeing the Cooper Tool Group move six unionized plants from the North (Plumb, Crescent, Wiss, Lufkin, Weller, and Nicholson saw and the Nicholson file plant) to the South, there must be significant cost improvement. Craft productivity can be measured and improved. In addition, if we can do it for in-house maintenance, we can surely do it for maintenance contractors as well. Therefore, what your wrench time or CU is can be shown in Figure 6.2. Figure 6.3 illustrates craft labor resources and key factors for keeping them as our most important maintenance resource.

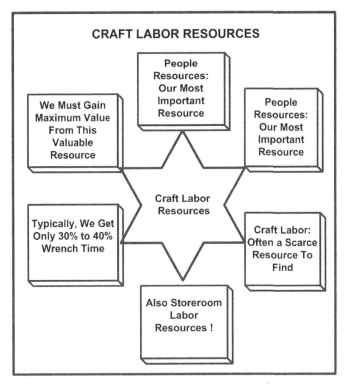

Figure 6.3 Craft labor resources—our most important maintenance resource.

With a good planning, estimating, and scheduling process, significant gained value is available from increased wrench time or hands-on tools time. An improvement in actual wrench time from 40% to 50% represents a 25% net gain in craft time available equivalent and a significant gained value. When we are able to combine gains in wrench time with increased craft performance (CP) when doing the job, we increase our total gain in craft productivity. Measuring and improving overall craft productivity should be a key metric/KPI to justify effective planning and scheduling, CMMS, and other investments for maintenance improvement. Increased reliability that increases asset uptime and throughput is likewise very important.

Before further addressing overall craft effectiveness, I want to cover three types of cost improvements:

1. Direct cost savings. They come from direct savings in standard labor and materials cost. For maintenance, this can come via staff reductions, by using contractors, or parts procurement of like quality parts at reduced costs.
2. Cost avoidances. A good maintenance example is when PM/PdM/CBM detects a problem when it can be a planned repair before a catastrophic, high consequence failure occurs. With a little estimating between cost of planned work versus possible cost of major unexpected failure, this extra can give you valid cost avoidances and return on investment (ROI) for PM/PdM/CBM activities.
3. Gained Value. This one is the value of (1) gained labor productivity or (2) value of gained production. It can be said simply as doing more with existing resources, i.e., gained value.

Planners, pay attention, as we will now look at a new term, *overall craft effectiveness* (OCE). As we saw previously, OCE has the same basic factors as (OEE; availability, performance, and quality) but applied to craft productivity and people resources as compared to OEE, focused upon critical physical assets and physical asset productivity.

The profit and customer-centered maintenance leader (in house or contractor) must consider total asset management in terms of improvement opportunities across all maintenance resources. There are many questions to be asked about how we can improve the contribution that each of these six resources (from Chapter 5) makes toward your goal for maintenance excellence:

- Physical resources: equipment and facilities.
- People resources: craft labor and equipment operators.
- Technical skill resources: craft labor that is enhanced by effective training.
- Material resources: maintenance repair operations (MRO) parts and supplies.
- Information resources: useful planning, scheduling, and reliability information, and not a sea of useless data.
- Hidden resources: The seemingly magical synergy of teamwork as a true people asset multiplier.

Measuring and improving OCE must be one of many components to continuous reliability improvement process and total asset management. OCE includes three key elements very closely related to the three elements of the OEE Factor.

Overall Equipment Effectiveness

We must clearly understand the elements of OCE and how the OCE factor relates to better use of our craft workforce. Most everyone recognizes and understands the world-class metric OEE that measures the combination of three elements for the physical asset: equipment asset availability, performance, and quality output. OEE is about measuring asset productivity. The calculation of OEE is shown in Figure 6.4.

The OCE factor focuses upon craft labor productivity and measuring/improving the value-added contribution that people assets make. Just like OEE, there are three elements to the OCE factor:

- Effectiveness factor: CU for OCE *and Asset availability for OEE.*
- Efficiency factor: CP for OCE *and Asset performance for OEE.*
- Quality factor: Craft service quality (CSQ) for OCE *and Quality of asset output for OEE.*

The OEE Factor = % Availability(A) x % Performance(P) x % Quality(Q)

An OEE Factor of 85% is recognized as world-class

Therefore OEE of 85% requires at least the 95% level for each of the 3 elements:

So if OEE = A x P x Q then if each factor is 95% or .95

OEE = .95 x .95 x 95 ≅ 85%

Figure 6.4 Overall craft effectiveness.

All three elements of OCE can be as well defined as all three of the OEE factors. We will now review the three key elements for measuring OCE and see how they very closely align with the three elements for determining the OEE factor for equipment assets. Figure 6.5 provides a comparative summary and Figure 6.6 defines how OCE is calculated.

OCE focuses upon your craft labor resources: I strongly believe in basic maintenance best practices as the foundation for maintenance excellence. This is what I call continuous reliability improvement (CRI). CRI is about maintenance business process improvement that includes opportunities across all maintenance resources: equipment and facility assets, as well as people resources—our crafts workforce and equipment operators. CRI must also include MRO materials management assets, maintenance informational assets, and the added value resource of synergistic team-based processes. CRI improves the total maintenance operation and can start with measuring and improving OCE.

The Maintenance Excellence Institute International advocates, supports, and clearly understands the need for reliability-centered maintenance (RCM) and total productive maintenance (TPM) types of improvement processes. But out on the shop floor, we see today's trend toward forgetting about the basics of "blocking and tackling" while going for the long touchdown pass with some new "analysis paralysis" scheme. RCM and root cause analysis (RCA) are not really analysis paralysis when done correctly with true information and when they are not based upon "precisely inaccurate" data.

Build upon the basics: Your approach must be built upon the basics and then include, but go well beyond, the traditional RCM/TPM approaches to CRI (Figure 6.7).

Maintenance excellence can start with PRIDE in maintenance: Do not take a piecemeal approach that focuses only RCM-type processes on physical assets and equipment resources. Often the maintenance information resource piece, among others, is a missing link for the successful RCM-type process. RCM alone can often become analysis paralysis with no data or bad data. Your approach should be about improvement opportunities across all maintenance resources. There of course must be priorities as to where we start and where we make investments. I have been asked numerous times what would I want first in starting a new maintenance operation? I reply that after finding the best craft people available (with PRIDE in maintenance) I would establish effective planning, estimating and scheduling, an effective PM/PdM program, and a responsive parts storeroom and procurement process.

Overall Craft Effectiveness (OCE)	Overall Equipment Effectiveness (OEE)	Elements of OEE and OCE
1. Craft Utilization or Pure Wrench Time (CU)	Asset Availability/Utilization (A)	Effectiveness
2. Craft Performance (CP)	Asset Performance (P)	Efficiency
3. Craft Service Quality (CSQ)	Quality of Asset's Output (Q)	Quality

Figure 6.5 Summary comparisons of overall craft effectiveness (OCE) and overall equipment effectiveness (OEE).

The OCE Factor = % Craft Utilization (CU) x % Craft Performance (CP) x % Craft Service

Quality (CSQ)   Therefore OCE = % CU x % CP x % CSQ

Typically CU and CP can be easily measured.

Craft Service Quality (CSQ) is somewhat harder to measure and can be more subjective.

NOTE: Later we will see how all three elements of OCE can be measured and

how all three contribute to increased craft productivity

Figure 6.6 Calculating overall craft effectiveness (OCE).

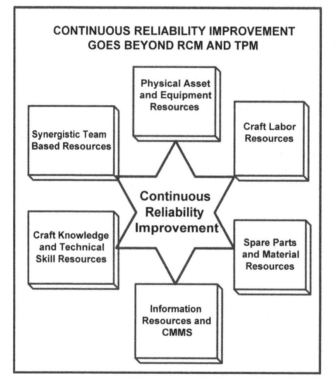

Figure 6.7 Continuous Reliability Improvements (CRI).

For example, with the crafts labor resource, we can easily measure the three elements of OCE, as we will see later. However, we can start the journey toward maintenance excellence by just helping to achieve PRIDE in maintenance from within the crafts workforce and among maintenance leaders at all levels and PRIDE in maintenance around the world as illustrated in Figure 6.8.

A very important question for the crafts workforce: I have seen maintenance operations and talked to crafts people all around the world, starting in South Vietnam in 1970. Almost all attendees at hundreds of my workshops also agree that their crafts

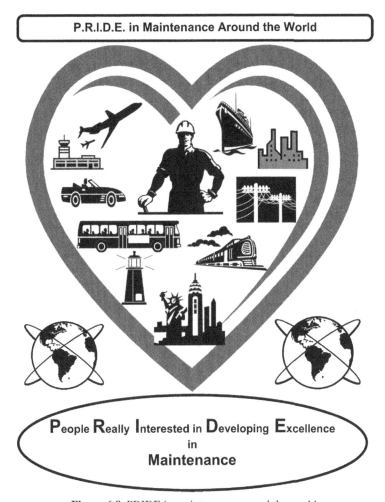

Figure 6.8 PRIDE in maintenance around the world.

people want to do a good job and be appreciated for what they do to help achieve the mission of the organization. In addition, when asked the following question, I have received positive responses from almost 100% of the thousands of dedicated professional crafts people I have interviewed. Dedicated union leaders and union crafts people are also included in my personal sample.

The question I always ask is, **"How would you do this job, lead this crew, or lead maintenance if it were in fact your own maintenance business?"** What if we could get attitudes that are more positive plus action from *all our crafts* focused on this important question? I think you as maintenance leaders and especially planner/schedulers can do it with a combination of many things from this book. So again I ask you to apply what you think will work and then test other ideas that you may not fully agree with right now from previous chapters.

Your own internal people can add greater value to your maintenance operation with a profit and customer-centered attitude about their job and the profession of maintenance. We feel strongly that maintenance excellence begins with PRIDE in maintenance. We will later see "How OCE Impacts Your Bottom Line" and how we can measure and improve the productivity of two important maintenance resources: craft labor as well as contractor performance expectations.

Remember that craft labor is a terrible thing to waste: How to improve OCE is a very key question we need to answer. Getting maximum value from craft labor resources and higher craft productivity requires measurement and as well as the big picture of knowing where you are now with a Scoreboard assessment. Maintenance operations that continue to operate in a reactive, run-to-failure, firefighting mode and disregard implementation of today's best practices will continue to waste their most valuable asset and very costly resource—craft time. Typically, due to no fault of the craft workforce, surveys and baseline measurements consistently show that only about 30–40% of an 8-h day is devoted to actual, hands-on wrench time. It is very important to understand, "How your valuable craft time can slip away," as illustrated in. Best practices such as effective maintenance planning/scheduling, preventive/predictive maintenance, more effective storerooms, and parts support all contribute to proactive, planned maintenance and more productive hands-on "wrench time." Measuring and improving OCE must be one of many components to CRI process and total asset management. OCE includes three key elements very closely related to the three elements of the OEE factor (Figure 6.9), which we will discuss next in more detail).

Craft Utilization

CU: The first element of the OCE Factor is CU or pure wrench time. This element of OCE relates to measuring how **effective** we are in planning and scheduling craft resources so that these assets are doing value-added, productive work (wrench time). Effective planning/scheduling within a proactive maintenance process is one important key to increased wrench time and CU. It is also about having an effective storeroom with the right part, at the right place in time to do scheduled work with minimal nonproductive time on the part of the crafts person or crew assigned to the job. Your valuable craft time can just slip away to a level of 30–40% or below when crafts service a large geographic area—a large plant, refinery, or facilities complex.

Overall Craft Effectiveness (OCE)	Overall Equipment Effectiveness (OEE)	Elements of OEE and OCE
1. Craft Utilization or Pure Wrench Time (CU)	Asset Availability/Utilization (A)	Effectiveness
2. Craft Performance (CP)	Asset Performance (P)	Efficiency
3. Craft Service Quality (CSQ)	Quality of Asset's Output (Q)	Quality

Figure 6.9 The three elements of overall craft effectiveness (OCE).

Pure wrench time is just that and does not include time caused by the following:

1. Running/traveling from emergency to emergency in a reactive, firefighting mode.
2. Waiting on parts issue, actually finding parts, or getting parts information.
3. Waiting on other asset info, asset drawings, repair instructions, documentation, etc.
4. Waiting for the equipment or asset to be shut down to begin work with all necessary risk assessments.
5. Waiting on rental equipment or contractor support to arrive at job site.
6. Waiting on other crafts or crews to finish their part of the job.
7. Traveling to/from job site.
8. All other make-ready, put-away, or shop cleanup time.
9. Meetings, normal breaks, training time, and excessive troubleshooting due to lack of technical skills.
10. Lack of effective planning and scheduling.

CU (or wrench time) can be measured and expressed simply as shown in Figure 6.10 as the ratio of:

Improve wrench time first Go on the attack to increase wrench time in your operation even if you do nothing to improve the other two OCE factors: CP and the CSQ level. As we will see in the following examples, very dramatic and significant tangible benefits can be realized with just focusing on increasing wrench time. Improvement of 10–30% points from your current baseline of wrench time can typically be expected. In addition, often this can be achieved just from more effective maintenance planning and scheduling. Let us now look at several examples showing the value of CU improvement within a 20-person workforce with an average hourly rate of $18.00 and see the significant benefits that a 10% increase in CU can provide.

Gained value of 10% in wrench time: What if through better planning and scheduling, good parts availability, and having equipment available to fix it on a scheduled basis, we are able to increase actual wrench time by just 10% from a baseline of 40%? What is the gained value to us if we get a wrench time increase across the board for a 20-person crew being paid an average hourly rate of $18/h? First let us look at what it is really costing us at various levels of wrench time.

Total Craft Hours Available and Annual Craft Labor Costs for Crew of 20 Crafts

$$20 \text{ Crafts} \times 40 \frac{h}{week} \times 52 \text{ weeks/year} = 41,600 \text{ Craft Hours Available}$$

41,600 Craft Hours @ $18/h = $748,800 Craft Labor Cost/Year

CU% =	Total Productive (Wrench Time)
	Total Craft Hours Available & Paid x 100

Figure 6.10 Craft utilization (CU) calculation.

Wrench Time and Actual Costs Per Hour at Various Levels of CU

Example: What if baseline for wrench time is 40%? With effective planning and scheduling we can achieve at a minimum of a 10-point improvement in CU from our current baseline. Starting from a baseline of 40% and increasing wrench time up to a level of 50%, we in effect get a 25% increase in craft capacity for doing actual work.

- **Total hours gained in wrench time: 4160 h**

  20,800 h @ 50% − 16,640 h @ 40% = 4160 h gained

- **Total gain in equivalent number of crafts positions: 5**

$$\frac{4160 \text{ Hours Gained}}{832 \text{ Average Wrench Time @ } 40\% \text{ Craft Utilization}} = 5 \text{ Equivalent Craft Positions}$$

- **Total gained value of 5 equivalent positions: $187,200**

$$5 \text{ equivalents} \times 40\frac{h}{week} \times 52\frac{weeks}{year} \times \frac{\$18.00}{h} = \$187,200 \text{ Gained Value}$$

Valuable craft time can be regained: For the 20-person craft workforce, just a 10% improvement up to 50% wrench time is 4160 h of added wrench time. This gain represents a 25% increase in overall craft labor capacity. The maintenance best practice for planning and scheduling requires a dedicated planner(s). An effective maintenance planner can support and plan for 20 to 30 crafts positions. With only a 10% increase in CU for a 20-person craft workforce, we can get more than a 5 to 1 return to offset a maintenance planner position.

Example B: What if wrench time is 30%? For many operations, wrench time is only about 30% and sometimes below 30%. Again, with effective planning and scheduling, good PM/PdM and parts availability we can eliminate excessive non-wrench time. An improvement of at least 20 points in CU is very realistic. If we begin from a baseline of 30% up to a level of 50%, we are in effect getting a 67% percent increase in craft capacity for actual hands-on work.

The gained value of going from 30% up to 50% wrench time:

- **Total hours gained in wrench time: 8320 h**

  20,800 h @ 50% − 12,480 h @ 30% = 8320 h gained

- **Total gain in equivalent number of crafts positions: 13**

$$\frac{8320 \text{ Hours Gained}}{624 \text{ Average Wrench Time Hours @ } 30\%} = 13.3 \text{ Equivalent Craft Positions}$$

- **Total gained value of 13.3 equivalent positions: $497,952**

$$13.3 \text{ equivalents} \times 40\frac{h}{\text{week}} \times 52\frac{\text{weeks}}{\text{year}} \times \frac{\$18.00}{h} = \$497,952 \textbf{ Gained Value}$$

Valuable craft time can be regained: Tremendous opportunities are available for the 20-person craft workforce with wrench time currently in the 30–40% range. Just a 10–20% improvement up to 50% wrench time can be from 4000 to 8000 h of added wrench time. This gain represents a 25–67% increase in overall net craft labor capacity. There is one important best practice needed to help you regain valuable craft resources. The maintenance best practice for planning and scheduling requires dedicated planners that are effective maintenance planners/schedulers that can support and plan for 20 to 30 crafts positions.

Use your CMMS/EAM as a mission-essential information technology tool that supports planning and scheduling, better MRO materials management, and effective preventive/predictive maintenance. They are three best practices for improving craft wrench time. Bottom line results that give us 5–13 more equivalent craft positions and up to $500,000 in gained value of more wrench time with existing staff can be dramatic proof that internal maintenance operations can be profit centered.

Craft Performance

CP: The second key element affecting OCE is CP. This element relates to how **efficient** we are in actually doing hands-on craft work when compared to an established planned time or performance standard. CP is expressed as the ratio of:

$$CP\% = \frac{\text{Total Planned Time (Hours)}}{\text{Total Actual Craft Hours Required}} \times 100$$

CP is directly related to the level of individual craft skills and overall trades experience as well as the personal motivation and effort of each craftsperson or crew. Effective craft skills training, technical development, and PRIDE in maintenance all contribute to a high level of CP.

CP calculation: For example, the planned time for a minor overhaul or planned repair is 10 h based on a standard procedure with parts list, special tools, permits required, etc.

- If the job is completed in 12 h, then $CP = \dfrac{10}{12} \times 100 = 83\%$
- If the job is completed in 9 h, then $CP = \dfrac{10}{9} \times 100 = 111\%$

An effective planning and scheduling function requires reasonable repair estimates. They are what I call reliable planning times needed for scheduling, and they should be established for as much maintenance work as possible. Because maintenance work is not highly repetitive, the task of developing reliable planning times is more difficult. However, there are a number of methods for establishing planning times for maintenance work including:

- **Reasonable estimates**: A knowledgeable person, either a supervisor or planner, uses their experience to provide their best estimate of the time required. This approach does not scope out the job in much detail to determine method nor outline job tasks or special equipment needed. However, it can have value when there is no reliable work measurement method in place.
- **Historical data**: The results of past experience are captured via the CMMS or other means to get average times to do a specific task. Over time, a database of estimated time is developed that can be updated with a running average time computed for the tasks. I often call this "hysterical data" due to inconsistencies in reporting time, actual location of work, and delays in getting parts or equipment access, etc.
- **Predetermined standard data**: Standard data tables for a wide range of small maintenance tasks have been developed. Standard data represents the building blocks that can then be used to estimate larger, more complex jobs. Each standard data table provides what the operation is, what is included in the time value, and the table of standard data time for the variables that are included. The universal maintenance standards (UMS) method used back in the 1970s and developed by the U.S. Navy and H. B. Maynard Inc. represents a predetermined standard data method. We used this method in the early 1970s to train 65 fleet maintenance planners for the North Carolina Department of Transportation's Division of Highways. UMS is a very detailed method when using standard data for every motion and turns of all types of tools used by a mechanic to complete a "benchmark job." Benchmark jobs were put on a "spreadsheet" for "slotting" to get wrench time, upon which allowances were added to create the planning time for scheduling. Once "spreadsheets" were developed for all trade areas and types of work, the relatively small number of benchmark jobs analyzed for putting on "spreadsheets" allows a good planner to estimate wrench time for a wide number of incoming jobs. In addition, the accuracy of the slotting process is 95% within the respective UMS time ranges.
- **The ACE team benchmarking process**: As a means to overcome many of the inherent difficulties associated with developing maintenance performance standards, the ACE team benchmarking process was developed. This process is detailed in Chapter 15, with complete forms in Appendix H. This process was developed back in 1978 by Ralph W. "Pete" Peters, founder of The Maintenance Excellence Institute International (TMEII) as part of a Masters program in management information systems at North Carolina State University. This method is based upon UMS slotting and range-of-time concept principles plus the principles of the Delphi technique to gain a consensus. It relies primarily on the combined experience and estimating and methods improvement ability of a group of skilled crafts personnel, planners, and supervisor. The objective is to determine reliable planning times for a number of selected "benchmark" jobs. This team-based process that uses skilled crafts people places a high emphasis on continuous maintenance improvement to reflect improvements in performance and repair methods plus repair quality.

Generally, the ACE team benchmarking process parallels the UMS approach in that the "range-of-time concept" and the "slotting" technique is used once the work content times for a representative number of "benchmark jobs" have been established. The ACE team benchmarking process focuses primarily on three key areas: (1) repair methods improvement to reduce mean time to repair (MTTR); (2) repair quality and safety; and (3) on the development of work content times for representative "benchmark jobs" that are typical types of craft work being performed.

Once a number of benchmark job times have been established, these jobs are then categorized onto spreadsheets by craft and task area and according to work groups that represent various ranges of times. Spreadsheets are then set up with four work groups/ sheet with each work group having a time slot or "range of time." For example, work group E would be for benchmark jobs ranging from 0.9 h up to 1.5 h and assigned a standard time (slot time) of 1.2 h. Likewise, work group F would be for benchmark jobs ranging from 1.5 h up to 2.5 h and assigned a standard time of 2.0 h. Spreadsheets include brief descriptions of the benchmark jobs and represent pure wrench time. Work content comparison is then done by an experienced person, typically a trained planner, to establish planning times within the 95% confidence range. A users guide complete with step-by-step procedures and forms for establishing the ACE team benchmarking process is available in Appendix G. Figure 6.11 below illustrates the ACE team time ranges for work groups A to T with a time range of up to 30 h, ranging from 28 to 32 h.

Planning times are essential: Planning times provide a number of key benefits for the planning/scheduling process. First, they provide a means to determine existing workloads for scheduling by craft areas and backlog of work in each trade area. Planning times allow the maintenance planner to balance repair priorities against available craft hours and to establish realistic repair schedules that can be accomplished as promised. Secondly, planning times provide a target or goal for each job that allows for measurement of CP. Due to the variability of maintenance type work and the inherent sensitivity toward measurement, the objective is not so much the measurement of individual CP. The planner must always remember that one real objective

The ACE System Time Ranges

	ACE SYSTEM TIME RANGES		
WORK GROUP	FROM	STANDARD TIME (Slot time)	TO
A	0.0	.1	.15
B	.15	.2	.25
C	.25	.4	.5
D	.5	.7	.9
E	.9	1.2	1.5
F	1.5	2.0	2.5
G	2.5	3.0	3.5
H	3.5	4.0	4.5
I	4.5	5.0	5.5
J	5.5	6.0	6.5
K	6.5	7.3	8.0
L	8.0	9.0	10.0
M	10.0	11.0	12.0
N	12.0	13.0	14.0
O	14.0	15.0	16.0
P	16.0	17.0	18.0
Q	18.0	19.0	20.0
R	20.0	22.0	24.0
S	24.0	26.0	28.0
T	28.0	30.0	32.0

Figure 6.11 Illustrates the ACE (a consensus of experts) team time ranges for work groups A to T.

Range of Overall Craft Effectiveness Element Values			
OCE Elements	Low	Medium	High
1. Craft Utilization (CU)	30%	50%	70%
2. Craft Performance (CP)	>80%	90%	95%
3. Craft Service Level CSQ)	>90%	95%	98%
The OCE Factor	22 %	43%	65%

Figure 6.12 Range of overall craft effectiveness (OCE) element values.

is measurement of the overall performance of the craft workforce as a whole. While measurement of the individual crafts person is possible, CP measurement is intended to be for the maintenance craft labor resources and getting maximum value as shown in Figure 6.12.

CSQ: The third element affecting OCE relates to the relative quality of the repair work. This element includes quality of the actual work, where certain jobs possibly require a callback to the initial repair, thus requiring another trip to fix it right the second time. Recall the last time you had to call back a repair person to your home for a job not performed to your satisfaction. This is not a good thing for a real profit-centered maintenance company! However, CSQ can be negatively impacted within a plant or facility complex due to no fault of the crafts person when hasty repairs, patch jobs, or inferior repair parts/materials create the need for a callback.

We can measure callbacks via the CMMS with special coding of callback work orders. Typically, the CSQ element of OCE is a more subjective value and therefore it must be viewed accordingly in each operation. However, the CSQ level does affect overall craft labor productivity and the bottom line results of the entire maintenance process. When reliable data are present for all three elements of OCE, then the OCE factor can be determined by multiplying each of these three elements:

$$\text{OCE} = \begin{array}{c} \text{CU\%} \\ \text{Craft} \\ \text{Utilization} \end{array} \times \begin{array}{c} \text{CP\%} \\ \text{Craft} \\ \text{Performance} \end{array} \times \begin{array}{c} \text{CSQ\%} \\ \text{Craft Service} \\ \text{Quality} \end{array}$$

What OCE can you expect? Since OCE is a rather new concept, there are actually a limited number of case studies outside the real worldwide experiences of MEI staff and alliance members. Some organizations try to measure just wrench time, and it is accepted that 30–40% is typical and 70% is great. Other organizations may measure and track CP if a sound planning process and reliable planning times are in place. Also, other good consulting firms shy away from the often sensitive issue of measuring craft labor at all, especially within a union environment. The Maintenance Excellence Institute International does recommend measurement even for a union shop. This is because at some point high wages, growing continuously without productivity improvement, will make the operation a candidate for contract maintenance.

The Maintenance Excellence Institute International (TMEII) feels strongly that measuring and improving productivity of craft labor resources is essential to profit-centered maintenance and continuous reliability improvement. Measuring and improving OCE must be addressed by today's in-house maintenance operation. Likewise, we feel that the range of OCE element values shown in Figure 6.13 represents

Level of Craft Utilization	Total Wrench Time (Hours)	Actual Hands On Cost Per Hour	Average Wrench Time Hours Per Craft Position
30%	12,480	$60.00	624
40%	16,640	$45.00	832
50%	20,800	$36.00	1040
60%	24,960	$30.00	1248
70%	29,120	$25.71	1456
80%	49,920	$22.50	1664
*85%	35,360	$21.18	1768
90%	37,440	$20.00	1872
100%	41,600	$18.00	2080

Figure 6.13 Actual cost per hour at various levels of wrench time. Note: Maximum possible CU is 85% (as shown previously in Figure 6.2 above considering paid holidays, vacation time, breaks, but not shop cleanup, employee meetings, craft training, etc.).

the high, medium, and low combinations for OCE. Successful operations can expect an OCE factor in the high range of 65% or more.

All three elements of OCE are important: Maintenance craft labor may be very efficient with 100% CP and still not be effective if CU is low and craft service quality is poor.

- Overall Craft Effectiveness = Craft Utilization × Craft Performance × Craft Service Quality
  30% × 100% × 70%
- Overall Craft Effectiveness = .3 × 1.00 × .7 = .21 × 100

Overall Craft Effectiveness = 21% OCE

Note: This is the prime example of having high efficiency but doing low quality work and having low actual wrench time putting out fires from emergency, nonplanned work.

The nature of determining the value of CSQ can be accomplished and requires a "good call" by the respective supervisor responsible for work execution. This element typically can used for calculating OCE if callbacks are recorded correctly. Plus, there should not be many callbacks as a percentage of work orders completed. The element of quality of repairs is a very important element of OCE. It is an important part of effective planning and scheduling related to monitoring and control of work. Another key part of planning is determining the scope of the repair job and the special tools or equipment that are required for a quality repair. That is also why Category 27, *Maintenance and Quality Control*, is included as a best practice category on the Scoreboard. A continuing concern of the maintenance planning function should be on improving existing repair methods whether by using better tools, repair procedures, or diagnostic equipment and using the right skills for the job. Providing the best possible tools, special equipment, shop areas, repair procedures, and craft skills can be a key contributor to improving CSQ. And CSQ is a key performance indicator that is determined from periodic review of callbacks, customer complaints, and customer surveys. Therefore, The Maintenance Excellence Institute International (TMEII) feels that the OCE factor is best determined by using all three elements for the OCE factor calculations:

OCE = Craft Utilization × Craft Performance × Craft Service Quality

The impact of improving both craft utilization and performance: Improved CU through more effective planning of all resources will increase available wrench time. Improved performance results from the fact that work is planned and the right tools, equipment, and parts are available made by planning the right craftsperson or crew for the job with the type of skills needed. Improving CP is a continuous process with a program for craft skills training and methods improvement to do the job right the first time in a safe and efficient manner. The ACE team benchmarking process mentioned earlier provides reliable planning times based upon "ACE" and a tremendous repair methods improvement effort as benchmark jobs are analyzed.

Example C: What if We Increase Wrench Time from 30% to 50% and CP from 80% to 90%

When we look at the combination of improving both craft utilization and performance, we see an even greater opportunity for a return on investment. Let us now look at a very realistic 20-point improvement in CU and a 10-point increase in CP for the same 20-person craft workforce shown in Figure 6.14, *actual cost per hour at various levels of wrench time*, and having an average hourly rate of $18.00.

Example C Details

- **Baseline Cost per Direct Maintenance Hour @ 30% Utilization and 80% Performance**

20 crafts × 40 h/week × 52 weeks/year × 0.30 (CU) × 0.80 (CP)
= 9984 Direct Craft Hours/Year
= 499 Direct Hours/Craft Position

Baseline Cost: $748,800 ÷ 9984 Direct Hours = $75 Cost per Direct Craft Hour

Examples	Baseline	Improve To:	Craft Labor Gain	Gained Value
A	CU @40%	CU@50%	5	$187,200
B	CU@30%	CU@50%	13	$497,952
C	CU@30% & CP@80%	CU@50% & CP@90%	17	$655,200

Figure 6.14 Summary comparisons of previous examples.

- **Improved Cost per Hour with 50% CU and 90% CP**

  20 crafts × 40 h/week × 52 weeks/year × 0.50 (CU) × 0.90 (CP)

  = 18,720 Direct Craft Hours/Year

  = 936 Direct Hours/Craft Position

  Cost per Direct Craft Hour @ 50% CU and 90% CP = $40

  $748,800 ÷ 18,720 Direct Hours = $40 Cost per Direct Hour.

- **Total Direct Craft Hours Gained = 8736 Total Direct Hours Gained (87% Increase)**

  18,720 direct hrs − 9984 direct hrs @ Baseline = 8736 Direct Craft Hours Gained

  8736 direct hrs Gained ÷ 9984 direct hrs @ Baseline = 0.87.5 × 100 = 87% Gain in Direct Craft Hours

- **Total Gain in Equivalent Number of Craft Position: 17 Equivalent Craft Positions**

  20 crafts × 0.87 (% Hours Gained) = 17.4 Equivalent Craft Positions

  8736 h Gained ÷ 499 h/Craft Baseline Average = 17.4 Equivalent Craft Positions

- **Total Gained Value = $655,200**

  Gain of 8736 Direct Craft Hours × $75 Baseline Cost/Direct Hour = $655,200 Gained Value

  $655,200 ÷ $748,800 = 87% Gain from a Baseline of 30% Wrench Time and 80% CP

  Summary of our previous examples: The previous examples have illustrated that increasing OCE provides greater craft capacity and gained value from increased wrench time. Improving CP in combination with improving CU simply compounds our return on investment, an astronomical amount of 87% as shown in Figure 6.15.

  Where can we apply OCE gained value? Maintenance operations that continually fight fires and react to emergency repairs never have enough time to cover all the work (core requirements) that needs to be done. Over time, more crafts people or more contracted services typically seem to be the only answers. Improving CU provides additional craft capacity in terms of total productive craft hours available. In relation to OEE, OCE is increased people asset availability and capacity. It is gained value that can be calculated and estimated and then measured. The additional equivalent craft hours can then be used to reduce overtime, devote to PM/PdM, reduce the current backlog, and attack deferred maintenance, which does not go away. Previously, Figure 6.2 showed, "How your valuable craft time can just slip away."

  Indiscriminate cutting of maintenance is bloodletting: Typically, operations that gain productive craft hours desperately need them to invest the time elsewhere. Likewise, we cannot automatically and indiscriminately reduce head count when we improve overall craft productivity. Indiscriminate cutting of maintenance is killing the goose that lays the golden egg. If an organization is not achieving core requirements for maintenance, the cutting of craft positions to meet budgets is like using bloodletting as a new cure for a heart attack. It will not work. Just like the high

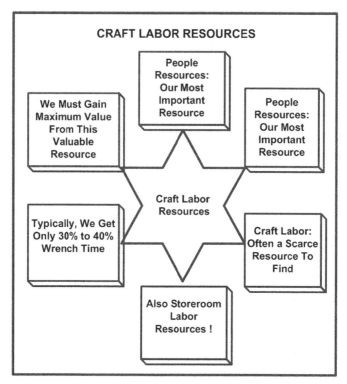

Figure 6.15 Craft labor resources.

cost of low-bid buying, gambling with maintenance costs can be fatal. Long-term stabilization and reduction of head count can occur. Attrition can absorb valid staff reductions that may result over the long term. We also may regain our competitive edge and get back some of the contract work we lost previously to low performance and productivity. We cannot indiscriminately cut craft labor resources when we increase OCE. The planner/scheduler helps ensure greater craft productivity and then maintains the total maintenance requirements from backlog management to use the gained craft capacity.

Think profit centered: Today's maintenance leaders (supervisors and planners included) and crafts people must develop the "maintenance-for-profit" mindset that the competition uses to stay in business. Measuring and improving OCE and the gained value received from improving our craft productivity is an important part of successful total asset management. Profit-centered in-house maintenance in combination with the wise use of high quality contract maintenance services will be the key to the final evolution that occurs. There will be revolution within organizations that do not fully recognize maintenance as a core business requirement and establish the necessary core competencies for maintenance. The bill will come due for those operations that have subscribed to the "pay me later syndrome" for deferred maintenance. It will be revolution within those operations that have gambled with maintenance and have lost with no time left before profit-centered contract maintenance provides the best financial option for a real solution.

Maintenance is forever: Contract maintenance will be an even greater option and business opportunity in the future. Again, we must remember that maintenance is forever! Some organizations today have neglected maintaining core competencies in maintenance to the point that they have lost complete control. The core requirement for maintenance still remains, but the core competency is missing. In some cases, the best and often only solution may be value-added outsourcing. Maintenance is a core requirement for profitable survival and total operations success. If the internal core competency for maintenance is not present, it must be regained. Neglect of the past must be overcome. It will be overcome with a growing number of profit-centered maintenance providers that clearly understand OCE and providing value added maintenance service at a profit. Your planner/schedulers can help your organization take a profits-centered approach.

Technology supports data collection for measuring OCE: We can easily collect data with today's technology (which, in fact, has been available for a long time). Bar coding can support asset tagging and identification, parts identification, as well as the work order itself. Parts charged to a work order that has a bar code is fast and accurate. But one challenge is linking a work order to an asset of a subcomponent of an asset. Often we see companies that have not finished the task of numbering all of their assets. The planner must make sure that asset numbering is complete. One way to get complete asset numbering is to do it "correctly the second time." During this second time numbering activity, many important things can occur:

1. Existing asset number can be confirmed. Many times migration of data to a new CMMS is done hastily and not really validated.
2. Missing asset numbers can be established. Missing data from an old system stays missing upon data migration unless it is validated.
3. Parent–child relationships can be reviewed. For example, one rule of thumb is that if a critical subsystem or component has a need for reliability improvement and its related failure data, then it should have an asset number.
4. A complete list of asset numbers to correct items 1, 2, and 3 above is the ideal time to affix appropriate bar codes to item outside and those inside the confines of the facility.
5. With assets now bar coded along with a bar-coded work order we can link the work performed to the asset, a subsystem, or a critical component with all the labor charged back to a work center or department account number.
6. The next step forward can be handheld devices that can wirelessly receive new work orders, contain complete PM task lists, collect time to a work order, and have clock on and off of a job as required. Prompts can be in place for defining type failure and cause of failures as well.
7. With all of the above in place, the once tedious task of using special codes for non-wrench time can be eliminated. Since we are most interested in two key areas of OCE, CP and CU (wrench time), they can now be easily accounted for due to bar coding on the asset.
8. The crafts person is assigned a job and "the total job clock starts." They clock on at the asset to begin their wrench-time work (hopefully planned with all items needed for the repair). If they must clock off to get parts, they clock off by swiping the bar code. When they come back, they clock on the job to finish it and clock off again.

9. All of the time clocked on the job is wrench time and all other non-wrench activities, such as travel to/from the site, personal fatigue and delays (PF&D), hopefully have been designated as job allowances for scheduling purposes.

10. When the crafts person reviews and provides all required input for the work order, they do it and then close the job to maintenance before required approvals are given and before final review by the planner.

Summary: So what does all this application of handheld devices and bar coding give us?

1. First, we can have time reporting for pay purposes with simple clock in and clock out at the site entrance.

2. We can easily track available hours charged to work orders where the goal can approach 100%. From the tracking of total planned time accrued to work orders, we can get actual time worked and therefore CP that we want to measure, across the entire workforce.

3. By clocking on and clocking off for actual hands on time, we are getting a much better picture of wrench time to give a better measure of CU.

4. We are also able to see the impact of emergency work when a person is pulled off and put back on a planned job.

5. Scope changes can be readily be analyzed for additional planned times or extremes when planned time is much more than originally planned.

Overall, the use of bar coding is not new for some maintenance operations. Some I have seen even included assigning a hardened laptop for each crafts person. Some make use of the phone system approach, while others use a handheld data collection device. The return on investment to enhance craft productivity and reliable data for RCM and root cause analysis (RCA) can be high when all elements we have discussed are in place.

What to Look for When Hiring a Reliable Planner/Scheduler

7

A. **Typical Roles and Responsibilities of a Planner/Scheduler**: Let us first look at the overall roles and responsibilities of a planner/scheduler as a baseline for upcoming topics related to each area of the planner position. Taking a fairly complete picture of the roles and responsibilities of planner/schedulers is a good first step. In this chapter, I provide some hints, observations, and recommendations that can assist you as maintenance leader, those who are in a planner/scheduler position, or someone that has just been selected for such a position. I also strive to illustrate how the planner/scheduler can make a significant impact on improving reliability and be a key leader for continuous improvement.

1. The title of this book, *Reliable Maintenance Planning, Estimating, and Scheduling*, indicates that this book will address a planner's additional responsibilities for supporting the site's reliability improvement process. The planner's unique position in the organization provides many natural job-related opportunities to contribute.

2. Maintenance planners/schedulers typically report to the maintenance leader.

3. The primary scope and role of planning and scheduling is to improve craft labor productivity and quality through the elimination of unforeseen obstacles, such as potential delays, coronation parts, machine time, and available resources. Therefore, the planner/scheduler is responsible for planning, estimating, and scheduling of all maintenance work performed across his or her areas of responsibility. The planner/scheduler supports direct liaison and coordination between operations and the maintenance department, along with maintaining appropriate records and files that can lead to meaningful analysis and reporting of results from the execution of work.

4. Remember that small operations may have a combination planner/scheduler, whereas larger plants, such as a large refinery, will normally have separate planners and schedulers. In some cases, I have seen day-to-day planners plus shutdown turnaround outage planners, along with planners for your facilities maintenance, such as control room office buildings and testing laboratories.

5. As stated, this position acts as the principal contact and liaison between maintenance and the various plant departments/customers that are served by maintenance. This person ensures his or her areas of responsibility for functions receives prompt efficient and quality service from maintenance. The planner/scheduler also strives to ensure that maintenance can provide this service and operations allows equipment to be shut down for various types of maintenance, including preventive and predictive maintenance's inspections.

6. The planner serves as the central point where all work requests are received and reviews these before they are turned into official work orders. True emergency work will typically go directly to the first-line supervisor or crew leader of the trade area required for the work. The work order in this case may be completed after the fact and would come back through the planner through a supervisor or craftsperson.

7. Reviewing and screening of each work order is a very critical task performed by the planner/scheduler. First, the planner/scheduler defines whether the work order is filled

Reliable Maintenance Planning, Estimating, and Scheduling. http://dx.doi.org/10.1016/B978-0-12-397042-8.00007-3

out correctly or if additional clarification is needed for the work requests. Here, the planner plays a big role in training and encouraging the operations staff to request work properly. In large refining operations, each operating unit may have a maintenance coordinator, who supports direct coordination with the planner as well the craft workforce when they come to perform repairs. For any type of operation, it is good practice to have someone designated within the department to create work orders and be able to see the ongoing status (hopefully online and in real time).

8. Other key areas during the review of work orders include clearly defining the scope of work, as well as checking the requested work/priority and requested completion dates to see if they are realistic and provide realistic lead times. One key point about planning is that we must have lead times. Without lead times, everything is unplanned reactive maintenance,

9. The planner/scheduler makes sure that accounting codes and other coding for chargebacks are correct and that all required approvals have been received.

10. At times, the planner/scheduler may need to discuss the details of the job with the person requesting the job when appropriate. If a maintenance coordinator is assigned within each unit of a large refinery, then that person typically will have written up the job request and has knowledge of the location within the unit and the problems to be repaired. In many cases, this is an engineer or experienced operator who should be able to define complete requirements or a work request.

11. The planner/scheduler will also assemble drawings, PID diagrams, and make additional diagrams in order to clarify work included on the work order.

12. At times, the work requests may not be valid. The originator should be questioned and issues should be readily resolved with operations. If issues are not readily resolved, the maintenance leader can be contacted in the next step that the planner/scheduler takes for resolution.

13. Some work orders may require information from plant engineering, so the planner might refer some work orders for further review by engineering.

14. If planned work orders involve the participation of several shops or functional crews, they are crossed over to a planner in that area. However, a single planner/scheduler must plan and then coordinate various functional crews with their respective supervisors during the scheduling process.

15. Planners should examine jobs to be done during the scoping process and determine the best way or method to accomplish the work. Planners may consult with the requester, maintenance supervisors, and subject matter experts within the craft workforce.

16. For job packages for more complex jobs, the documentation required may be much more than normal. It may include blueprints, drawings, parts list, special procedures, or repair or instructional manuals for the repair—all that is needed to clarify the intent of the work order to the craftsperson doing the job.

17. Planners normally will identify parts and materials required. They may create the purchase requisition, monitor status of incoming parts not in inventory, and arrange for kitting or delivery to the job site, shop, or secure job boxes as appropriate throughout the site.

18. Planners should determine if critical spares/insurance items are in stock by verifying availability with stores or via direct online access. Note that planners are in an excellent position to recommend items to consider for including within the storeroom as critical spares. They see what is repetitively coming up for nonstock item purchasing as well as what is being repaired over and over again. Being active in this area can support improving reliability and uptime.

19. A top priority for the planner is ensuring that safety needs are met as well as all health, safety, security, and environmental issues.

20. For reliable estimations, the sequence of job steps should be well defined and come from doing a good scope of work. Also, for each step, the number of technicians that were required, man-hours for each step, and total duration of the job should be established. A parts list should be developed, along with any special tools or equipment that might be needed.

21. A reliable method to estimate times for scheduling must be in place. The Maintenance Excellence Institute International highly recommends use of the ACE (A Consensus of Experts) team benchmarking process. This method is based on the slotting concept but uses a consensus of experts via a cross-functional team to analyze benchmark jobs (for wrench time), which then are used in slotting with site-specific allowances added to determine a reliable time scheduling purposes. More details for applying the ACE team process are provided in Chapter 15.

22. Cost estimates may be required for selected jobs, including direct labor costs, material costs, and rental equipment costs, to reach a total estimated cost. As discussed previously, one good performance metric is to establish a reasonable cost variance, typically from 5% to 10%. My philosophy has always been to measure contractors just as we measure our in-house craft personnel related to costs and productivity. Therefore, an extreme cost variance by a contractor may become obvious early on before the total cost has exceeded all expectations and it is too late to make adjustments. Planners should be accountable for cost estimates that they make, and project engineers should be accountable for their cost estimates as well.

23. Maintaining the total maintenance requirements in the form of backlog is a critical role of the planner/scheduler. This is a backlog file of work orders waiting for scheduling in accordance with their priority limits with an estimated date for completion. Work orders without parts in hand are never ever included on a weekly schedule. Proactive planning occurs when a part comes in, the equipment is made available, and the repair is made with the needed parts.

24. The total maintenance requirement backlog consists of **total backlog plus ready backlog**. It should be accurate and readily available for the maintenance leader to see and to brief top leaders on staffing needs, use of overtime, and work that can be referred until shutdown occurs. However, deferred maintenance needing immediate attention that will be of greater cost if left unrepaired should be reported to top leaders at every opportunity.

25. Based upon approved schedules, the planner assembles all relevant work orders plus supporting documents for the area supervisors. At this time, any planning/job packages are discussed, along with any special instructions or considerations to be observed in the execution of the job. Any new upcoming jobs are also reviewed. All work orders, including emergencies for tracking purposes, come through the planner.

26. Follow-up to progress being made on the current schedule is important and can come from a number of sources. The key source is from the maintenance supervisor who monitors order progress and provides feedback to the planner. The planner then carefully reviews the completed schedule and corresponding work orders. If real-time labor reporting is being done for a big job, then good visibility of status is available. In this case, the planner, supervisor, and person requesting the job can track hours spent versus hours planned for a status. However, it is the supervisor's responsibility to inform the planner or customer of the true job status. With staffing adjustments, the supervisor may bring a job back on schedule. For jobs that will not meet their scheduled completion times, the supervisor must notify both the planner and the customer.

27. Based on current control procedures, the planner provides a final review of work orders and administrative support performs the final closeout in the CMMS (computerized maintenance management system) system. This is an important step toward work order closure, allowing the planner to ensure that key information is accurately charged to the work order, such as valid work completion, cause of failure, failure codes, and craft time.

28. Selected jobs may be charged to standing work orders/blanket work orders for minor repairs or for asset logbook-type jobs. The planner needs to be kept advised as to the status of these jobs. Generally, they are closed out monthly and reported, with a new blanket order created for the next month's work with this type of general-purpose work order. Note that *troubleshooting is an important skill. It can often be scheduled by a planner needing the true scope of work that is not visible until the item is disassembled and parts for replacement are determined. My recommendation is to consider this time as wrench time and track it as a work type if need be. This is also an area for additional training to ensure systematic troubleshooting practices are used by all.* There are a number of technical training companies that offer both electrical and mechanical troubleshooting courses.

29. Where there are multiple planners across a large site, each planner must ensure that they coordinate complex multiskilled jobs with the applicable shops served by other planners.

30. Planners should schedule other weekly meetings with operations supervisors and maintenance supervisors who work in relevant areas regarding facilities or equipment to be maintained and work requests generated.

31. A skilled planner can make excellent recommendations to the operations team related to long-term maintenance needs. In collaboration with production, the planner can prepare a weekly or monthly forecast of work expected to be scheduled. At times, the planner may solicit relatively simple corrective maintenance in order to avoid major repairs at a later date. As we discussed previously, finding small repairable items from PM tasks, which can be planned and scheduled before a catastrophic failure, provides direct cost avoidance and return on investments for PM and PdM. It is a good rule of thumb to track these work types with a separate repair type code, showing jobs that were generated as a result of PM/PdM. Even better is when the planner can extract related actual repair costs to projected costs of a catastrophic failure.

32. Planners should always maintain part of their ready backlog that can be scheduled when windows of opportunity occur as equipment is not being used for any reason.

33. When the ready backlog has all the resources necessary to schedule, the planner can file these by supervisor and by completion date. This can be readily accomplished by a computerized system to include visibility to the requester with regard to their specific work requests (now the form of work orders), status of work orders on the schedule, and completed work orders in their respective areas. However, the planner maintains the master backlog summary, which must be accurate and define all completed jobs without delay, jobs that are questionable, backlog status, and estimated times, especially for current backlog jobs ready for scheduling.

34. After the job is fully planned, with man-hours and job duration established, the planner verifies the availability of parts material and any special tools prior to scheduling. Always remember that a job does not appear on a schedule unless all parts, materials, and special tools/equipment are available.

35. With regards to PM/PdM, the planner should have knowledge of required resources within each department and attempt to level load this type of work to ensure that all annual PMs are spread out across the year and do not all come at the same time.

36. As a schedule is being firmed up for the next week, generally the planner will convene a scheduling meeting during the last days of the current week. Typically, the day may be Thursday (or Wednesday in the Middle East, where there is a Sunday to Thursday work week). At times, this may include planner participation as part of a regularly scheduled production/operations review meeting, with the planner scheduled on the agenda to review next week's schedule. It is at these scheduling meetings that commitment is gained for equipment to be available for PMs and other scheduled repairs. It is a good practice to have operations agree and formally sign off on next week's schedule.

37. For a continuous process operation, such as a refinery, gas plant, or petrochemical operation running 24/7, much can occur on Friday, Saturday, and Sunday if the scheduling meeting is held on Thursday afternoon, for example. Therefore, work generated after the schedule is put in place must be viewed in terms of priority. Whether or not it should be added to the schedule can be finalized after the weekend.

38. It is highly recommended that the first day of the scheduled week be as firmly established as is possible.

39. If anything is rigidly flexible, it must be the weekly schedule. Best practice is for operations to approve any break in an established schedule by signing off at the scheduling meeting. As stated previously, critical spares coming in during the schedule week that can be used for critical equipment repair are coordinated with operations, equipment is made available, and repairs are made by adjusting lesser priority items on the schedule. Again, this is very proactive planning process that does count toward customer service (i.e., schedule compliance).

40. As a planner begins to develop a schedule, craft labor availability must be known on a daily basis. It is generally a supervisor's responsibility to let the planner know about planned vacations and sick leave. This affects the labor resources that are available on a daily basis during the weekly schedule.

41. When developing a schedule, the planner considers the ready backlog in each supervisor's area and/or trade group. At this point, the planner must match labor requirements with labor availability and take into account any carryover work from the previous week's schedule. All known skill requirements are determined, and one planner may need to make arrangements with other planners. For example, a job such as on-site fabrication with welding may come from coordination with a central shop planner. The planner may also have visibility across a number of shops, monitoring backlog and seeing differences in workloads. Therefore, it is recommended that craft resources be cross-leveled to accomplish work in areas with greater backlog.

42. At times, there may be special skills involved that require planner knowledge of the workforce to designate specific crafts resources to these type of jobs. Normally, this is common knowledge between the planner and the supervisor. Nonetheless, the planner coordinates with the supervisor so that specific individuals can be assigned with the technical skills needed for a particular job.

43. While planners do not actually assign crafts to specific jobs, they allocate and coordinate these requirements through a maintenance supervisor and, in some cases, the maintenance leader.

44. When selecting jobs for scheduling, the planner strives to meet the deadlines established by the requesting department while maintaining the preventive maintenance schedule. This is essential. If any work orders cannot be scheduled within the requested priority lead time, the department management and the requester should be given prompt notification so that appropriate actions can be taken to get the work done in a satisfactory and timely manner. Here is where our organizational structure, with the planner as the direct

liaison, comes into effect by creating close communications when schedule complete dates cannot be met or equipment may not be released for repairs.

45. As discussed, the planner attends meetings with the operations planning department and participates actively in the overall plant scheduling for the following week's work. Here, the planner negotiates downtime during which preventive or corrective repair can be performed. At this point, the next week's schedule can be finalized to ensure that scheduled work balances with the labor resources available so that a full day's work is provided for each person. This relates directly to a term I first heard from George Smith: "keep a half a day ahead" (KAHADA). In general, this is accomplished via the supervisor. It can be as simple as putting in work orders and a slot for each craftsperson or as sophisticated as wireless communication of work orders to a handheld device used by the craftsperson for work orders and time reporting, as we discuss in Chapter 6.

46. Why is the KAHADA system a good thing to practice? It like knowing who your next opponent is for a game in sports. It gives you time to have a mental picture and think about the first job tomorrow morning—to get psyched up, so to speak. On the other hand, if the job looks to be tough, then a person might decide to take a vacation day—not really! Most maintenance people welcome a challenge, so KAHADA helps them while allowing the storeroom to have the right parts and the planner and supervisor to avoid a hectic shift startup.

47. One area directly related to reliability is for the planner to recommend equipment that might be included within the preventive maintenance program. Also, the practice of continuous renewal of work orders should be taken from feedback as the craft workforce completes each PM/PdM task. Here, the planner can help encourage craftspeople to make recommended changes and have the major responsibility for updating PM/PdM task descriptions, frequencies, and estimated times.

48. The planning and scheduling of PMs must be coordinated with operations and maintenance supervisors. There must be a well-defined timeframe for completion of each PM on the schedule. Normally, PM compliance is calculated on a monthly basis; however, PMs should be released on a weekly basis. This helps to eliminate last-minute completions at the end of the month. This method establishes a sense of discipline so that PMs are scheduled on a weekly basis and hopefully completed on a weekly basis, depending on the equipment availability. PdM that can be performed with the equipment running is ideal. This can be done by an individual with, for example, a vibration monitoring device, infrared device, or by continuous monitoring via wireless communications to the plant's control center.

49. With regard to man-hours required for a PM or PdM program, it is always good for the planners to know the total equivalent man-hours required to achieve 100% PM compliance. This can be done when reliable estimates are in place, simply by calculating the number of occurrences per year for weekly, monthly, semiannual, and annual requirements to get the total man-hours required, then dividing by average man-hours worked per year by a craft position. Figure 7.1 provides a structure for determining the equivalent number of craftspeople required for 100% PM/PdM compliance.

Top leaders must clearly understand the resources needed for an effective PM/PdM program that achieves 100% compliance to existing tasks. I have seen some organizations complete project work to the point that PMs were being almost totally neglected. If all PMs have estimated times and frequencies, this calculation is very straightforward to compile from the PM task database. Often, a trained planner can do it with just Microsoft Excel or Crystal Reports. Also, with SAP Business Objects Analysis edition for Microsoft Office, it is to easy analyze, digest, and share data from multidimensional sources in a Microsoft Office environment and use powerful analytics to discover, compare, and

VERY IMPORTANT: Define the Required Staffing for Your Preventive/Predictive Maintenance Program

Frequency of Routines	Craft Hours per Occurrence	Repetitions per Year	Craft Hours per Year
Daily	?	365	?
Weekly	?	52	?
Monthly	?	12	?
Quarterly	?	4	?
Semi-Annual	?	2	?
Annual	?	1	?
Other (Overhauls0)	?	.5 to .1	?
		Total/Year ➡	27,000 Hrs. (example)
		Required Staff @ 1800 Hrs/Yr ➡	15 Equivalent Crafts

Note: Do not develop this staffing estimate from history. The proper estimate is for achieving 100% PM/PdM Compliance; the current program, if you had the needed craft resources.

Figure 7.1 A method for defining total staffing requirements for PM/PdM needs.

forecast business drivers in Excel. SAP Crystal Reports that produces clear and customizable reports for all types of business insight that are easy to understand and act upon, delivered upon demand.

50. Just as planners can support reliability improvement, they can also review with the maintenance supervisors the actual hours required versus estimated labor and material used for completed jobs. This may be in order to determine any corrective action needed to improve the accuracy of estimating and for improving methods of doing work.

51. Planners can assist maintenance and operations management with periodic reviews of cost and where recommended corrective actions are needed to reduce maintenance costs. Planners are in an excellent position to see this type of information as it comes through via the completed work orders.

52. Another important responsibility of the planner is to keep the maintenance leader properly informed on all abnormal or critical situations and to seek advice on matters outside the planner's knowledge or authority. The planner should have the knowledge to make suggested improvements to the planning and scheduling process and other areas within maintenance, even some denoted from the Scoreboard for Maintenance Excellence. As stated in Chapter 5, the planner should be able to see the big picture across his or her total maintenance operation regarding best practices that could help their own organizations.

53. Another important database that the planner develops is the file of standard work orders for job plans. These are regularly occurring repair jobs seen from history that can simplify the planning process by these jobs becoming templates for other jobs.

54. A file of job packages for major jobs also may include templates taken from previously performed job plans. Therefore, maintaining a good record and file of job packages is also important for the planner, along with other records, files, and reports, to include preparation and distribution of meaningful and accurate control reports, such as The Reliable Maintenance Excellence Index.

55. This position also has the responsibility of performing other tasks and special assignments as needed by the maintenance leader.

B. Summary of Position Goals and Relationships: The primary goals of the planner/scheduler can be summarized as follows:

1. Ensuring that the production/operations areas are served and receive prompt efficient and quality service from the maintenance function, which operates at a high level of productivity
2. Ensuring that the maintenance function is given every opportunity to provide production/operations with the service that it requires
3. Accurately defining and providing reliable estimates for work requests
4. Properly preparing schedules and distributing them along with other meaningful control reports.

Relationships: The maintenance planner/scheduler coordinates with many areas within a plant. While reporting to the maintenance leader, this position works closely with maintenance supervisors, operational supervisors, stores, and procurement personnel, and maintains good working relationships with all other organizational units within the plant.

C. Selection Criteria: Requirements and Qualifications: The selection of the right person for a planner/scheduler position is an important decision process. Ideally, the following areas should be a part of the position requirements and qualifications:

1. It is necessary to have a mechanical/electrical background, with technical school training desired.
2. This position should have adequate trade knowledge to establish reliable estimated labor hours and materials as well as to visualize the job to be performed. Proper scoping of the work will ensure that the required job tasks are considered.
3. Good communication skills by oral/written means are very desirable and critical to the success of a planner/scheduler.
4. At all times, the planner/scheduler should display a high degree of respect for and be able to tactfully deal with both senior and subordinate employees.
5. This position must have the ability and willingness to handle and organize various types of paperwork and have higher than average administrative and mathematical skills.
6. Planners in today's modern technology world must have or acquire a good knowledge of personal computers and possess reasonable data entry and typing skills.
7. Planning and organizational skills must be well developed or acquired via concentrated formal and informal training.
8. Planners/schedulers must be the champions of a proactive maintenance strategy as they are the cornerstone for the overall improvement process within maintenance.
9. Good work instructions must be understood and developed by planners/schedulers who know and understand what good work instructions are.
10. The ability to read and understand blueprints and shop drawings is a necessary skill.
11. In turn, planners/schedulers should be able to produce easily understood sketches, etc., that can clarify work instructions when needed.
12. Planners/schedulers must thoroughly understand the overall work management process, including the use of work orders, priorities, planning methods, scheduling techniques, etc.
13. The ability to monitor multiple jobs in a controlled situation while simultaneously considering new jobs coming in and closing out completed jobs is needed.

14. The planner/scheduler must gain the respect of his or her peers, have tough skin during tough times, and be able to function well under pressure.
15. Maintenance at times can be in a state of chaos. Planner/schedulers in turn must be able to bring order to the situation at hand.
16. Planners/schedulers must always **Remember Who They Work For?**—the customer of maintenance working with a mindset and commitment to good customer service.
17. They must possess a personal leadership style and a demonstrated level of capability that will gain respect from within both maintenance and the production operations.
18. Lastly, they must enjoy their work, have fun, and always demonstrate a philosophy of positive expectations during all encounters with maintenance staff.

Summary: Planners/schedulers should always remember that they were selected for a very important purpose. Often, they may be near the pay level of a maintenance supervisor or foreman. We have just seen the extensive list of roles and responsibilities each person in this position is expected to fill. New or current planners/schedulers must realize the importance of their position and what they can do to improve the productivity of both people and physical assets. This position is an excellent stepping-stone to greater responsibilities. I have seen this happen firsthand with the planners we recruited, selected, and trained from 65 mechanics who became service managers within the North Carolina Department of Transportation. All of them became either a division superintendent, an equipment inspector, or a shop supervisor.

Planner Review of the Maintenance Business System—Your CMMS-EAM System

8

Planners today very seldom operate without a computerized maintenance management system (CMMS), which I like to refer to as the maintenance business management system. With a payroll module plus purchase order creation, it could fit that purpose. This chapter introduces the CMMS Benchmarking System as the second benchmarking tool and the improvement process the planner or others can take to gain better use of your CMMS and information technology for maintenance. This benchmarking tool is introduced as a means to help the planner/scheduler evaluate the effectiveness of their current CMMS, to define functional gaps, and to define how to enhance current use or to help upgrade functional gaps. It is also a methodology to help develop and justify a CMMS replacement strategy where many existing planners can play a critical role.

Often this is the data migration from a stand alone system into an enterprise asset management (EAM) system such as SAP, JD Edwards One World, and others. Often a change provides a company with a chance to get it right the second time. It has been said many times that computerizing a bad process workflow and procedures only increases the magnitude of the problem. Therefore, the key to successful implementation is to ensure the correct best practices are in place such as planning and scheduling, and a good preventive and predictive maintenance (PM/PdM) program and well-planned storeroom and parts inventory is in place. In the 1980s, I developed this from a concept used by the late Oliver Wight, the guru for materials requirements planning (MRP II) for shop floor production control systems that were emerging. Our plant was the pilot for all of Cooper Industries divisions and plants. A little background from Oliver Wight International would be very appropriate here on your journey toward maintenance excellence.

Journey to Business Excellence.

The journey to excellence is never ending. It encompasses every part and every process in your company. The will to sustain such a journey depends on visible results delivered continually to all your stakeholders. For this reason, the journey begins with short projects that produce results quickly and grows into a longer business improvement program that assures success in the future.

How do you measure your progress? The Oliver Wight Business Maturity Map enables you to understand the maturity of your business and the projects within your business improvement program that will deliver the best real gain now. This is addressed in Oliver Wight education, which is arranged both publicly and privately in all parts of the world.

Reliable Maintenance Planning, Estimating, and Scheduling. http://dx.doi.org/10.1016/B978-0-12-397042-8.00008-5

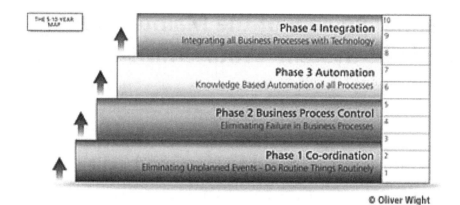

© Oliver Wight

Is the journey worth the effort? Only the excellent companies consistently win in business, and the standard of excellence is raised every day as customers expect more value and shareholders expect more return. An excellence program unites your people, your customers, and your suppliers by engaging them in a common set of goals, and achieves a pace of change to outperform competition and differentiate your company in the marketplace.

"Our new Sixth Edition Class A Checklist excellence standard will be more demanding upon companies and may require a longer period of improvement. These higher standards reflect the realities of today's increasingly competitive markets. While the journey may be long, the rewards arrive relatively quickly. You can expect that each project within your business improvement program will offer significant and sustainable benefits with rapid payback to your business. I encourage you to sign up for their site and use the excellent free knowledge base of manufacturing info available at http://www.oliverwight.com/checklist.htm .The Checklist raises the standard of demonstrated superior business performance to qualify for the coveted Class A award. No longer an MRP checklist or a manufacturing excellence standard, the Sixth Edition Checklist is a standard of excellence for all businesses. Take up the challenge and begin a journey to Class A with Oliver Wight International, and use the Scoreboard and CMMS benchmarking for your journey to maintenance.

The CMMS Benchmarking System has nine benchmark categories and 50 benchmark items, is easily adaptable, and can be specifically tailored to all CMMS systems for their intended application and operating context. The CMMS system is an internal benchmarking tool like the Scoreboard that is becoming a model process for benchmarking effective use of CMMS. It is designed as a methodology for developing a benchmark rating of your current CMMS into four classifications (Class A, B, C, or D) to determine how well this tool is supporting best practices and the total maintenance process. It is not designed to compare the functionality of various CMMS systems nor is it intended to compare vendors. The CMMS Benchmarking System provides a methodology for developing a benchmark rating of your existing CMMS to determine how well this tool is supporting best practices and the total maintenance process in all operations, large or small.

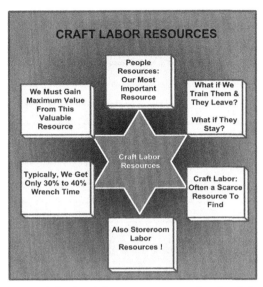

Figure 8.1 Effective computerized maintenance management system is key to continuous reliability improvement.

It can also be used as a method to measure the future success and progress of a CMMS system implementation that is now being installed. Maintenance best practices are the key and the CMMS is the information technology tool that links it all together. Maintenance information is one of the key maintenance resources (Figure 8.1) and must be a part of your approach to continuous reliability improvement, which covers all six maintenance resources.

A summary of The CMMS Benchmarking System is shown in Figure 8.2 with the nine assessment categories that include a total of 50 benchmark items for benchmarking your CMMS installation. The CMMS Benchmarking System rating scale is shown in Figure 8.3.

Conducting the CMMS Benchmark Evaluation?

The CMMS benchmark evaluation can be conducted internally by the planner/scheduler, maintenance leader, or via an internal team effort of knowledgeable maintenance people. Other options include using support from an independent resource to provide an objective maintenance benchmarking resource. The Scoreboard for Maintenance Excellence™ process in combination with the CMMS Benchmarking System provides powerful tools to help achieve greater value from all the six types of maintenance resources across all types of maintenance operations.

The CMMS Benchmarking System provides a means to evaluate and classify your current installation as either "Class A, B, C or D." Nine major categories are included along with 50 specific benchmark items. Each benchmark item that is rated as being

The CMMS Benchmarking System	
CMMS BENCHMARK CATEGORIES	**Benchmark Items**
1. CMMS Data Integrity	6
2. CMMS Education and Training	4
3. Work Control	5
4. Budget and Cost Control	5
5. Planning and Scheduling	7
6. MRO Materials Management	7
7. Preventive and Predictive Maintenance	6
8. Maintenance Performance Measurement	4
9. Other Uses of CMMS	6
TOTAL CMMS BENCHMARK ITEMS	**50**

Figure 8.2 Summary—the computerized maintenance management system (CMMS) Benchmarking System best practice categories.

CMMS BENCHMARKING SYSTEM RATING SCALE	
Class A	180 - 200 points (90% +)
Class B	140 - 179 points (70% to 89%)
Class C	100 - 139 points (50% to 69%)
Class D	0 - 99 points (up to 49%)

Figure 8.3 The computerized maintenance management system (CMMS) Benchmarking System rating scale.

accomplished satisfactorily receives a maximum score of 4 points. If an area is currently being "worked on," a score of 1, 2, or 3 can be assigned based on the level of progress achieved. For example, if spare parts inventory accuracy is at 92% compared to the target of 98%, a score of 3 points is given. A maximum of 200 points is possible. A benchmark rating of "Class A" is within the 180 to 200 point range. The complete CMMS Benchmarking System is included as an Appendix, and a sample is included below in Figure 8.4 for Category A-CMMS Data Integrity; Items #1 and #2.

Developing a future "Class A" CMMS installation requires that each organization start early in the implementation phase with establishing how they will determine the overall success of their installation. The CMMS Benchmarking System provides the framework for internal benchmarking of the CMMS installation as it matures. It is recommended that a team process be used for the CMMS benchmarking evaluation and that it be included as part of the CMMS evaluation team's initial work. Appendix

CMMS BENCHMARKING CATEGORIES and ITEM DESCRIPTIONS	YES (4 Points)	NO (0 Points)	WORKING ON IT (1, 2 or 3 Points)
A. CMMS DATA INTEGRITY			
1. Equipment (asset) history data complete and accuracy 98% or better			
2. Spare parts inventory master record accuracy 98% or better			

Figure 8.4 Example Category A—computerized maintenance management system (CMMS) data integrity from the CMMS Benchmarking System

D includes a charter format for a Leadership Driven, Self-Managed Team at GRIDCO Ghana for first evaluating, selecting, and implementing CMMS with concurrent best practices of which planning and scheduling was a critical need across all six regions of Ghana's power transmission grid.

Establishing a "Class A" CMMS requires that a number of key databases be established and that a number of maintenance best practices be in place. Data integrity, accuracy, and continuous maintenance of the key databases provide the foundation for a "Class A" CMMS installation. There are a number of other factors related to the CMMS and to maintenance best practices that in combination produce a future "Class A" installation.

We will now review each of the nine major categories from the CMMS Benchmarking System and provide key recommendations and examples for each of the 50 benchmark items to get your CMMS implementation started on the right track from day one.

A. Computerized maintenance management system data integrity

1. Equipment (asset) history data complete and accuracy 98% or better
2. Spare parts inventory master record accuracy 98% or better
3. Bill of materials for critical equipment includes listing of critical spare parts
4. Preventive maintenance tasks/frequencies data complete for 95% of applicable assets
5. Direct responsibilities for maintaining parts inventory database are assigned
6. Direct responsibilities for maintaining equipment/asset database are assigned

---

1. **Accuracy of equipment history database**: The equipment database represents one of the essential databases that must be developed or updated as part of implementing a new CMMS. It requires that a complete review of all equipment be made to include all parent/child systems and subsystems that will be tracked for costs, repairs performed, etc. The work to develop or update this database should begin as soon as possible after the data structure of the equipment master file for the new CMMS is known. A number of examples from a refinery's use of SAP are included.

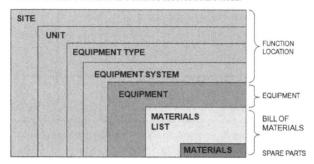

SAP Structure and Function Location

Function Locations are used as a detailed asset structure model

The equipment master information for a piece of equipment (parent/child), manufacturer, serial number, equipment specs, and location will all need to be established. If the installation and removal of components within certain process type operations requires tracking by serial number and compliance to process safety management requirements, these equipment items will have to be designated in the equipment database.

If an equipment database exists as part of an old CMMS, now is the time to review the accuracy of the old equipment database prior to conversion to the new system. Conversion of the new equipment master database into the new system should be done only after a thorough and complete update of the old database has occurred. Once the new equipment master database has been converted to the new CMMS, a process to maintain it at an accuracy level of 98% or above should be established.

SAP Structure and Operation

0424-ACID	ACID PLANT
0424-ADMN	REFINERY ADMIN
0424-ALK2	ALKY #2
0424-ALK2-A	Safety
0424-ALK2-C	Compressors
0424-ALK2-E	Electrical
0424-ALK2-F	Furnaces, Heaters, Boilers
0424-ALK2-G	General
0424-ALK2-H	Exchangers, Air Coolers, etc.
0424-ALK2-I	Instrumentation
0424-ALK2-K	Buildings
0424-ALK2-L	Lines, Piping, Valves
0424-ALK2-N	Non Maintenance
0424-ALK2-P	Pumps/Other Rotating Equipment
0424-ALK2-R	Reactors, Regenerators
0424-ALK2-S	Systems
0424-ALK2-V	Vessels, Towers, Drums, Tanks
0424-ALK2-X	Relief Valves
0424-ALK3	ALKY #3
0424-ARUA	AROMATICS RECOVERY A
0424-ARUB	AROMATICS RECOVERY B

Typical Rotating Equipment

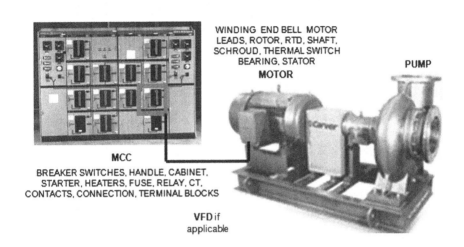

WINDING END BELL MOTOR
LEADS, ROTOR, RTD, SHAFT,
SCHROUD, THERMAL SWITCH
BEARING, STATOR
MOTOR

PUMP

MCC
BREAKER SWITCHES, HANDLE, CABINET,
STARTER, HEATERS, FUSE, RELAY, CT,
CONTACTS, CONNECTION, TERMINAL BLOCKS

VFD if
applicable

SAP Structure and Operation

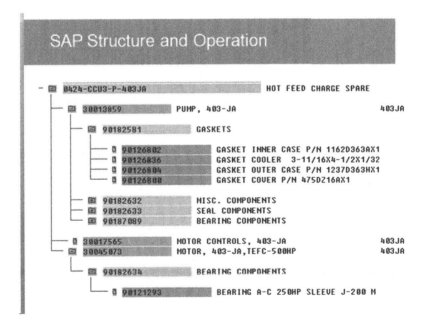

```
- ▣ 0424-CCU3-P-403JA                   HOT FEED CHARGE SPARE
  ├─ ▣ 30013859           PUMP, 403-JA                         403JA
  │   ├─ ▣ 90182581           GASKETS
  │   │   ├─ ▯ 90126802           GASKET INNER CASE P/N 1162D363AX1
  │   │   ├─ ▯ 90126836           GASKET COOLER   3-11/16X4-1/2X1/32
  │   │   ├─ ▯ 90126894           GASKET OUTER CASE P/N 1237D363HX1
  │   │   └─ ▯ 90126800           GASKET COVER P/N 475D216AX1
  │   ├─ ▣ 90182632           MISC. COMPONENTS
  │   ├─ ▣ 90182633           SEAL COMPONENTS
  │   └─ ▣ 90187089           BEARING COMPONENTS
  ├─ ▯ 30017565           MOTOR CONTROLS, 403-JA                 403JA
  └─ ▣ 30045073           MOTOR, 403-JA,TEFC-500HP               403JA
      └─ ▣ 90182634           BEARING COMPONENTS
          └─ ▯ 90121293           BEARING A-C 250HP SLEEVE J-200 M
```

Typical Instrument Equipment

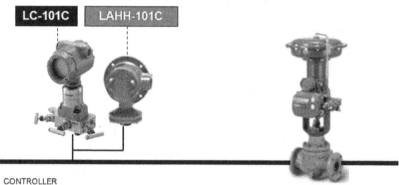

INSTRUMENT FUNCTION LOCATION

0424-SRUC-I-L101C

LC-101C   LAHH-101C

CONTROLLER
ELEMENT
GAUGE
as required

SAP Structure and Operation

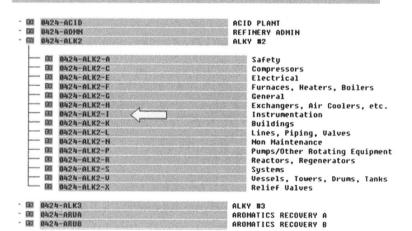

0424-ACID	ACID PLANT
0424-ADMN	REFINERY ADMIN
0424-ALK2	ALKY #2
0424-ALK2-A	Safety
0424-ALK2-C	Compressors
0424-ALK2-E	Electrical
0424-ALK2-F	Furnaces, Heaters, Boilers
0424-ALK2-G	General
0424-ALK2-H	Exchangers, Air Coolers, etc.
0424-ALK2-I	Instrumentation
0424-ALK2-K	Buildings
0424-ALK2-L	Lines, Piping, Valves
0424-ALK2-N	Non Maintenance
0424-ALK2-P	Pumps/Other Rotating Equipment
0424-ALK2-R	Reactors, Regenerators
0424-ALK2-S	Systems
0424-ALK2-U	Vessels, Towers, Drums, Tanks
0424-ALK2-X	Relief Valves
0424-ALK3	ALKY #3
0424-ARUA	AROMATICS RECOVERY A
0424-ARUB	AROMATICS RECOVERY B

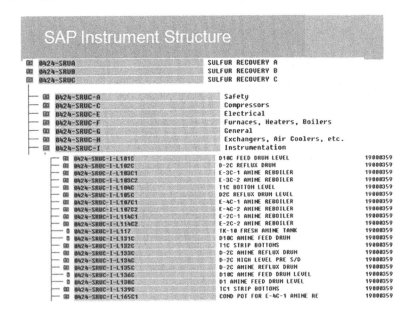

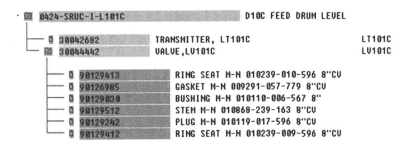

2. **Accuracy of spare parts database**: The spare parts database represents another key database that must be developed or updated as part of implementing a new CMMS. For operations not having a parts inventory system, this will require doing a complete physical inventory of spare parts and materials. All inventory master record data for each item will need to be developed based on the inventory master record structure for the new CMMS and loaded directly to the inventory module.

Win-Win Control of Materials, Tools and Equipment

Operations that have an existing spare parts database should take the time to do a complete review of it prior to conversion. Typically, this will allow for purging the database of obsolete parts and doing a complete review of the inventory master record data. This can be a very time-consuming process, but it allows the operation an excellent opportunity to revise part descriptions, review safety stock levels, reorder points and vendor data, and start the new CMMS with an accurate parts inventory database.

Personal Observations of Storerooms and MRO Materials Management

- Most storerooms are not organized, are inefficient, and are costly to operate.
- Few companies know what parts or tools they have on-hand or where they are.
- Parts databases are inaccurate, not up-to-date and contain duplications.
- Parts databases often never get set up correctly with new CMMS
- Parts-related downtime is very costly and reduces profit
- Parts storage areas represent a major source of corporate maintenance savings.
- Organized storerooms can correct inefficiencies of the past
- **Successful planning and scheduling depends on having an effective storeroom**

3. **Bill of materials**: One key functional capability of CMMS is to provide a spare parts listing (bill of materials) within the equipment module. This requires researching where spare parts are used and linking inventory records with equipment master records that are component parts of an equipment asset. This function would also add, change, or delete items from an established spares list or copy a spares list to another equipment master record. In addition, this feature would copy all or part of a spares list to a work order job plan and create a parts requisition or pick list to the storeroom.

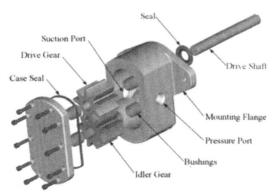

The process of establishing a spares list is time-consuming and would involve only major spares that are currently carried in stock. Most CMMS systems have the capability to build the spare parts list as items are issued to or purchased for a piece of equipment. It is recommended that equipment bill of materials be established, but the conversion of equipment master data can take place without this information being available. Because bill of materials for spare parts is so beneficial for planning purposes, it is recommended that the process to identify and code key critical spares in the equipment master database be a priority to complete.

4. **PM tasks/frequencies**: The PM/PdM database is another key database necessary for establishing a "Class A" installation. If a current PM/PdM database is present, it is recommended that the existing procedures be reviewed and updated prior to conversion to the new system. If the existing PM/PdM database has been updated continuously on the old system, conversion can probably occur directly from the old to the new PM/PdM database; this, however, will depend on the PM/PdM database structure of the new system.

It is recommended that in the very early stages of a new CMMS benchmark/selection process, the status of the current PM/PdM program be evaluated; this is an excellent time to establish a team for applying reliability-centered maintenance. This process provides the best maintenance strategy for the failure within the equipment's operating context. If a process for the review/update of PM/PdM procedures has not been in place, then it is very important to get something started as soon as possible. This provides an excellent opportunity to establish a team of experienced craft people, engineers, and maintenance supervisors to work on PM/PdM procedures to review and update task descriptions, frequencies, and making sure that all equipment is covered by proper procedures. It is also a good idea to know the equivalent staffing needed to achieve 100% PM compliance. The format below is a good guideline for a summary of overall calculations.

VERY IMPORTANT: Define the Required Staffing for Your Preventive/Predictive Maintenance Program

Frequency of Routines	Craft Hours per Occurrence	Repetitions per Year	Craft Hours per Year
Daily	?	365	?
Weekly	?	52	?
Monthly	?	12	?
Quarterly	?	4	?
Semi-Annual	?	2	?
Annual	?	1	?
Other (Overhauls0	?	.5 to .1	?
		Total/Year ➡	27,000 Hrs. (example)
		Required Staff @ 1800 Hrs/Yr ➡	15 Equivalent Crafts

Note: Do not develop this staffing estimate from history. The proper estimate is for achieving 100% PM/PdM Compliance; the current program, if you had the needed craft resources.

5. **Maintaining parts database**: After a new CMMS is installed, it is highly recommended that one person be assigned direct responsibility for maintaining the parts database. This person would have responsibility for making all additions and deletions to inventory master

records, changing stock levels, reordering points and safety stock levels, and changing any data contained in the inventory master records. This person could also be designated responsibility for coordinating the development of the spares list if this information is not available. This person would be responsible for recommending obsolete items based on monitoring of usage rates or due to equipment being removed from the operation. The practice of having one primary person assigned direct responsibility for the inventory master records can help ensure that parts database accuracy is 98% or greater.

How Planning & Scheduling Depends on Effective Stores and Procurement

6. **Maintaining equipment database**: It is also highly recommended that one person also be assigned direct responsibility for maintaining the equipment database. This person would be responsible for making all changes to equipment master records. Information on new equipment would come to this person for setting up parent–child relationships of components in the equipment master records. Information on equipment being removed from the operation would also come to this person to delete equipment master records.

Coordination between this person and the person responsible for the parts database would be required to ensure that obsolete parts are identified and/or removed from the inventory system due to removal of equipment.

B. Computerized maintenance management system (CMMS) education and training

7. Initial CMMS orientation training for all maintenance employees
8. An ongoing CMMS training program for maintenance and storeroom employees
9. Initial CMMS orientation training for operations employees
10. CMMS systems administrator (and backup) designated and trained

---

7. **Initial CMMS training**: One of biggest roadblocks to an effective CMMS installation is the lack of initial training on the system. Many organizations never take the time up front to properly train their people on the system. Shop-level people must gain confidence in using the system for reporting work order information and knowing how to look up parts information. The CMMS implementation plan should include an adequate level of actual hands-on training on the system for all maintenance employees prior to the "go live" date. It is important to invest the time and expense to "train the trainers" who can in turn can assist with the training back in the shop. Many organizations set up "conference room pilots" where the CMMS software is set up and training occurs with actual data using CMMS vendor trainers or in-house trainers. It is highly recommended that competency-based training be conducted so that each person can demonstrate competency in each function they must perform on the system. See Appendix L–An SAP Planner Training Checklist. Planners should be validated in their competency during the vendor training provided, whether it is SAP or any other system.

8. **Ongoing CMMS training**: The CMMS implementation plan must consider having an ongoing training program for maintenance and storeroom personnel. After the initial training, there must be someone in the organization with the responsibility for ongoing training.

If a good "trainer" (often the planner as a "super user") has been developed within the organization prior to the "go live" date, this person can be the key to future internal training on the new system. Ongoing training can include one-on-one support that helps to follow up on the initial training. *Note*: Very important in my opinion and experience is to select a person that can serve as a back up to the trainer.

9. **Initial CMMS training for operations personnel**: The customers of maintenance must gain a basic understanding of the system and know how to request work, check status of work requested, and understand the priority system. During implementation, operations personnel need to get an overview of how the total system will work and the specific things they will need to do to request work. If the organization has a formal planning and scheduling process, they will also need to know the internal procedures on how this will work.

10. **CMMS systems administrator/backup trained**: It is important that each site have one person trained and dedicated as the systems administrator with a backup trained whenever possible. Typically, this person will be from information services and have a complete knowledge of system software, hardware, database structures, interfaces with other systems, and report writing capabilities. The systems administrator will also have responsibility for direct contact with the CMMS vendor for debugging software problems and for coordinating software upgrades.

C. Work control

---

11. A work control function is established or a well-defined documented process is being used
12. Online work request (or manual system) used to request work based on priorities
13. Work order system used to account for 100% of all craft hours available
14. Backlog reports are prepared by type of work to include estimated hours required
15. Well-defined priority system is established based on criticality of equipment, safety factors, cost of downtime, etc.

---

11. **Work control function established:** A well-defined process for requesting work, planning, scheduling, assigning work, and closing work orders should be established. The work control function will depend on the size of the maintenance operation. Work control may involve calls coming directly to a dispatcher who creates the work order entry and forwards the work order to a supervisor for assignment. The work request could also be forwarded directly to an available crafts person by the dispatcher for execution of true emergency work.

Work control can also be where work requests are forwarded manually or electronically to a planner who goes through a formal planning process for determining scope of work, craft requirements, and parts requirements to develop a schedule. PM/PdM work would be generated and integrated into the scheduling process. The status of the work order would be monitored, which might be in progress, awaiting parts, awaiting equipment, awaiting craft assignment, or awaiting engineering support, etc. A work order backlog would also be maintained to provide a clear picture of work order status. Effective work control provides systematic control of all incoming work through to the actual closing of the work order. The work control process should be documented with clearly defined written procedures unique to each maintenance operation.

9. WORK MANAGEMENT AND CONTROL: MAINTENANCE AND REPAIR (M/R)

ITEM #	Rating: Excellent – 10, Very Good – 9, Good – 8, Average – 7, Below Average – 6, Poor – 5	RATING
1.	A work management function is established within the maintenance operation generally crafted along functionality of the CMMS.	
2.	Written work management procedures which governs work management and control per the current CMMS is available.	
3.	A printed or electronic work order form is used to capture key planning, cost, performance, and job priority information. 10=Bar coded assets, parts and work order.	
4.	A written procedure which governs the origination, authorization, and processing of all work orders is available and understood by all in maintenance and operations.	
5.	The responsibility for screening and processing of work orders is assigned and clearly defined.	
6.	Work orders are classified by type, e.g. emergency, planned equipment repairs, building systems, PM/PdM, project type work, planned work created from PM/PdM's.	
7.	Reasonable "date-required" is included on each work order with restrictions against "ASAP", etc.	
8.	The originating departments are required to indicate equipment location and number, work center number, and other applicable information on the work orders.	
9.	A well-defined procedure for determining the priority of repair work is established based on the criticality of the work and the criticality of equipment, safety factors, cost of downtime, etc.	
10.	Work orders are given a priority classification based on an established priority system.	
11.	Work orders provide complete description of repairs performed, type labor and parts used and coding to track causes of failure.	
12.	Work management system provides info back to customer; backlogs, work orders in progress, work completed, work schedules and actual cost charge backs to customer. 10= real time system	
	9. Work Management and Control: Maintenance and Repair SUBTOTAL SCORE POSSIBLE:	**120**

10. WORK MANAGEMENT AND CONTROL: SHUTDOWN, TURNAROUNDS, AND OUTAGES (STO)

ITEM #	Rating: Excellent – 10, Very Good – 9, Good – 8, Average – 7, Below Average – 6, Poor – 5 or less	RATING
1.	Work management and control is established for major overhaul repairs, shutdowns, turnarounds and outages (STO) and includes effective work management and control by in house staff and contracted resources.	
2.	Work management and control of major projects provide means for monitoring project costs, schedule compliance and performance of both in house and contracted resources with a robust project management system.	
3.	Work orders are used to provide key planning info, labor/material costs and performance info for major all STO and overhaul work.	
4.	Equipment history is updated with info from work orders generated from major overhaul repairs, and SATO work.	
5.	The responsibility for screening and processing of work orders for major repairs is assigned to one person or unit.	
6.	Change order procedures and control are clear to all and approved at the appropriate level based on company requirements.	
7.	Change orders are reviewed by planners just as they review all jobs; scope of work, key job steps, equipment required and total additional cost and impact on total STO duration and appropriate approvals received before work execution.	
8.	Work orders for major repairs, shutdown, and overhauls are monitored for schedule compliance, overall costs and performance info including both in house staff and contracted services.	
9.	Cost variances are measured at key milestone with cost variance info so extreme variance can be investigated sooner than later when it is too late. A 5%-10% variance is set as maximum with clear reason for increased scope of work.	
10.	Has the current level of plant maintenance/asset management achieved the desired reliability to make to make an STO a) needed at longer timeframe than normal b)needed at a shorter timeframe or c) needed the appropriate period based on age and state of asset capabilities in their operation context.? a)= 10,9,8; b)=5; c=7,6	
11.	The organization has the capability to manage the turnaround program and be cost effective as compared to the best in the sector, has a strategy for reducing costs in the face of an ageing plant and rising manpower and material costs and where can we get high level technical advice?	
12.	The organization knows what manpower is available in-house, the competence levels where to get additional resources, who will lead the site team, who will do the work plus the cross functional team to design, monitor and control the event organization?	
13.	Have STO's received significant level of attention companies, have a history of tolerating higher than necessary downtime have older age of plants, see STO as a "necessary evil" are striving to lengthen the STO intervals from 12 – 24 months and even4, 5 and sometimes 8 years	
14.	The organization's history of planning & preparation for STO's has been: carried out more carefully, alignment of capital programs, has been scrutinizing and challenging scope of work	

#	Description		
15.	A process of assessing plant equipment deterioration is in place, the likely impact on reliability is known, the planning stocks and safeguarded and has partnering with major plant overhaul engineering contractors and have a learning organization from past history to manage STO's effectively		
16.	The plant beginning an STO has personnel available when required and capable of performing design specifications economically and a) safely for life of plant, b) knows sum of activities performed to protect reliability of the plant, c) helps provide consistent means of production, d) help generate profit all with e) reducing the Total Cost of Ownership (TOC)		
17.	Top Leaders clearly understand that STO is a significant maintenance and engineering event during which new plant is installed, existing plant overhauled, and redundant plant removed which has a direct connection between successful accomplishment and the company's profitability		
18.	The company includes profit lost during period of STO is considered part of turnaround cost because they know the total true cost of event and the real impact can be assessed		
19.	All involved with STO's realize the potential hazard to plant reliability or can diminish or destroy reliability if not properly planned, prepared, executed, poor decisions by managers and engineers, bad workmanship, use of incorrect materials and damage done while plant is being shut down, overhauled, restarted		
20.	Technical uncertainty due to occurrence of unforeseen problems can be accurately reported, knowing when cost estimates are being exceeded, event's duration must be extended, how both cost and duration increases be justified, Are reasonable cost and time contingencies built into an STO plan with accurate loss of revenue/profit considered.		
21.	Have Top Leaders created their business strategy to manage the STO basic objectives to eliminate STO's all together unless proven it is absolutely necessary,		
22.	If an STO is proven to be necessary, the Top Leader ensures that it will align with maintenance objectives, production requirements, business goals.		
23.	When beginning as STO the Top Leader has formed a chartered leadership driven, self-managed (not a committee) forming a cross functional staff to help a committed company get the best STO value.		
24.	The STO team has senior managers, responsible for long-term strategy and meet at regular intervals throughout year to review current performance and formulate high-level strategies for management of events such as a long-term STO program		
25.	Is an STO truly aligned to overall business strategy which include an evolution of asset management's driven search for change to preventive/predictive maintenance, being driven by technical considerations and a philosophy of maintenance prevention and continuous reliability improvement.		
26.	STO's are driven by business needs and question every maintenance practice to determine if it can be eliminated by addressing cause that generated the need and examines the largest maintenance initiatives first during an STO.		
	10. Work Management and Control: Shutdowns, Turnarounds and Outages (STO) **SUBTOTAL SCORE POSSIBLE:**		260

12. Online work request based on priorities: Requesting work online represents an advanced CMMS functional capability where the customer enters the work request directly into the system on a local area network or via e-mail. Online work requests would include basic information about work required, equipment location, date work is to be completed by, name of requestor, and priority of the work. This information would go to the work control function where the jobs would be planned, scheduled, and assigned based on the overall workload. The requestor would have the capability to track the status of their jobs online and even give final approval that the work was completed satisfactorily.

13. Work order system accounts for 100% of craft hours: Handled devices can have a high return on investment when applied to time reporting to work orders as well as wireless transfer of work orders to the crafts person. This method of data collection uses barcodes on work orders, assets, and on all parts. All craft work should be charged to a work order of some type. Accountability of labor resources is an important part of managing maintenance as an internal business. Quick reporting to standing work orders can be established for jobs of short duration within a department or for the reporting of non-craft time such as meetings, delays in getting the equipment to work on, training, and chasing parts.

14. Backlog reports: Maintaining good control of the work to be done is essential to the maintenance process. Having the capability to visually see the backlog helps to effectively plan and schedule craft resources.

The CMMS reporting system should provide the capability to show the backlog of work in a number of ways, some of which include:

• By type of work	• By overdue work orders
• By craft	• By parts status
• By department	• By priority

WEEK NO.	START DATE	A — ACTUAL TOTAL M.H. WORKED	B — M.H. SPENT ON SCH. WORK	C — M.H. SPENT ON UNSCH. WORK	BREAKDOWN OF SCHEDULED WORK				BREAKDOWN OF UNSCHEDULED WORK		BACKLOG		% PERFORMANCE
					D — ROUTINE AND STANDING	E — P.M.	F — FIRM SCH.	G — OTHER SCH. WORK	H — CRITICAL PRODUCTION EQUIPMENT BREAKDOWNS	I — OTHER UNSCH. WORK	J — MAN-HOURS	K — WEEKS	L — EST. SCH. M.H. STD HRS.
			M.H. %	M.H. %	M.H. %	M.H. %	M.H. %	M.H. %	M.H. %	M.H. %			%
													B
Total To Date This Qtr.													
Weekly Avg. This Qtr.													
Goal													

An effective report to collect and display data

15. **Priority system:** A "Class A" CMMS installation will have in place a priority system that allows the most critical repairs to get done first. An effective priority system adds professionalism to the maintenance operation and directly supports effective planning and scheduling. There are two basic systems for establishing priorities:

   a. Straight numeric – Priority 1, 2, 3, 4, 5 etc. where each priority level is defined by a definition, such as Priority 1: A true emergency repair that affects safety, health, or environmental issues.

   b. Ranking Index of Maintenance Expenditures (RIME) system – The RIME is a system that combines the criticality index of the equipment (10 highest to 1 lowest) with criticality of the work type (10 highest to 1 lowest importance) to compute the RIME priority number. The RIME priority number equals the equipment criticality index multiplied by the criticality number of the work type. Many CMMS systems will compute the RIME number when assets are assigned critical values and work types designed the same way as shown below.

RIME System Example (Distribution Center)

Work Type Priority → → / Asset Criticality ↓↓		Safety & True Emergency (SAF)	Preventive and Predictive Maintenance (PM)(PDM)	Project Work (PRJ)	Corrective Maintenance and Warranty Work (CMA)	Operations Service (OPS)	Routine Normal Safety (RNS)	Tenant Improvement (TIM)	Inspections (INS)	Miscellaneous (MIS)	Housekeeping (HSK)
		10	9	8	7	6	5	4	3	2	1
SORTER	10	100	90	80	70	60	50	40	30	20	10
Conveyor	9	90	81	72	63	54	45	36	27	18	9
Utilities	8	80	72	64	56	48	40	32	24	16	8
Turret Truck	7	70	63	56	49	42	35	28	21	14	7
Lift Truck	6	60	54	48	42	36	30	24	18	12	6
Stock Picker	5	50	45	40	35	30	25	20	15	10	5
Miscellaneous Mobile Equipment	4	40	36	32	28	24	20	16	12	8	4
Office Facilities	3	30	27	24	21	18	15	12	9	6	3
Miscellaneous support To Other Asset	2	20	18	16	14	12	10	8	6	4	2
Buildings, Roads and Grounds	1	10	9	8	7	6	5	4	3	2	1

Ranking Index of Maintenance Expenditures

D. Budget and cost control

16. Craft labor, parts, and vendor support costs are charged to work order and accounted for in equipment/asset history file
17. Budget status on maintenance expenditures by operating departments is available

---

D. Budget and cost control—Cont'd

---

18. Cost improvements due to computerized maintenance management system and best practice implementation have been documented
19. Deferred maintenance and repairs are identified to management during budgeting process
20. Life cycle costing is supported by monitoring of repair costs to replacement value

---

16. **Craft labor, parts, and vendor support costs**: The equipment history file should provide the source of all costs charged to the asset. Here it is important to ensure that all labor is charged to the work orders for each asset and that parts are charged to the respective work orders.

17. **Budget status–operating departments**: Operating departments should be held accountable for their respective maintenance budgets. With an effective work order system in place for charging of all maintenance costs, the accounting process should allow for monitoring the status of departmental budgets. One recommended practice is for maintenance to be established as a zero-based budget operation and that all labor and parts be charged back to the internal customer. This practice helps ensure accountability for all craft time, parts, and materials to work orders.

18. **Cost improvements due to CMMS**: The impact of a successful CMMS installation should be reduced costs and achieving gained value in terms of greater output from existing resources. The CMMS team should be held accountable for documenting the savings that are achieved from the new CMMS and the maintenance best practices that evolve. The areas that were used to justify the CMMS capital investment such as reduced parts inventory, increased uptime, and increased craft productivity should all be documented to show that improvements did occur.

19. **Deferred maintenance identified**: It is important that maintenance provides management with a clear picture of maintenance requirements that require funding for the annual budget. Deferred maintenance on critical assets can lead to excessive total costs and unexpected failures. Benefits from CMMS will provide improved capability to document deferred maintenance that must be given priority during the budgeting process each year.

Beware of the Maintenance Iceburg

20. Life cycle costing supported: Complete equipment repair history provides the base for making better replacement decisions. Many organizations often fail to have access to accurate equipment repair costs to support effective replacement decisions and continue to operate and maintain equipment beyond its economically useful life. As a result, the capital justification process then lacks the necessary life cycle costing information to support replacement decisions.

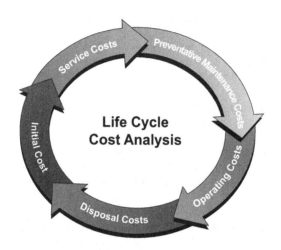

E. Planning and scheduling

21. A documented process for planning and scheduling has been established
22. The level of proactive, planned work is monitored and documented improvements have occurred
23. Craft utilization (true wrench time) is measured, and documented improvements have occurred
24. Daily or weekly work schedules are available for planned work
25. Status of parts on order is available for support to maintenance planning process
26. Scheduling coordination between maintenance and operations has increased
27. Emergency repairs, hours, and costs are tracked and analyzed for reduction

---

21. Planning and scheduling: This maintenance best practice area is essential to better customer service to operations and for greater utilization of craft resources. For most maintenance operations with 25–30 crafts people, a fulltime planner can be justified. The CMMS system functionality must support the planning process for control of work orders, backlog reporting, status of work orders, parts status, craft labor availability, etc. The planning and scheduling function supports changing from a "run-to-failure strategy" to one for proactive, planned maintenance.

11. SHOP LEVEL RELIABLE PLANNING, ESTIMATING AND SCHEDULING (M/R)

ITEM #	Rating: Excellent – 10, Very Good – 9, Good – 8, Average – 7, Below Average – 6, Poor – 5 or less	RATING
1.	A formal maintenance planning function has been established and staffed with qualified planners in an approximate ratio of 1 planner to 20-25 crafts people.	
2.	The screening of work orders, reliable estimating of repair times, coordinating of repair parts and planning of repair work is performed as a support service to the supervisor.	
	Planner/Schedulers realize their primary scope and role of planning and scheduling is to improve craft labor productivity and quality through the elimination of unforeseen obstacles such as potential delays coronation parts machine time and available resources.	
3.	Planner/Schedulers clearly understand the scope of their defined roles and responsibilities within your organization and are in an organization structure that promotes close coordination, cooperation and communications with their customer in operations.	
3.	The planner uses the priority system in combination with parts and craft labor availability to develop a start date for each planned job to be scheduled	
4.	A daily or weekly maintenance work schedule is available to the supervisor who schedules and assigns work to crafts personnel with multiple week "look a heads" if required.	
5.	The maintenance planner develops reliable and well accepted estimated times for planned repair work and includes on work order for each craft to allow performance reporting, backload levels and even documentation of work competency for selected jobs.	
6.	A day's planned work is available for each crafts person with at least keeping a half a day ahead (KAHADA) during the working day known in advance.	
7.	A master plan for all repairs is available indicating planned start date, duration, completion date, and type crafts required to define "total maintenance requirements".	
8.	The master plan is reviewed and updated by maintenance, operations, and engineering as required with project type work expected from maintenance. Care is taken not to overload maintenance with project work that causes PM/PdM and other work to be neglected.	
9.	Total maintenance requirements are a total of Total Backlog + Ready Backlog that has all resources (except labor or equipment availability) available to be scheduled.	
10.	A firm rule of thumb is never to put anything on the schedule without parts in house. But have contingency plan if needed parts arrive for critical equipment.	

11.	When parts arrive for critical equipment and can be inserted to the current schedule this is very proactive maintenance cooperation with operations.		
12.	Scheduling/progress meetings are held periodically with operations to ensure understanding, agreement and coordination of planned work, backlogs, and problem areas.		
13.	Operations cooperate with and support maintenance to accomplish repair and PM schedules.		
14.	Operations staff signs off the agreed upon schedule and are responsible to approve change in schedules and are accountable to TOP Leaders for adverse results.		
15.	Set-ups and changeovers are coordinated with maintenance to allow scheduling of selected maintenance repairs, PM inspections, and lubrication services during scheduled downtime or unexpected "windows of opportunity" to insert Ready Backlog jobs into the schedule.		
16.	Planned repairs are scheduled by a valid priority system, completed on time and in line with completion dates promised to operations and measured accordingly.		
17.	Deferred maintenance is clearly defined on the master plan and increased costs are identified to management as too the impact of deferring critical repairs, overhauls, etc.		
18.	Maintenance planners and production planners work closely to support planned repairs, to adjust schedules and to ensure schedule compliance in a mutual goal.		
19.	The planning process directly supports the supervisor and provides means for effective scheduling of work, direct assignment of crafts and monitoring of work in progress by the supervisor.		
20.	Planners training has included formal training in planning/scheduling techniques, super user training on the CMMS, report generating software or via Excel and on the job training to include developing realistic planning times for craft work being planned. Understand use of MS Project or the company's larger project management system such as Primavera 6.		
21.	Benefits of planning/scheduling investments are being validated by various metrics that document areas such as reduced emergency work, improved craft productivity, improved schedule compliance, reduced cost and improved customer service.		
22.	Planning and scheduling procedures have been established defining work management and control procedures, the planning/scheduling process, the priority system, etc.		

23.	A reasonable number of backup planner/schedulers are selected and properly trained and used to cover for the full time staff. The number is based on the size and type work being planned. Ideally just like the full time planner should have good shop experience and sound technical experience.	
24.	If a maintenance coordinator is assigned within a unit of a large refinery or any production the planner will coordinate with that person about the job request, location within the unit, the problems to be repaired and related risks. In many cases this is an engineer or experienced operator who should be able to define complete requirements or a work request and in some case prepare a risk assessment for the planner to use for the job.	
25.	If issues and of any nature arise and are not readily resolved by the planner and operations, the maintenance leader should be the next step that the planner/scheduler takes for resolution.	
26.	If planned work orders involve participation by several shops or functional crews they are crossed over to a planner in that area. However a single planner/scheduler must plan and then coordinate various functional crews with the respective supervisors during the scheduling process.	
27.	Planners are in an excellent position to ensure critical spares by asset are accounted for as well as to recommend items to consider for including within the storeroom as critical spares.	
29.	Planners see what is repetitively coming up for non-stock item purchasing as well as what is being repaired over and over again, Are your planners active in this area can support improving reliability and uptime.	
30.	Planners help ensure that warranted parts or equipment is denoted in the equipment file and that work orders for warranted parts or equipment are flagged during the planning process to document supplier reimbursements.	
	11. Shop Level Reliable Planning, Estimating and Scheduling M/R SUBTOTAL SCORE POSSIBLE:	**300**

12. STO AND MAJOR PLANNING/SCHEDULING WITH PROJECT MANAGEMENT

ITEM #	Rating: Excellent – 10, Very Good – 9, Good – 8, Average – 7, Below Average – 6, Poor – 5 or less	RATING
1.	The planning and scheduling function includes major repairs, overhauls and shutdown, turnaround and outage (STO) type work not considered as part of normal maintenance work and any work requiring an STO event,	
2.	The planner team is a resource (or member) for the STO team of senior managers and the Maintenance Leader. Planners should meet at regular intervals to review current jobs awaiting a planned STO event.	
3,	Are your Total Backlog jobs coded and planned effectively to await an STO event. In large plants and refineries planners support the plant schedulers with normally detailed job packages for estimates of all required resources for an STO job	
4.	Schedulers from Item #3 coordinate parts/materials develop daily or weekly schedules, monitor status of work along with onsite observations, from the supervisor input and from a planner's job package which could include several crews, defined job steps and estimated time for each step. With real time reporting to a project management system or CMMS status including costs can be readily determined from progress reporting.	
5,	Current planning/scheduling manpower is available with the competency levels needed to support the site team during an STO.	
6.	All planners and schedulers involved with STO's must realize the potential hazard to plant reliability or can diminish or destroy reliability if not: properly planned, prepared and execute.	

7.	There may be even poor decisions by managers and engineers, bad workmanship, use of incorrect materials and damage done while plant is being shut down, overhauled or restarted. Planners realize that properly planned, well prepared work and work executed to all HSSE requirements is essential.	
8.	The use of work orders, estimating of repair times, coordinating and staging of repair parts/materials and planning/scheduling of internal resources and contractor support is also included for major work and STO work not considered day to day maintenance and repair.	
9.	A project work schedule or formal project management system is used to manage status and cost variance for STO work.	
10.	The current CMMS is integrated and linked to the project management system in real time when STO work orders or a change order work is approved.	
11.	Estimated labor and materials are established prior to project start using work orders with effective labor and material reporting to track overall cost, work progress, schedule compliance, etc.	
12.	The master plan for all major STO repairs, overhauls and new installation is available indicating planned start date, duration, completion date, and type crafts required.	
13.	Resources required for day to day maintenance work are not compromised by having to perform major repair type work, installation, modifications etc, consuming in house resources required for PM's and other day to day type work.	
14.	Scheduling/progress meetings are held periodically with operations to ensure understanding, agreement and coordination of major work and problem areas such as asset being ready for scheduled work.	
15.	Major work performed by contractors is preplanned, scheduled and includes measuring performance of contracted services.	
16.	Planning and scheduling procedures have been established for project type work.	
	12.STO and Major Maintenance Planning/Scheduling with Project Management SUBTOTAL SCORE POSSIBLE:	**160**

22. **Planned work increasing**: The bottom line results for the planning process are to actually increase the level of planned work. Percent planned work should be monitored and included as one of the overall maintenance performance metrics. In some organizations with effective PM/PdM programs, the level of planned work can be in the 90% range or more.

Planned vs. Unplanned

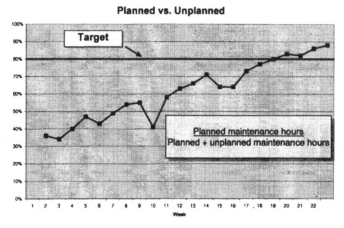

23. **Craft utilization measured and improving**: Effective planning and scheduling is essential to increasing the level of actual hands-on wrench time of the craft workforce. Improving craft utilization allows more work to get done with current staff by eliminating non-craft activities such as waiting for equipment, searching for parts, and scheduling the right sequence for different crafts on the job.

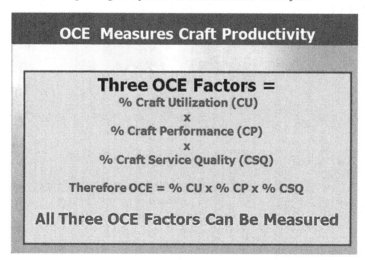

24. **Work schedules available**: One key responsibility of the planning process is to establish realistic work schedules for bringing together the right type craft resources, the parts required, the equipment to be repaired or serviced, along with having the time available to complete the job right the first time. The actual schedule may only start with a one-day schedule and gradually work up to scheduling longer periods of time. Work schedules provide a very important customer service link with operations that helps to improve overall coordination between maintenance and operations.

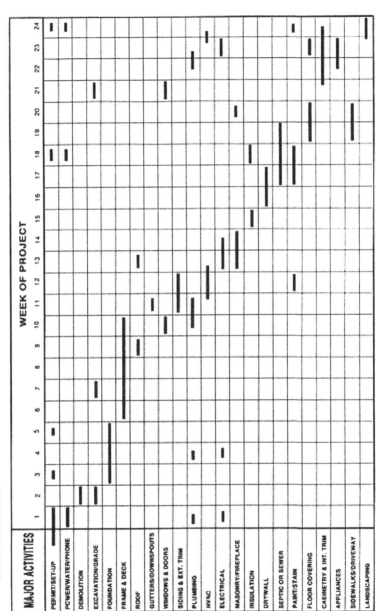

25. **Spare parts status is available**: One of the most essential areas to support effective planning is the maintenance storeroom and the accuracy of the parts inventory management system. Jobs should not be put on the schedule without parts being on hand. The planner must have complete visibility of inventory on-hand balances, parts on order, and the capability to reserve parts for planned work.

✓ IC: Insufficient craft capacity; lack of cross training contributes to shortage of the right skills

✓ SO: Stock outs are frequent
  - ✓ Inaccurate inventory control
  - ✓ Excessive time finding what is in storeroom
  - ✓ Parts not requisitioned

✓ PP: Planned jobs do not reflect reality
  - ✓ Scope expands
  - ✓ Plan for parts is incorrect
  - ✓ Job steps incomplete
  - ✓ Lockout, specifications and regulations not documented

26. **Scheduling coordination with operations**: As the planning function develops, there will be improved coordination with operations to develop and agree upon work schedules. This may involve coordination meetings near the end of each week to plan weekend work or to schedule major jobs for the upcoming week. Direct coordination with operations allows maintenance to review PM/PdM schedules or to review jobs where parts are available to allow the job to be scheduled based on operations scheduling equipment availability.

27. **True emergency repairs tracked**: Many organizations really focus on reducing true emergency repairs that create uncertainty for operations scheduling and contribute to significantly higher total repair costs than planned work. Improved reporting capabilities of an effective CMMS will allow for better tracking of emergency repairs, document causes for failures, and assist in the elimination of the root causes for failures.

Maintenance repair operations (MRO) materials management is the overall area of MRO parts and materials procurement, storage, inventory management, and issues represents another best practice area that often needs major work when implementing a CMMS and developing a "Class A" installation. Many organizations never take the time to set up a well-planned and controlled storeroom operation and often find out that their parts database is a weak link needing major updates before it can truly be used effectively.

F. Maintenance repair operations (MRO) materials management

28. Inventory management module fully utilized and integrated with work order module
29. Inventory cycle counting based on defined criteria is used, and inventory accuracy is 95% or better
30. Parts kitting and staging is available and used for planned jobs
31. Electronic requisitioning capability available and used

F. Maintenance repair operations (MRO) materials management

32. Critical and/or capital spares are designated in parts inventory master record database
33. Reorder notification for stock items is generated and used for reorder decisions
34. Warranty information and status is available

---

28. **Inventory management module**: The work order module must be fully integrated with an accurate parts inventory management module to charge parts back to work orders, to check parts availability status for planned work, to reserve parts, and to check status of direct purchases. A "Class A" CMMS installation will develop, maintain, and fully utilize the inventory module and ensure that it is fully integrated with the work order module.

29. **Inventory cycle counting established**: Inventory accuracy should be one of the key metrics for MRO materials management, and it can best be accomplished by cycle counting rather than annual physical inventories. Most CMMS systems will allow for your own criteria to be developed such as doing an ABC analysis of inventory items (based on either usage value or frequency of issue) and then scheduling of periodic counts for each classification of inventory item that you want to cycle count. For example, A items would be counted more frequently than B and C items. The real value of cycle counting is that it is a continuous process that creates a high level of discipline and allows for inventory problems and adjustments to be made throughout the year rather than once following the annual inventory.

Annual Physical Inventory versus Cycle Counting Best Practices

Definition of Cycle Counting

- Periodic inventory system audit-practice in which different portions of an inventory are counted or physically checked on a continuous schedule.

- Each portion is counted at a definite, preset frequency to ensure counting of each item at least once in an accounting period (usually an year).

- Fast-moving or more expensive items are counted more often than slower moving or less expensive ones, and certain items are counted every day. Also called cycle inventory.

30. **Parts kitting and staging**: This best practice area is key to the planning process and can evolve over time as the planning process matures to the point of being able to give the storeroom prior notification on the parts required for planned jobs. Controlled staging areas are set for parts that are either pulled from stock or received from direct purchases.

31. **Electronic parts requisitioning**: This functional capability can provide paperless work flow for requisitioning of parts directly from maintenance to the storeroom for creation of a pick list for the item or go to purchasing to create a purchase order for a stock item or direct purchase. In some cases, electronic requisitioning might go directly to the vendor using e-commerce capability.

32. **Critical spares identified**: Critical spares (or insurance items) that may be a one-of-a-kind, high cost spare are often in the parts inventory system. It is recommended that these items be classified and identified in the item master record as such. This practice will help separate these items from the regular inventory management process and identify them as a separate part of the total inventory value that has been fixed. Critical spares should also be identified in the spares list for the equipment they have been purchased for.

Solving Special Storage Challenges

- Electronic Components
- Pipe & Metal for Fab'n
- Project Materials
- Motors
- Rebuild Kits
- Belts
- Filters
- Large Pumps
- Other Special Storage Challenges?
- Space for Controlled Shop Stock

228

33. **Reorder notification process**: The capability to determine when and what to reorder based on a review of stock level reorder points is an important feature for a "Class A" installation. A recommended reorder report should be generated periodically and reviewed for validity as well as for any future needs that may not be reflected in current on-hand balances. Based on final review of the recommended reorder report, electronic requisitioning then could occur directly to purchasing.

34. **Warranty information**: Many organizations fail to have a process in place to track warranty information and in turn incur added costs by not being able to get proper credit for items under warranty. Tracking specific high value parts or components and specific equipment under warranty should be a CMMS functionality of the equipment master or the inventory item master database. The system should provide a quick reference and alert to the fact that the item is still under warranty and that a follow-up claim to the vendor is needed.

G. Preventive/predictive maintenance (PM/PdM)

35. PM/PdM change process is in place for continuous review/update of tasks/frequencies
36. PM/PdM compliance is measured and overall compliance is 98% or better
37. The long-range PM/PdM schedule is available and level loaded as needed with computerized maintenance management system (CMMS)
38. Lube service specifications, tasks, and frequencies included in CMMS database
39. CMMS provides mean time between failures, mean time to repair, failure trends, and other reliability data
40. PM/PdM task descriptions contain enough information for new crafts person to perform task

35. **PM/PdM change process**: This best practice area simply ensures that PM/PdM procedures are subject to a continuous review process and that all changes to the program are made in a timely manner. The CMMS system should provide an easy method to update task descriptions and task frequencies and allow for mass updating when the procedure applies to more than one piece of equipment.

36. **PM/PdM compliance is measured**: One key measure of overall maintenance performance should be how well the PM/PdM program is being executed based on the schedule. Measuring PM/PdM compliance ensures accountability from maintenance and from operations. Normally, a scheduling window of a week will be established to determine compliance. A goal of 98% or better for PM/PdM compliance should be expected.

37. **Long range PM/PdM scheduling**: As a PM/PdM schedule is loaded to the system, peaks and valleys may occur for the actual scheduling due to frequencies of tasks coming due at the same time period. The CMMS system should provide the capability to level load the actual PM/PdM schedule and to view upcoming PM/PdM work loads to assist in the overall planning process.

38. **Lubrication services**: Ideally, lubrication services, tasks, frequencies, and specifications should be included as part of the PM/PdM module. A continuous change process for this area should also be put in place as well as an audit process established to ensure all lube and PM/PdM tasks are being performed as scheduled.

39. **CMMS captures reliability data**: The elimination of root causes of problems is the goal rather than just more PM/PDM. One important feature of a "Class A" installation is being able to capture failure information that can in turn be used for reliability improvement. This requires development of a good coding system for defining causes for failures and accurately entering this information as the work order is closed.

40. **Complete PM/PDM task descriptions**: PM/PdM task descriptions often provide vague terminology to check, adjust, inspect, etc. and do not provide clear direction for specifically what is to be done. Task descriptions should be reviewed periodically and details added that to the level that a new crafts person would understand exactly what is to be done and be able to adequately perform the stated task.

20. RELIABILITY CENTERED MAINTENANCE (RCM)

ITEM #	Rating: Excellent – 10, Very Good – 9, Good – 8, Average – 7, Below Average – 6, Poor – 5	RATING
1.	A process such as RCM is used to determine the maintenance requirements of any critical physical asset in its operating context.	
2.	Criticality analysis to define top candidates for review has been conducted those factors in production consequence (pd), safety/ env. cons (sf), service level (sl), redundancy(rf), frequency of failure (ff) and down time (dt)"	
3.	The maintenance team strives to ensure any physical asset continues to do whatever its users want it to do in its present operating context?	
4.	Operations context is clearly defined and understood by maintenance and operations to include the type of process which is the most important feature of the "operating context". Some areas must be very clear before starting the RCM process. ✓ Continuous processing where failure may stop entire plant or failure could significantly reduce output ✓ Batch processing or discrete manufacturing	
5.	Other factors related to operating context of the asset include: Redundancy factors, Quality standards, Environmental standards, Actual physical location, Safety hazards, Shift arrangements , Work in process levels, Repair time & overall Mean Time to Repair (MTTR), Spares availability, Market demand, Raw material supply.	
6.	Strategies and operating principles are defined and characterize the key steps which are: ✓ A focus on the preservation of system function ✓ The identification of specific failure modes to define loss of function or functional failure ✓ The prioritization of the importance of the failure modes, because not all functions or functional failures are equal ✓ Failure consequences ✓ The identification of effective and applicable PM/PdM tasks for the appropriate failure modes ✓ Applicable means that the task will prevent, mitigate, detect the onset of, or discover the failure mode Effective means that among competing candidates the selected PM/PdM task is the most cost-effective option	
7.	Maintenance can define ways the asset can fail to fulfill its function where a functional failure is defined as the inability of any asset to fulfill a function to a standard of performance which is acceptable to the user.	
8.	Maintenance understands that it is more accurate to define failure in terms of a specific function rather than the asset as a whole. Categories of functional failures may include Partial and total failure, Upper and lower limits, Gauges and indicators along with the operating context. The RCM process looks to define/record all functional failures with each function.	
9.	The organization's work order system captures functional failures, cause of failures and frequencies.	

10.	When defining functions the function statement should consist of a verb, object and a desired standard of performance: "To pump water from Tank X to Tank Y at not less than 300 gallons per minute.
11.	Performance standards are defined in two ways: Desired Performance: What the user wants the asset to do and Design Capability: What the asset can do?
12.	Function statements may include the appropriate but different types of performance standards depending on the asset such as Multiple, Quantitative, Qualitative, Absolute, Variable or Upper and lower limit types of standards.
13.	A failure mode that causes functional failures is being documented by the work order system there is by helping maintenance define the options for a maintenance strategy or development of caring equipment plans.
14.	Based upon an analysis of work completed a reactive maintenance strategy of dealing with failure events after they occur is prevalent.
15.	Proactive maintenance that deals with events before they occur or deciding, planning and scheduling how they should be dealt with is a solid strategy.
16.	Failure Mode Categories are designated when a) Capability falls below desired performance b)Desired performance rises above capability and when Asset not capable from the start per design or operating procedures.
17.	Failure Mode Categories Capability falling below desired performance can be identified as deterioration (all forms of "wear and tear, Lubrication failure, Dirt/dust, Disassembly where integrity of the assemblies (welds, rivets, bolts) decline due to fatigue or corrosion and human errors.

18.	When desired performance rises above capability the following can be identified:
	  a) Sustained, deliberate overloading
	  b) Sustained, unintentional overloading and
	  c) Incorrect process materials

19.	The Level of Detail in Defining Failure a Mode provides enough detail to select a suitable failure management strategy.
20.	Failure modes which might reasonably be expected to occur in the current operating context are defined and will include; Failures that have occurred on the same or similar assets, Failure modes which are already under PM/PdM, Any other failure modes that are considered real possibilities and Where consequences are very severe if a failure occurs.
21.	Failure effects and failure consequences' are not the same. Failure effects answer the question, "what happens", whereas failure consequences answer the question "how does it matter?" Failure consequences are factored as one element of component/equipment criticality rating method being used.

22.	Describing the effects of a failure strives to answer this question "What evidence is there that the failure occurred" part of which may be answered by a maintenance planner?
	  ⟩ In what way did the failure pose a threat to safety & the environment?
	  ⟩ In what ways did the failure affect production or operations?
	  ⟩ What physical damage is caused by the failure?
	  ⟩ What must be done to repair the failure?
	Is this reported correctly on a work order, is part of a standard procedure or becomes an after action report

	completed for Total Cost of Down Time (TCDT).

23.	Completing the Failure Modes and Effects Analysis (FMEA) utilizes steps 1 to 20 above that point's maintenance toward the best maintenance strategy to use.
24.	Failure consequences answer the question "How does it matter?" They're many consequences of failure that: ✓ Impact on output, quality and customer service ✓ Personal safety and environmental issues ✓ Increase in operating cost, energy consumption ✓ Related to Nature and severity of effects govern and whether the users of the asset really believe that a failure matters

25.	The focus on consequences starts the RCM process of task selection
	✓ Assessment of the effects of each failure mode
	✓ Classifying into 4 basic categories of consequences
	  1. Safety and environmental consequences
	  2. Operational consequences
	  3. Non-operational consequences
	  4. Hidden failure consequences

26.	What can be done to predict or prevent each failure? ✓ Proactive tasks: Tasks undertaken before a failure occurs to prevent asset from going into a failed state. They include: preventive, predictive, scheduled restoration/overhaul, scheduled discard/replacement, condition monitoring and proactive task is worth doing if it reduces the consequences of failure enough to justify the direct and indirect cost of doing the task
27.	The organization uses the RCM process to make the best failure management decision for critical assets by considering: ✓ Age-related failures ✓ Non-age related failures (operator error) ✓ Cost factors for scheduled restoration/overhaul ✓ Cost factors for scheduled discard/replacement ✓ Identifying potential failures and the P-F Interval which are time that a potential failure begins and the time that a functional failure actually occurs and the question "Is the P-F interval enough time to deal with failure/consequences" and what condition monitoring options available?
28.	Decisions are made on what should be done if a suitable proactive task cannot be found? ✓ For hidden functions that cause multiple failures look for a possible failure finding task; if one is not available maybe redesign? ✓ If safety or environmental issue cannot be resolved by proactive task; redesign or change the process. ✓ For operational consequences, if no proactive tasks available and costs are less then no scheduled maintenance (run to failure). Look to redesign if costs too high. ✓ For non-operational consequences (same as above)
29.	There is a very clear and solid understanding about the P-F Interval Curve and the high cost of gambling with

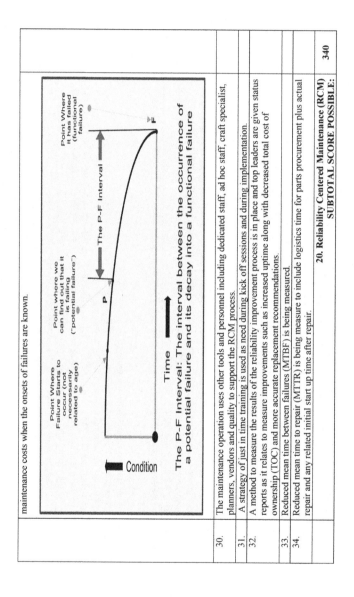

maintenance costs when the onsets of failures are known.

The P–F Interval: The interval between the occurrence of a potential failure and its decay into a functional failure

30.	The maintenance operation uses other tools and personnel including dedicated staff, ad hoc staff, craft specialist, planners, vendors and quality to support the RCM process.	
31.	A strategy of just in time training is used as need during kick off sessions and during implementation.	
32.	A method to measure the results of the reliability improvement process is in place and top leaders are given status reports as it relates to measure improvements such as increased uptime along with decreased total cost of ownership (TOC) and more accurate replacement recommendations.	
33.	Reduced mean time between failures (MTBF) is being measured.	
34.	Reduced mean time to repair (MTTR) is being measure to include logistics time for parts procurement plus actual repair and any related initial start up time after repair.	
	20. Reliability Centered Maintenance (RCM) **SUBTOTAL SCORE POSSIBLE:**	**340**

H. Maintenance performance measurement

41. Downtime (equipment/asset availability) due to maintenance is measured and documented improvements have occurred
42. Craft performance against estimated repair times is measured and documented improvements have occurred
43. Maintenance customer service levels are measured and documented; schedule compliance improvements have occurred
44. The maintenance performance process is well established and based on multiple indicators compared to baseline performance values

---

41. **Equipment downtime reduction**: Another key metric for measuring overall maintenance performance is increased equipment uptime. The improvement in this metric is a combination of many of the previously mentioned best practices all coming together for improved reliability. Downtime due to maintenance should be tracked and positive improvement trends should be occurring within a "Class A" installation.

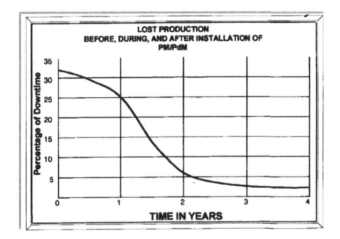

42. **Craft performance:** Two key areas affecting overall craft productivity are craft utilization (wrench time) and craft performance. Measurement of craft performance requires that realistic planning times be established for repair work and PM tasks. A standard job plan database can be developed for defining job scope, sequence of tasks, special tools listing, and estimated times. The goal is measurement of the overall craft workforce and not individual performance. Planning times are also an essential part of the planning process for developing a more accurate picture of workload and to support scheduling of overtime and staff additions.

43. **Maintenance customer service**: The results of improved maintenance planning must be improved customer service. The overall measurement process should include metrics such as compliance to meeting established schedules and jobs actually completed on schedule.

44. Maintenance performance measurement process: In this area, it is important to have a performance measurement process that includes a number of key metrics in each of the following major categories:

a. Budget and cost
b. Craft productivity
c. Equipment uptime
d. Planning and scheduling
e. Customer service
f. MRO materials management
g. Preventive and predictive maintenance

The overall maintenance performance process should be established so that it clearly validates the benefits being received from the CMMS and maintenance best practice implementation.

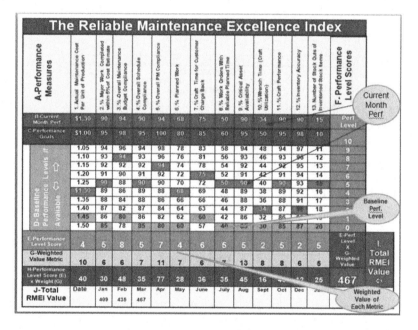

I. Other uses of computerized maintenance management system (CMMS)

45. Maintenance leaders use CMMS to manage maintenance as internal business
46. Operations staff understand CMMS and use it for better maintenance service
47. Engineering changes related to equipment/asset data, drawings, and specifications are effectively implemented
48. Hierarchies of systems/subsystems used for equipment/asset numbering in CMMS database
49. Failure and repair codes used to track trends for reliability improvement
50. Maintenance standard task database available and used for recurring planned jobs

45. Maintenance managed as a business: One true indicator for a successful CMMS installation is that it has changed the way that maintenance views its role in the organization. It should progress to the point that maintenance is viewed and managed as an internal business. This view requires greater accountability for labor and parts costs, greater concern for customer service, better planning, and greater attention to reliability improvement and increased concern for the maintenance contribution to the bottom line.

46. Operations understands benefits of CMMS: There is direct evidence that operations understands that an improved CMMS is a contributor to improved customer service. The scheduling process is continuously improving through better coordination and cooperation between maintenance and operations within a "Class A" installation.

47. Engineering changes: Accurate engineering drawings are essential to maintenance planning and to actually making the repairs. Asset documentation must be kept up to date based on a formal engineering change process. Feedback to engineering must be made on all changes as they occur on the shop floor. Engineering must in turn ensure that master drawings are updated and that current revisions are made available to maintenance. Appendix I: Management of Change (MOC) Procedure – Courtesy of Peru LNG is provided as an MOC guideline.

48. Equipment database structure: To provide equipment history information in a logical parent–child relationship, the equipment database structure has been developed using an identification of systems and subsystems. Accessing the equipment database should allow for drill down from a parent level to lower level child locations that are significant enough for equipment master information to be maintained. This definitely includes instrument loops within refineries and all continuous processing plants.

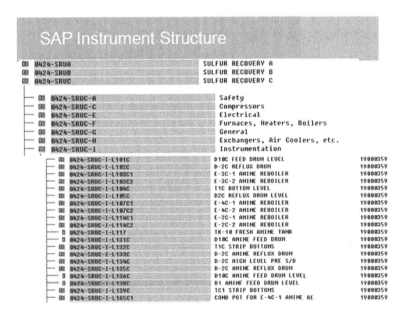

49. Failure and repair codes: The reporting capability of the CMMS should provide good failure trending and support analysis of the failure information that is entered from completed work orders. Improving reliability requires good information that helps to pinpoint root causes of failure.

50. **Maintenance standard task database**: Developing the maintenance standard task database (or standard repair procedures and detailed job packages) for recurring jobs is an important part of a planner's job function. This allows for determining scope of work, special tools and equipment, and for estimating repair times. Once a standard repair procedure is established, it can then be used as a template for other similar jobs, resulting in less time for developing additional repair procedures. This database is an excellent source of benchmark jobs for review by the ACE team discussed in Chapter 16, Developing Improved Repair Methods plus Reliable Maintenance Planning Times with the ACE Team Process™.

Summary: Developing a "Class A" CMMS installation requires the combination of good system functionality and improved maintenance practices. The CMMS team should begin very early during implementation with how it will measure the success of the installation. The recommendations provided here for using the CMMS Benchmarking System can help your organization achieve maximum return on its CMMS investment.

   Understand the power of CMMS/EAM to support potential savings: The evaluation of your CMMS using the CMMS Benchmarking System will identify improvement opportunities that translate into direct savings. It is important that these areas be highlighted and that the future process for performance measurement is focused upon these specific areas that may have been used initially for CMMS/EAM capital project justification. The opportunities to realize both quantifiable and qualifiable benefits are numerous. Maintenance must be given the best practice tools, the people resources, and capital investments to address the improvement opportunities, and in turn are held accountable for results. As summarized in Part 2, there are 12 key areas where direct savings, cost avoidances, and gained value can be established and documented. Effective CMMS/EAM will contribute to all of them and help to increase:

1. Value of asset/equipment uptime providing increased capacity and throughput
2. Value of increased quality and service levels due to maintenance

3. Value of facility availability or cost avoidance from being nonavailable
4. Value of increased direct labor utilization (production operations)
5. Gained value from increased craft labor utilization/effectiveness via gains in wrench time
6. Gained value from increased craft labor performance/efficiency
7. Gained value of clerical time for supervisors, planners, engineering, and administrative staff
8. Value of MRO materials and parts inventory reduction
9. Value of overall MRO materials management improvement
10. Value of overall maintenance costs reductions with equal or greater service levels
11. Value of increased facility and equipment life and net life cycle cost reduction
12. Other manufacturing and maintenance operational benefits, including improved reliability and other reduced costs

Use CMMS to develop your maintenance operation as a profit center: A fully utilized CMMS is your business management system to support the business of maintenance. It is a mission-essential information technology tool, and effective physical asset management and maintenance is also mission essential and a core requirement for success. Often we see the CMMS being purchased as "the solution" and never really integrated with the business system, or the necessary basic best practices are not initiated to really make the IT investment work. Often maintenance is only viewed as a "necessary evil" and not as a valid "profit center" and internal business. Many times the maintenance leaders cannot sell management on doing maintenance the right way or able to convince them that the right thing to do is to shut down for preventive maintenance. Conversely, when maintenance is viewed as a "profit center," the opportunities to realize both quantifiable and quantifiable benefits are numerous. In turn, maintenance support to the profit optimization process continues when CMMS is used effectively to develop your maintenance operation as a profit center.

Defining Maintenance Strategies for Critical Equipment With Reliability-Centered Maintenance (RCM)

This chapter is included to expose the planner/scheduler to the concept of reliability-centered maintenance (RCM). Some professionals believe that the so-called classical RCM has been made much more complicated than it needs to be. Nonetheless, RCM was first introduced in the field of commercial aviation. It made its way to the nuclear industry in the mid-1980s, then spread to other organizations. In all, it is been estimated that more than 60% of all RCM programs initiated have failed to be successfully implemented. According to Neil Bloom (2005) in his book *Reliability Centered Maintenance—Implementation Made Simple,* RCM became overly complicated in its transfer from the airlines and history. Also, it is his belief that the successful application of the process is inversely proportional to the complexity it has acquired. Bloom also stated that some consultants employ an ***elixir of obfuscation*** to allow them sole possession of understanding the process, and hence a continued income stream.

> *Bloom (2005) clearly stated that RCM is not a preventive maintenance (PM) reduction program. It is a reliability program. RCM will indeed identify those unnecessary PMs that may become candidates for deletion. RCM is almost always described as a process of identifying critical components whose failure would result in an unwanted consequence to one's facility. As a planner, your organization may have experienced some of the following reasons for lack of success or it may have a successful. You may be what is needed to get the RCM analysis on the right track. Many times, one or more of the following occurrences cause a lack of success:*

1. Lost of in-house control
2. An incorrect mix of personnel performing the analysis
3. Unnecessary and costly administrative burdens
4. Fundamental RCM concepts are not understood
5. Confusion determining system functions
6. Confusion concerning system boundaries and interface
7. Divergent expectations
8. Confusion regarding convention
9. Misunderstanding hidden failures and redundancy
10. Misunderstanding run-to-failure
11. Inappropriate component classifications
12. Instruments were not included as part of the RCM analysis

Reliable Maintenance Planning, Estimating, and Scheduling. http://dx.doi.org/10.1016/B978-0-12-397042-8.00009-7

My goal for this chapter is for the planner/scheduler to understand item 4—the fundamental RCM concepts. Let us start with a definition and then a question. First, RCM can be defined as a process that is used to determine the maintenance requirements of any physical asset in its operating context. It provides a detailed process to answer this question: What must be done to ensure that any physical asset continues to do whatever its users want it to do in its present operating state?

RCM can be defined as the seven key elements as shown in Figures 9.1–9.17.

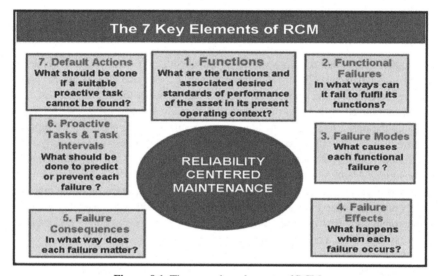

Figure 9.1 The seven key elements of RCM.

Reliability-Centered Maintenance (RCM)

The principles which define and characterize RCM are:
✓ A focus on the preservation of system function

✓ The identification of specific failure modes to define loss of function or functional failure

✓ The prioritization of the importance of the failure modes, because not all functions or functional failures are equal and

✓ The identification of effective and applicable PM/PdM tasks for the appropriate failure modes

    ✓ Applicable means that the task will prevent, mitigate, detect the onset of, or discover the failure mode

    ✓ Effective means that among competing candidates the selected PM/PdM task is the most cost-effective option

Figure 9.2 The principles that define and characterize RCM.

Reliability-Centered Maintenance (RCM)

These RCM principles, in turn, are typically implemented in a seven step process:

1. The objectives of maintenance with respect to any particular item/asset are defined by the <u>functions</u> of the asset and its associated desired <u>performance standards</u>.

2. <u>Functional failure</u> (the inability of an item/asset to meet a desired standard of performance) is identified. This can only be identified after the functions and performance standards of the asset have been defined.

3. Failure modes (what causes each functional failure) are identified.

4. Failure effects (describing what will happen if any of the failure modes occur) are documented. (Consequences of failures)

Figure 9.3 RCM is typically implemented in seven steps.

Reliability-Centered Maintenance (RCM)

5. Failure consequences are quantified to identify the criticality of failure. RCM not only recognizes the importance of the failure consequences but also classifies these into four groups:
   - ✓ Hidden failure
   - ✓ Safety and environmental
   - ✓ Operational
   - ✓ Non-operational

6. Functions, functional failures, failure modes are criticality analyzed to identify opportunities for improving performance and/or safety.

7. Proactive tasks are established which may include:
   - ✓ Scheduled on-condition tasks (which employ condition-based or predictive maintenance)
   - ✓ Scheduled restoration/overhaul
   - ✓ Scheduled discard tasks/replacement

Figure 9.4 RCM is typically implemented in seven steps.

1. Functions and Performance Standards

What are the functions and associated performance standards of the asset in its present operating context?

Defining the Operating Context: **The type of process is the most important feature of the "operating context". It must be very clear before starting the RCM process.**
✓ **Continuous processing**
    ✓**Failure may stop entire plant**
    ✓**Failure could significantly reduce output**
✓ **Batch processing or discrete manufacturing**

Other factors related to operating context of the asset:
✓ **Redundancy**	✓ **Quality standards**
✓ **Environmental standard**	✓ **Safety hazards**
✓ **Shift arrangements**	✓ **Work in process**
✓ **Repair time**	✓ **Spares**
✓ **Market demand**	✓ **Raw material supply**

Figure 9.5 Functions and performance standards.

2. Functional Failures

In what way does the asset fail to fulfill its function?

Functional Failure: A functional failure is defined as the inability of any asset to fulfill a function to a standard of performance which is acceptable to the user.

✓ **More accurate to define failure in terms of a specific function rather than the asset as a whole.**

✓ **Categories of functional failures**
    ✓ **Partial and total failure**
    ✓ **Upper and lower limits**
    ✓ **Gauges and indicators**
    ✓ **The operating context**

✓ **The RCM process looks to define/record all functional failures with each function.**

Figure 9.6 Functional failures.

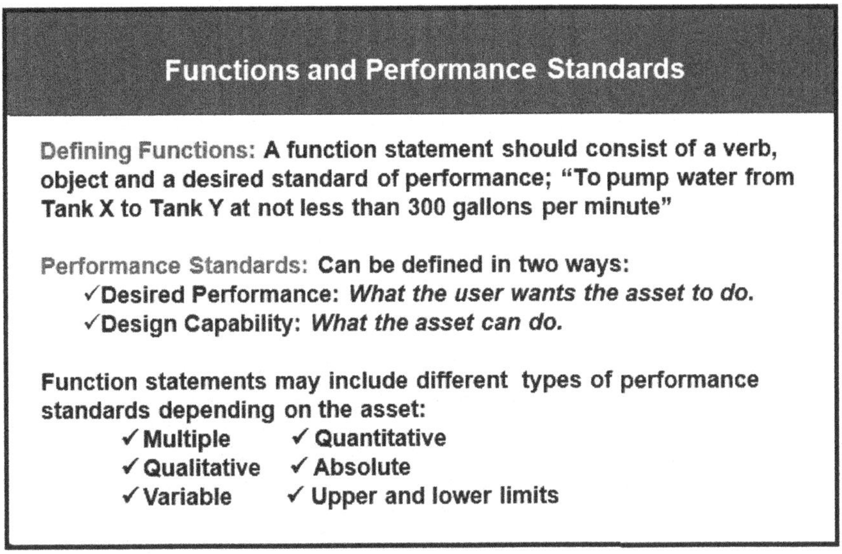

Functions and Performance Standards

Defining Functions: A function statement should consist of a verb, object and a desired standard of performance; "To pump water from Tank X to Tank Y at not less than 300 gallons per minute"

Performance Standards: Can be defined in two ways:
  ✓ Desired Performance: *What the user wants the asset to do.*
  ✓ Design Capability: *What the asset can do.*

Function statements may include different types of performance standards depending on the asset:
  ✓ Multiple   ✓ Quantitative
  ✓ Qualitative   ✓ Absolute
  ✓ Variable   ✓ Upper and lower limits

Figure 9.7 Functional failures.

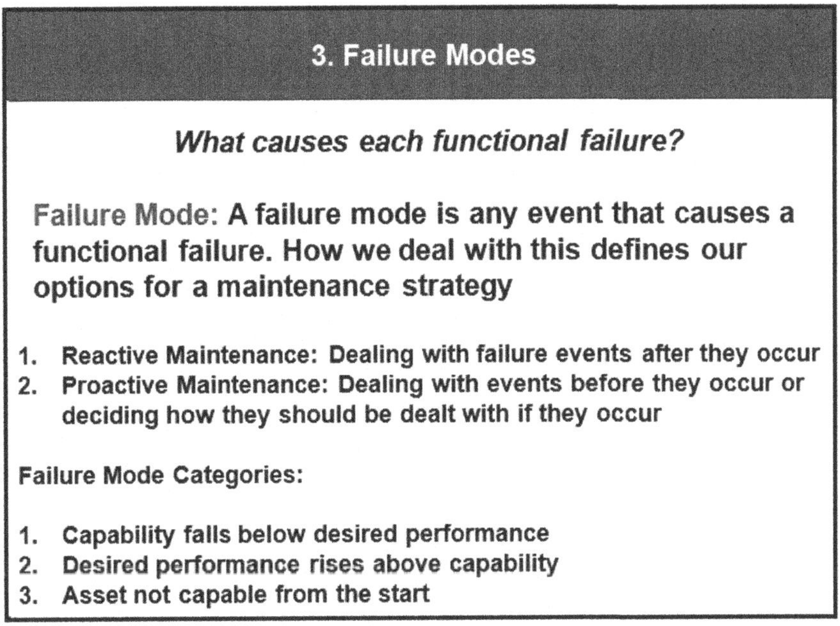

3. Failure Modes

What causes each functional failure?

Failure Mode: A failure mode is any event that causes a functional failure. How we deal with this defines our options for a maintenance strategy

1. Reactive Maintenance: Dealing with failure events after they occur
2. Proactive Maintenance: Dealing with events before they occur or deciding how they should be dealt with if they occur

Failure Mode Categories:

1. Capability falls below desired performance
2. Desired performance rises above capability
3. Asset not capable from the start

Figure 9.8 Failure modes.

4. Failure Effects

What happens when each failure occurs?

Functional Effects: Failure effects are not the same as failure consequences. Failure effects answer the question, "what happens", whereas failure consequences answers the question "how does it matter?"

Describing the effects of a failure strives to answer these questions: What evidence is there that the failure occurred?

✓ In what way did the failure pose a threat to safety & the environment?

✓ In what ways did the failure affect production or operations?

✓ What physical damage is caused by the failure?

✓ What must be done to repair the failure?

Items #1 to # 4 all lead now to completing the Failure Modes and Effects Analysis (FMEA)

Figure 9.9 Failure effects.

System: 5MW Gas Turbine	Sub-System: Exhaust System	EXAMPLE: FAILURE MODES AND EFFECTS ANALYSIS (FMEA)	
FUNCTION	FUNCTIONAL FAILURE (Loss of Function)	FAILURE MODE (Cause of Failure)	FAILURE EFFECT What happens when it fails?
1. To channel all the hot turbine air without restriction to a fixed point 10m above the roof of the turbine building.	A. Unable to channel gas at all.	1. Silencer mountings corroded away.	Silencer assembly collapses and falls to bottom of stack. Back pressure causes the turbine to surge violently and shut down on high exhaust gas temperature. Downtime to replace silencer up to four weeks.
	B. Gas flow resticted.	1. Part of silencer falls off due to fatigue.	Depending on the nature of blockage, exhaust temperature may rise to where it shuts down the turbine. Debris could damage parts of the turbine. Downtime to repair silencer up to four weeks.
	C. Fails to maintain the gas.	1. Hole in flexible joint from corrosion.	The joint is inside turbine hood, so leaking exhaust gases would be extracted by the hood extraction system. Fire and gas detection equipment inside hood is unlikely to detect an exhaust gas leak, and temperatures are unlikely to rise enough to trigger the fire wire. A severe leak may cause gas demister to overheat, and may melt control wires near the leak with unpredictable effects. Pressure balance inside the hood are such that little or no gas is likely to escape from a small leak, so a small leak is unlikely to be detected by smell or hearing. Downtime to replace joint is 3 days.
		2. Gasket in ducting improperly fitted.	Gas escapes into turbine hood and ambient temperature rises. Building ventilation system would expel gases through louvers to atmosphere. So concentration of gases is unlikely to reach noxious levels. A small leak at this point would be audible. Downtime to repair up to 4 days.
		3. Hole in upper bellows due to corrosion.	The upper bellows are outside the turbine building, so a leak here discharges to the atmosphere. Ambient noise levels may rise. Downtime to repair, a few days to several weeks.
2. To reduce exhaust noise to ISO Rating 30 at 50m.	A. Noise level exceeds ISO Rating 30 at 50m.	1. Silencer material retaining mesh corroded away.	Most of the material would be blown out, but some might fail to the bottom of the stack and obstruct the turbine outlet causing high EGT and possible turbine shutdown. Noise levels would rise gradually. Downtime to repair about 2 weeks.

Figure 9.10 Failure modes and effects analysis (FMEA).

5. Failure Consequences

In what way does each failure matter?

Failure Consequences: **Failure consequences answers the question "how does it matter". They're many consequences of failure:**
- ✓ **Impact on output, quality and customer service**
- ✓ **Personal safety and environmental issues**
- ✓ **Increase in operating cost, energy consumption**

Nature and severity of effects govern whether users of the asset really believe that a failure matters

The focus on consequences starts the RCM process of task selection
- ✓ **Assessment of the effects of each failure mode**
- ✓ **Classifying into 4 basic categories of consequences**
    1. **Safety and environmental consequences**
    2. **Operational consequences**
    3. **Non-operational consequences**
    4. **Hidden failure consequences**

Figure 9.11 Failure consequences.

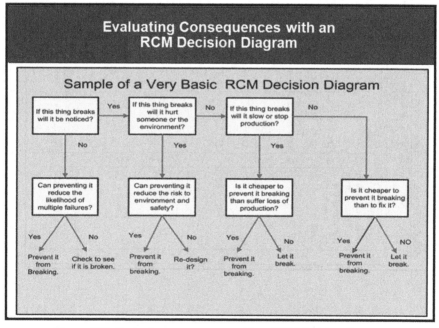

Figure 9.12 Evaluating consequences with an RCM decision diagram.

6. Proactive Maintenance

Step 6 in the RCM process strives to make the best failure management decision by considering:

- ✓ **Age-related failures**
- ✓ **Non-age related failures (operator error)**
- ✓ **Cost factors for scheduled restoration/overhaul**
- ✓ **Cost factors for scheduled discard/replacement**
- ✓ **Identifying potential failures and the P-F Interval which is:**
  - ✓ **The time that a potential failure begins and**
  - ✓ **The time that a functional failure actually occurs**
  - ✓ **Is the P-F interval enough time to deal with failure/consequences?**
  - ✓ **Condition monitoring options?**

Figure 9.13 Proactive maintenance.

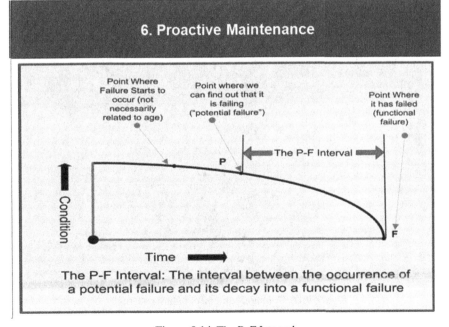

Figure 9.14 The P–F Interval.

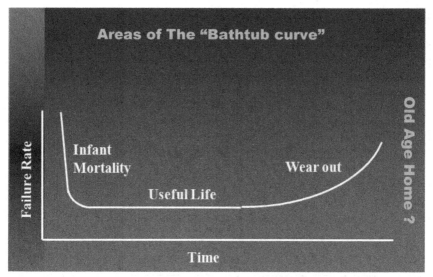

Figure 9.15 Areas of the "bathtub curve "

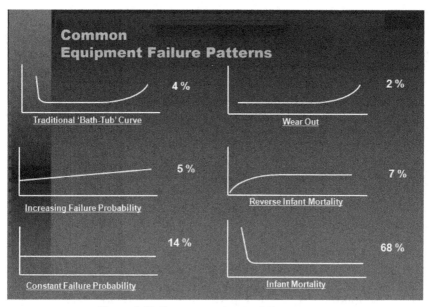

Figure 9.16 Common equipment failure patterns.

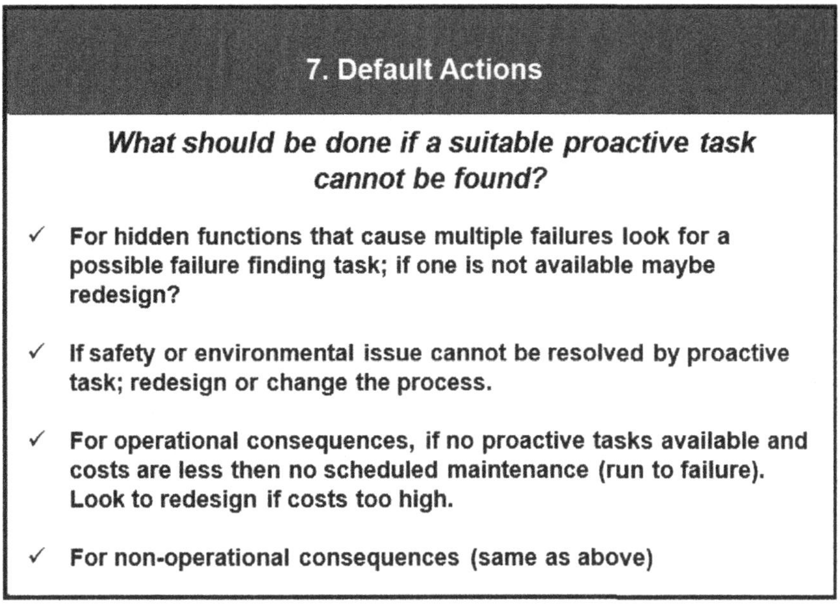

---

7. Default Actions

What should be done if a suitable proactive task cannot be found?

✓ **For hidden functions that cause multiple failures look for a possible failure finding task; if one is not available maybe redesign?**

✓ **If safety or environmental issue cannot be resolved by proactive task; redesign or change the process.**

✓ **For operational consequences, if no proactive tasks available and costs are less then no scheduled maintenance (run to failure). Look to redesign if costs too high.**

✓ **For non-operational consequences (same as above)**

Figure 9.17 Default actions.

Further Reading

Bloom, Neil B., 2005. Reliability Centered Maintenance; Implementation Made Simple. McGraw-Hill Companies Inc.
Moubray, John, 1997. Reliability-centered Maintenance. Industrial Press.
Smith, Anthony M., Hinchcliffe, Glenn, 2004. RCM-gateway to World Class Maintenance. Elsevier-Butterworth-Heinemann.

Defining Total Maintenance Requirements and Backlog

This book strives to answer many important questions, including the who, what, when, where, why, and "how to" of effective planning, estimating, and scheduling. So far, I have tried to answer some of these questions in previous chapters and appendices.

1. **Why?** Chapters 1, Chapter 2, and 6.
2. **Who?** Appendix C.
3. **What?** Chapter 7.
4. **When?** Generally, you apply reliable maintenance planning estimating scheduling (RMPES) when there is a desire to move from reactive, firefighting-type maintenance to a proactive, planned maintenance strategy.
5. **Where?** This book focuses on the oil, gas, and petrochemical sectors, but much of the information is applicable to all types of operations.
6. **How much?** Chapter 10.
7. **How to?** Chapters 3, 4, 5, 8, and 9, and Appendix A.

In this chapter, we look at some of the items related to the "how much?" question regarding total maintenance requirements (TMR) and current backlog. In the simplest terms, backlog is the work that has not yet been completed. Trying to manage maintenance without managing the backlog can be a losing situation. However, you must have a certain level of backlog in order to plan and schedule work. Backlog provides a picture of resources and whether it is possible to achieve work with in-house staff, overtime, staff additions, or outsourcing to contract maintenance providers.

As a work request is generated and becomes a work order, it contains key information, of which one is a priority code. In this way, the planner can determine when the job will be planned, scheduled, and performed in relation to other jobs being requested. As a planner completes all the planning functions for each work order, the job then becomes available to schedule. If the planner and scheduler are two positions, the scheduler then receives the available planned job. Likewise, if the planning process is not completed, the job is not ready for the schedule and is unavailable. This may be due to a number of reasons, such as awaiting parts because of out-of-stock items or nonstocked items that must be purchased.

I have always considered TMR as one of the key items that a maintenance leader, supported by the planner/scheduler, should always be ready to review with top leaders. We can think of TMR as all of the work required during a period of time, such as on an annual basis. TMR includes everything that needs to be done, including (most importantly) the identified deferred maintenance, which can cost more if not repaired in a reasonable timeframe. TMR also can be minor project work that maintenance typically is staffed to perform and has done in the past.

Reliable Maintenance Planning, Estimating, and Scheduling. http://dx.doi.org/10.1016/B978-0-12-397042-8.00010-3

When the maintenance leader requires additional staff resources, a valid estimation of TMR plus the validation of current craft productivity (OCE) is available. In this case, using these current facts on workload increases the chance that a request using documented needs will be successful. If not successful, then the maintenance leader will have provided due diligence by informing top leaders of critical deferred maintenance. TMR also includes the equivalent staffing needed to achieve 100% preventive maintenance compliance.

In most cases, the planner may have this information in a master plan, broken down into two areas:

1. Total backlog, which includes work waiting to be scheduled (unavailable) due to a number of reasons.
2. Ready backlog, which is work that is ready to go on to a schedule, with all parts and materials being ready (available).

In this case, what I call *total maintenance requirements* is the total backlog plus ready backlog. TMR includes measured labor hours required and translated into labor weeks. The labor hours that the work is estimated to take are the available resource. The normal range for backlog is typically based on 80–90% of jobs being planned. Most companies set their own criteria for total backlog.

- Ready available backlog is equal to 2–4 weeks of labor hours.
- Total backlog.
- TMR—the total backlog (unavailable) plus the ready backlog (available)—for the schedule Total backlog is equal to 4-6 weeks of labor hours.

Backlog "is what it is," so to speak, if maintenance is clearly working at a documented high level of productivity and deferred maintenance is clearly being reported correctly to top leaders.

Now, we will take a look at some of the formats for joining backlog, coding backlog, and computing weeks of backlog. Again, all of this is a critical role and responsibility of the planner/scheduler. First, as a minimum, the planner should maintain a continuous TMR list of total backlog plus ready backlog, with the following items summarized in an end-of-the-month report:

- Total weeks of total backlog and ready backlog.
- Total TMR jobs/labor hours for all open work orders.
- Total backlog jobs/labor hours for work orders unavailable to be scheduled.
- Total ready backlog job/labor hours ready to be scheduled.
- Backlog aging
  - <1 month
  - <1–2 months
  - <3–6 months
  - >6 months
- Total backlog summary by reason code for being unavailable.
- Trend charts
  - Total labor hours completed versus number of work orders completed.
  - Total available labor hours.
  - Total scheduled overtime hours. (Figures 10.1–10.6).

Backlog Management

- Types of Backlog
  - Ready Backlog: Jobs Ready to Go
  - Total Backlog = Ready Backlog + Other open work orders
  - Ready Backlog: 2 to 4 Weeks
  - Total Backlog: 4 to 8 Weeks
- Establish a Valid Priority System
- Establish Valid Job Status Codes
- Maintain Backlog Integrity. It must be;
  - Complete
  - Current
  - Pure
  - Reliable

Figure 10.1 Key elements of backlog management.

Valid Priority System

- Numerical
  - 1= An immediate true emergency, a life safety or regulatory issue
  - 2= Required within 24 hours
  - 3= Required in 2-5 days
  - 4= Required in 1-2 weeks
  - 5= Required in over 2 weeks
- RIME System
  - Ranking Index for Maintenance Expenditures (RIME)
  - Based on Criticality of the Asset/Equipment
  - Based on Criticality of the Work Type

Figure 10.2 Two types of priority systems.

RIME System

RIME System: Ranking Index for Maintenance Expenditures and Calculated by:

Criticality of the Asset/Equipment (Rated 10 down to 1)

X

Criticality of the Work Type (Rated 10 down to 1)
= RIME Number for the Work Order

Figure 10.3 The Ranking Index for Maintenance Expenditures (RIME) system.

RIME System Example (Distribution Center)

Work Type / Priority → / Asset Criticality ↓↓		Safety & True Emergency (SAF)	Preventive and Predictive Maintenance (PM)(PDM)	Project Work (PRJ)	Corrective Maintenance and Warranty Work (CMA)	Operations Service (OPS)	Routine Normal Safety (RNS)	Tenant Improvement (TIM)	Inspections (INS)	Miscellaneous (MES)	Housekeeping (HSK)
		10	9	8	7	6	5	4	3	2	1
SORTER	10	100	90	80	70	60	50	40	30	20	10
Conveyor	9	90	81	72	63	54	45	36	27	18	9
Utilities	8	80	72	64	56	48	40	32	24	16	8
Turret Truck	7	70	63	56	49	42	35	28	21	14	7
Lift Truck	6	60	54	48	42	36	30	24	18	12	6
Stock Picker	5	50	45	40	35	30	25	20	15	10	5
Miscellaneous Mobile Equipment	4	40	36	32	28	24	20	16	12	8	4
Office Facilities	3	30	27	24	21	18	15	12	9	6	3
Miscellaneous support To Other Asset	2	20	18	16	14	12	10	8	6	4	2
Buildings, Roads and Grounds	1	10	9	8	7	6	5	4	3	2	1

Figure 10.4 Example of Ranking Index for Maintenance Expenditures (RIME) system.

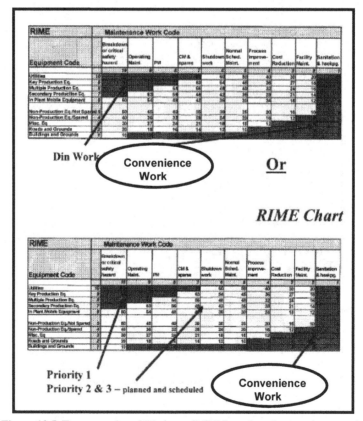

Figure 10.5 Two examples of "do it now" (DIN) work and convenience work.

Benefits of the RIME System

- Adds objectivity and professionalism into the planning and scheduling process
- Does not eliminate the need to apply individual judgment for final priorities for scheduling
- Lessens the impact of the "squeaky wheel"
- Provides better understanding about maintenance to the customers.
- Criticality of assets is determined in asset database
- Criticality of work types is established

Figure 10.6 Benefits of the Ranking Index for Maintenance Expenditures (RIME) system.

One advantage of using the Ranking Index for Maintenance Expenditures (RIME) system is that it provides another level of professionalism for the planning process. In other words, if the operations team understands that it is based on criticality of equipment and criticality of the work type, then the acceptance of work to be scheduled increases. The RIME system can also be used by the planner as a validation tool in determining the validity of a job priority assigned by the originator. Most often, the job priority is self-evident and validation by the planner is seldom required. It is been said that "the squeaky wheel gets the most grease," such as a persistent operations person who wants to get work done immediately. However, if there is a question about job priority assignments, the planner must always direct questions back to the work order approving authority for resolution. For a refinery unit, this authority may be granted to the unit maintenance coordinator, unit manager, or the maintenance leader. If procedures state that operations must approve all scheduled breaks, then the unit manager must sign off on work that changes the schedule significantly.

One of the things I distinctly remember about the scoreboard assessment at the Marathon Ashland Petroleum refinery in Robinson, Illinois (the 1999 North American Maintenance Excellence (NAME) Award winner) was a comment by their senior planner. He stated, "We have a very hard-and-fast rule about classifying work as an emergency work here at our refinery" Now here we are within a site with all types of critical failure consequences, but they wanted to be proactive and plan work to be as safe as possible. Not a bad idea! So let us consider what Kister and Hawkins defined as emergency ("E") work, which all companies should adapt:

"Emergency—*Must be performed immediately*. Higher priority than scheduled work, critical machinery down or in danger of going down until requested work is complete. 'E' to be used only if production loss, delivery performance, personnel safety (new and eminent), and equipment damage or material loss are involved and no bypass is available. Start immediately and work expeditiously and continuously to completion, including the use of overtime without specific further approval. Only personnel authorized to approve overtime can assign 'E' to work orders. Emergency work order reports will be sent to the plant manager for review."(Figures 10.7–10.13).

Maintaining an accurate backlog is an important responsibility of the maintenance planner. From this document, schedules based on priorities are created. This document in terms of TMR defines staffing needs for the valuable resource of craftspeople. It is a means to define the facts of workload to top leaders. In turn, the backlog must have accurate estimates of the time required for as many jobs as possible. It must also have reliable estimates of jobs that have not been completed through the planning process. Every company site is different, and requirements for backlog levels may vary. One constant is that there must be control of the backlog, regardless of what has been accrued. Backlog is necessary for having planned

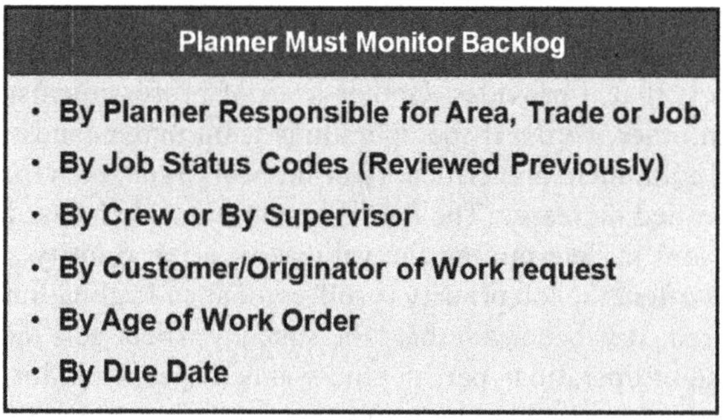

Planner Must Monitor Backlog

- By Planner Responsible for Area, Trade or Job
- By Job Status Codes (Reviewed Previously)
- By Crew or By Supervisor
- By Customer/Originator of Work request
- By Age of Work Order
- By Due Date

Figure 10.7 Areas where the planner should monitor backlog.

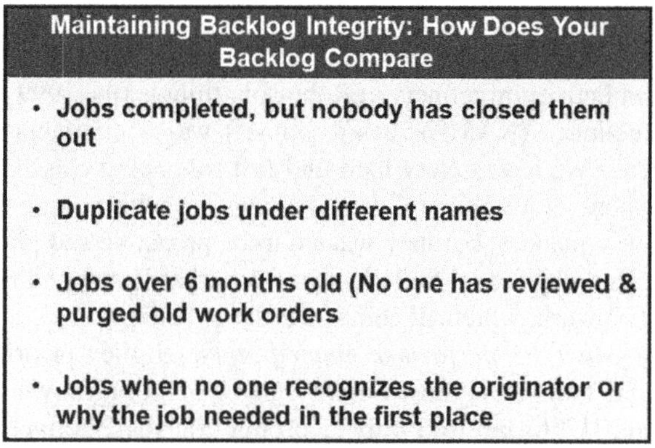

Maintaining Backlog Integrity: How Does Your Backlog Compare

- Jobs completed, but nobody has closed them out

- Duplicate jobs under different names

- Jobs over 6 months old (No one has reviewed & purged old work orders

- Jobs when no one recognizes the originator or why the job needed in the first place

Figure 10.8 Maintaining backlog integrity.

Maintaining Backlog Integrity: How Does Your Backlog Compare

- A poorly described job (no one can figure out what to do to what)

- Job status not filled in and nobody know status
  - Were parts required?
  - Were parts ordered?
  - Were they delivered-where are they now?

- Jobs that need to be done but not on the backlog

Figure 10.9 Maintaining backlog integrity.

Balancing Maintenance Resources with the Maintenance Workload Must;

- Ensure that expectations for backlog relief (reduction) are realistic
- Make allowances for all commitments to indirect activity
- Clarify the craft time to be allocated to true emergency work
- Clearly define labor hours for PM/PdM requirements

Figure 10.10 Balancing the maintenance resources with the backlog.

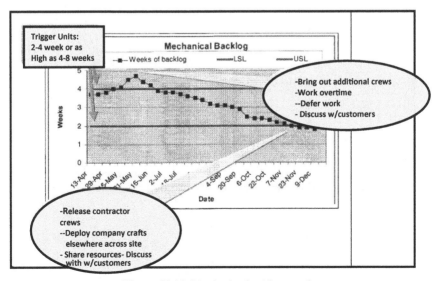

Figure 10.11 Monitoring backlog trends.

Backlog Job Status Codes

Status Code	Status Code Description	Type of Backlog
AP	Awaiting Planning	Total
AE	Awaiting Engineering Review	Total
AA	Awaiting Approval	Total
PF	Pending Funding	Total
DF	Deferred-Funding	Total
PO	Waiting PO to be Issued	Total
AM	Awaiting Receipt of Material	Total
FP	Further Planning Required	Total
DP	Requires Downtime-Programmed Shutdown	Total
PW	Requires downtime- Weekend	Ready
DS	Requires downtime-Not Scheduled	Ready
DO	Requires downtime- Await Window of Opportunity	Ready
R	Ready to be scheduled	Ready
FI	Ready for Fill In Assignment	Ready
S	On the current schedule	Not in Backlog
CO	All maintenance work complete	Not in Backlog
CM	WO pending material close out	Not in Backlog
CH	Closed to Equipment History	Not in Backlog
CP	Print Revision not received	Not in Backlog

Figure 10.12 Backlog status code examples.

Example: Maintenance Work Program to Address Backlog Relief (Reduction)

Work Program for Backlog Relief

Available Resources- Crew Size = 20 Crafts

Straight Time Man-hours Available Per Week	800
Planned Overtime Per Week	96
Man-Hours Contracted or Borrowed Per Week	0
Total Man-Hours Available Per Week	**896**

Less Indirect Commitment (Average)

Lunch (if paid)	0
Vacation	120
Absence	34
Training	56
Meetings	40
Special Assignments	40
Average Man-Hours Loaned to Other Areas	40
Other Indirect	10
Total Indirect Hours Projected Per Week	**330**
Total Hours Available Per Week for Direct Work (Wrench Time)	**566**

Commitments Other Than Backlog Relief

Emergency/Urgent (Unschedulable)	100
Routine PM/PdM	120
Other Fixed Routine Assignment	0
Sub Total	**220**
Net Resource Available for Backlog Relief	**346**

Backlog Data	Current	Backlog Weeks	Current	Target
Man Hours in Ready Status	3200	Ready Backlog	9.2 Weeks	2 to 4 Weeks
Total Man-Hours Of Backlog	4800	Total Backlog	13.9 Weeks	4 to 6

Figure 10.13 Calculating weeks of backlog.

work to develop effective scheduling. In Chapters 11–18, we will review the five key areas of the planning process:

1. Planning
2. Estimating
3. Scheduling
4. Monitoring
5. Controlling

Overview of a Reliable Planning-Estimating-Scheduling-Monitoring-Controlling Process

When we consider the overall range of planning and scheduling, there are five key phases:

1. Planning
2. Scheduling
3. Estimating
4. Monitoring
5. Controlling

As we discussed previously, the planner begins with a reliable backlog of total maintenance requirements (TMR). From that baseline, labor resources are considered. During annual budgeting, one of the main questions asked by top leaders is how many staff positions are required. Measurement of TMR provides the definition of inherent workload. It is the staff required to preserve an asset related to its size, replacement value, and usage. Without this investment, the asset will inevitably deteriorate. In many cases, there is a shortfall between what is needed and what gets accomplished, thereby creating deferred maintenance. Deferred maintenance is a key element for which the maintenance leader must clearly show that short-term savings will create higher long-term costs.

The maintenance function must do a better job in defining staffing needs. That is why I always say (1) define total maintenance requirements and (2) show maximum craft labor productivity. If you have done this, you have done due diligence in regard to your current state of maintenance. Chapter 10, Defining Total Maintenance Requirements and Backlog, is a very important part of the planner's job and one that can help directly define staff requirements. Without solid evidence and facts, relying on vague evidence and indirect measures of increased in work (such as rising customer complaints, increasing downtime, growing use of contractors, etc.) may be inadequate justification for additional staff. Other areas contributing to maintenance costs and staffing are the shown in Figure 11.1.

A good book for your maintenance library is *Maintenance Planning, Scheduling and Coordination* by Don Nyman and Joel Levitt. *Monitoring and controlling* is much like their term *coordination* that is shown in Figure 11.2.

One last illustration concept from their book is Figure 11.3, which illustrates a steady state of parts maintenance (PM) work and backlog relief work that is planned work. In addition, we see urgent response work, deferred maintenance work, plus capital program requirements across time periods showing a reduction in urgent response/emergency work. This in turn allows for deferred maintenance to be accomplished.

Reliable Maintenance Planning, Estimating, and Scheduling. http://dx.doi.org/10.1016/B978-0-12-397042-8.00011-5

• Processes and equipment used	• Single or multi-shift operation
• Frequency of process improvements or changes	• The product mix
• Frequency is set up and product changeover	• Sector within the industry
• Is purchasing based on life cycle cost and quality?	• Small, medium or large operation
• Are contracts awarded to the low bidder i.e. and at time the high cost of low bid buying	• The expectation for maintenance
• Policy and procedures for purchasing, engineering and quality	• History of the site
• Customer service measured by percent uptime, schedule compliance, delivery, safety and other measures	• Location of the site, unit and whether indoor/outdoor
• Scope of asset utilization	• Is this a new Greenfield, recently commissioned operation or an aging facility?
• The maintenance staff's knowledge and skill levels and dedication	• Age of processing equipment and control systems
• Expectations of internal customers as well as skills expected and dedication	• Type of equipment installation, materials and workmanship of facilities and equipment
• Availability of spare parts and vendors	• Quality of process design; latest technology applications?
• Availability of qualified contractors located near the site	• Accessibility to the equipment performing maintenance
• Size of the capital budget in relation to the overall replacement asset value (RAV)	• Size of the facility, layout and operating context
• Organizational change such as turnover within maintenance, operations and support staff	• Products and processes
• Business competition	• Hours of operation, one, two or three shift operation and weekends and 24/7
	• Regulated industries such as refining, nuclear, pharmaceutical etc.
	• ISO 55000 compliance goal?

Figure 11.1 Areas contributing to maintenance costs and staffing.

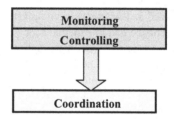

Figure 11.2 Monitoring and controlling is coordination.

All of this occurs when we have a complete shift from an environment that is reactive to a maintenance strategy that is proactive and well -planned.

Appendix E– Case Study: Process Mapping for a Refinery – Work Initiation to Completed Work Reliability Improvement Analysis is a very detailed process mapping within a refinery. Figure 11.4 illustrates the planning, estimating, and scheduling work process flow. Key to this illustration is the center block containing coordination and communication. I often call it the three C's: coordination, cooperation, and communication.

Let us now review how management can better understand the symptoms of ineffective planning in Figure 11.5, Figure 11.6, and Figure 11.7.

As it has been said before, "Nothing happens until somebody sells something!" This too applies to planning/scheduling, which we touched on in Chapter 1, Profit and Customer-centered Benefits of Planning and Scheduling. Figure 11.8, Figure 11.9, and Figure 11.10 provide areas where we must strive to sell the benefits of planning and scheduling.

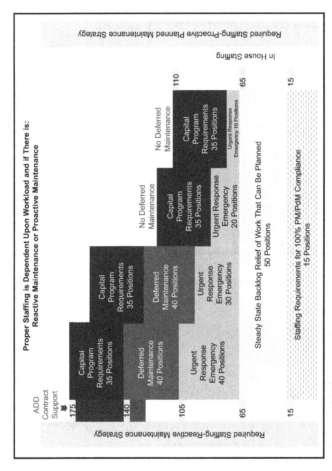

Figure 11.3 Illustrates a steady state of PM work and planned backlog relief work with reduction in urgent response/emergency work, which in turn allows for deferred maintenance to be accomplished.

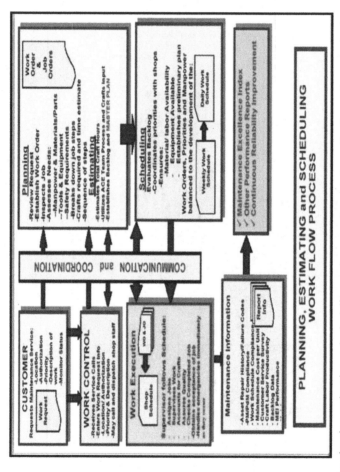

Figure 11.4 Illustrates the planning, estimating and scheduling work process integrated coordination and communication.

Management Must Understand Symptoms of Ineffective Job Planning

- **Delays encountered by our most valuable resources, the craft work force**
  - Gaining information about the job
  - Obtaining permits
  - Identifying and obtaining parts and materials
  - Identifying blue prints, tools and skills needed
  - Getting all of above to the job site
  - Waiting for required parts not in stock
- **Crafts waiting at job site for supervisor or operations to clarify work to be done**

Figure 11.5 Examples of ineffective planning #1.

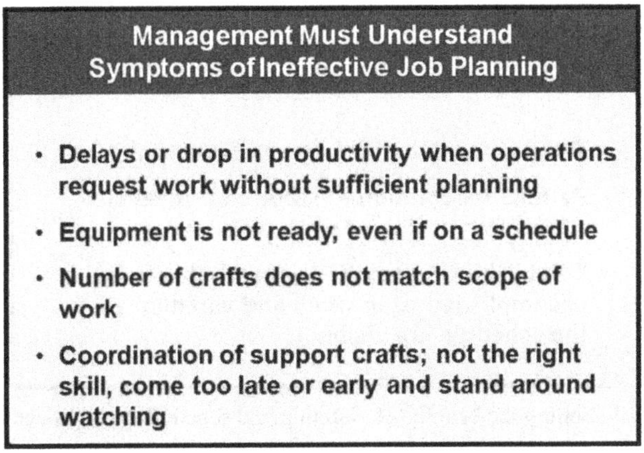

Management Must Understand Symptoms of Ineffective Job Planning

- **Delays or drop in productivity when operations request work without sufficient planning**
- **Equipment is not ready, even if on a schedule**
- **Number of crafts does not match scope of work**
- **Coordination of support crafts; not the right skill, come too late or early and stand around watching**

Figure 11.6 Examples of ineffective planning #2.

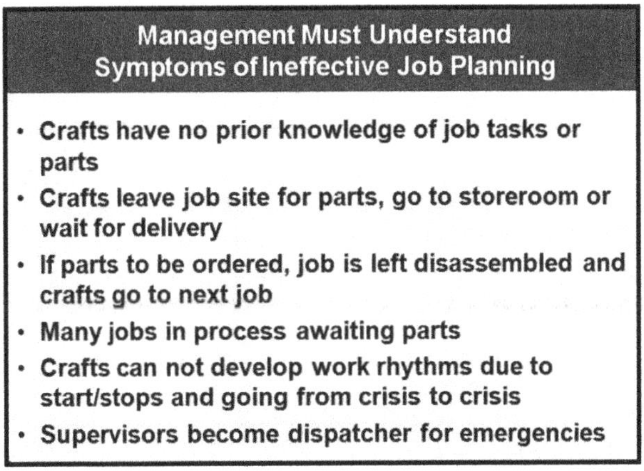

Management Must Understand Symptoms of Ineffective Job Planning

- **Crafts have no prior knowledge of job tasks or parts**
- **Crafts leave job site for parts, go to storeroom or wait for delivery**
- **If parts to be ordered, job is left disassembled and crafts go to next job**
- **Many jobs in process awaiting parts**
- **Crafts can not develop work rhythms due to start/stops and going from crisis to crisis**
- **Supervisors become dispatcher for emergencies**

Figure 11.7 Examples of ineffective planning #3.

Selling the Benefits of Planning and Scheduling to Management

- Provides central source of equipment condition, workload & resources available to perform it
- Improves employee safety & regulatory compliance
- Helps achieve optimal level of maintenance in support of long and short-term operational needs
- Challenges work request of questionable value

Figure 11.8 Selling the benefits of planning and scheduling to management #1.

Selling the Benefits of Planning and Scheduling to Management

- Provides forecast of labor and material needs
- Permits recognition of labor shortages and allows for leveling of peak workloads
- Establishes expectations for what is to be accomplished each week and variation from the schedule are visible

Figure 11.9 Selling the benefits of planning and scheduling to management #2.

Selling the Benefits of Planning and Scheduling to Management

- Improves productivity by anticipating needs and avoiding delays
- Increases productivity of both operations & maintenance
- Provides factual data; performance measurement, cost variations
- Provides info to identify problems that need focused attention

Figure 11.10 Selling the benefits of planning and scheduling to management #3.

Selling benefits of planning and scheduling to operations is essential. As they are the customers, they are the service receivers, and they are focused on production with the process that provides plant throughput to create profit. Within refineries, for example, it is important to understand the integration of complex processes and types of output that are possible. Figure 11.11 illustrates common processes found in a refinery.

Within operations shown in Figures 11.11–11.13 are many complex and hazardous maintenance challenges. They can be summarized as four major challenges in Figures 11.14 and 11.15.

Therefore, we must sell planning and scheduling to the operations management staff and operators. They must be trained in the CMMS/EAM as to information required for creating a work request. If a modern CMMS/EAM in place, they can do this electronically while monitoring their own backlog of work, schedule, and status of work in progress. Figures 11.16 and 11.7 define key benefits to the operations/production customers.

Just remember, the planner you must be a salesman too as shown in Figure 11.18! Many times this will be selling benefits to your own maintenance organization. Sometimes this can be very difficult if a planner is not well trained, respected, and viewed as a valuable asset supporting the maintenance mission (Figures 11.19–11.24).

Figure 11.25 illustrates most all of the "right things" that planner/scheduler must get right in terms of planning a maintenance job (Figure 11.26).

Figure 11.27 (from *Maintenance Planning, Scheduling and Coordination* by Don Nyman and Joel Levitt) shows how small reductions in all areas of nonproductive time can increase direct wrench time available from 35% up to 65%. Figures 11.28 and 11.29 include two case study examples. Figure 11.30 is a very good way to see the scope of possible gained value in your operation. First, provide a good estimate of true wrench time. This totals hours working directly on a job (tool time) compared to total labor hours paid. Second, estimate your goal for improving the craft utilization factor of OCE (overall craft effectiveness), which is wrench time. Then do the simple math based on number of pure crafts people and average annual wages (Figures 11.31 and 11.32).

Now is an appropriate time to discuss where the maintenance storeroom is included within the total operation. There are three main options: (1) it reports to maintenance, (2) it is part of finance/purchasing, and (3) it could be a contracted storeroom in Figures 11.33–11.36.

When we organize the planning process it should report directly to the maintenance leader and be structured so that there is open and cooperative communication with the operational side of the business. That type of organization is shown in Figure 11.37 and of course depends as much on relationships between planners, maintenance supervisors, and operational staff as much as any other factor. Figure 11.37 strives to illustrate a closed-loop-type organization with actions clearly displaying a service-oriented approach to operations. This relationship is further enhanced when production planning can also work hand in hand with maintenance planning. In contrast to Figure 11.38, this could be viewed as indirect liaison with planner reporting primarily to the maintenance leader.

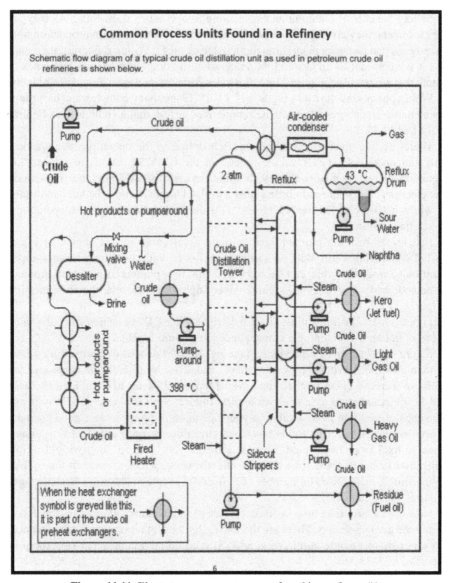

Figure 11.11 Illustrates common processes found in a refinery #1.

The maintenance supervisor is a key maintenance leader. In Figures 11.39–11.42 we will cover the important role played by the supervisor who supports the execution of work from the schedule and also the monitoring and controlling areas of the five steps in the overall planning process (Figures 11.43–11.55).

Now to add to Chapter 7, What to Look for When Hiring a Reliable Planner-Scheduler, we will see that good planning starts with a good planner having qualities as shown in Figure 11.56 and Figure 11.57. As a special note, I personally recommend having at

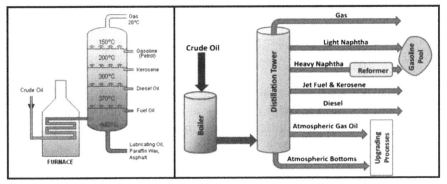

Figure 11.12 Illustrates common processes found in a refinery #2.

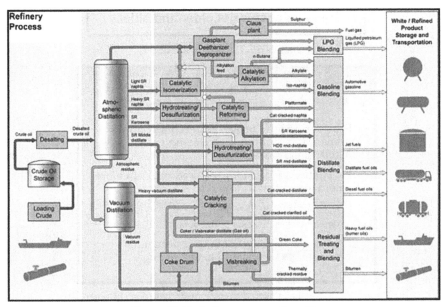

Figure 11.13 Refinery process flow #3.

Today's Very Real Challenges

<u>Challenge One:</u> Maintain existing facilities and equipment in safe and sound conditions.

<u>Challenge Two:</u> Improve, enhance and then maintain existing physical assets to achieve environmental, regulatory life safety/security standards & energy best practices.

Figure 11.14 Challenges #1 and #2.

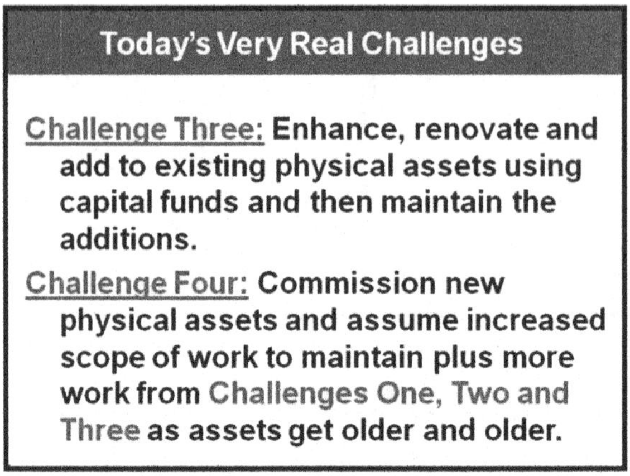

Today's Very Real Challenges

<u>Challenge Three:</u> Enhance, renovate and add to existing physical assets using capital funds and then maintain the additions.

<u>Challenge Four:</u> Commission new physical assets and assume increased scope of work to maintain plus more work from Challenges One, Two and Three as assets get older and older.

Figure 11.15 Challenges #3 and #4.

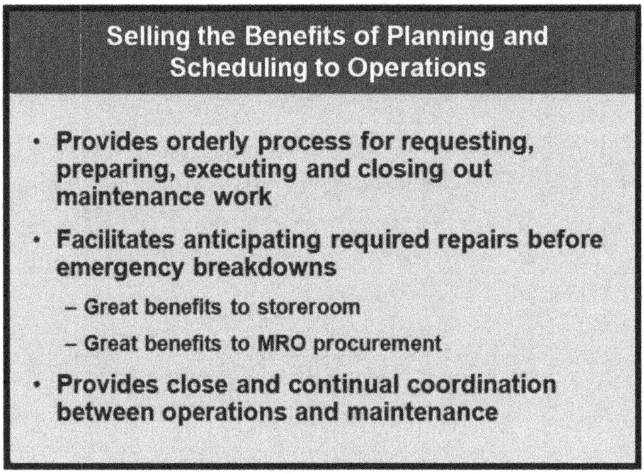

Selling the Benefits of Planning and Scheduling to Operations

- Provides orderly process for requesting, preparing, executing and closing out maintenance work
- Facilitates anticipating required repairs before emergency breakdowns
  - Great benefits to storeroom
  - Great benefits to MRO procurement
- Provides close and continual coordination between operations and maintenance

Figure 11.16 Key benefits to the operations/production customers #1.

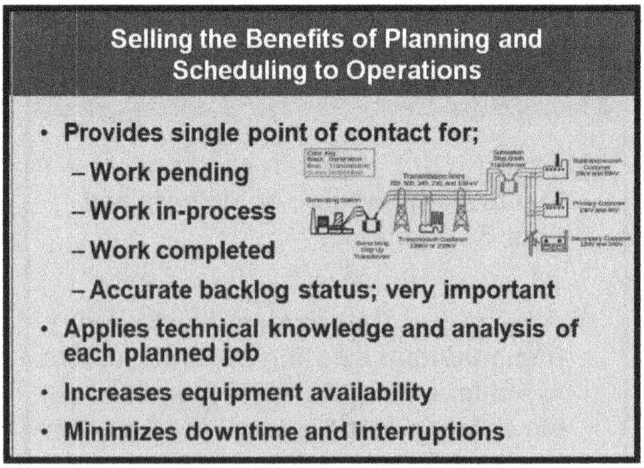

Selling the Benefits of Planning and Scheduling to Operations

- Provides single point of contact for;
  - Work pending
  - Work in-process
  - Work completed
  - Accurate backlog status; very important
- Applies technical knowledge and analysis of each planned job
- Increases equipment availability
- Minimizes downtime and interruptions

Figure 11.17 Key benefits to the operations/production customers #2.

Figure 11.18 Selling benefits to your own maintenance organization.

Selling the Benefits of Planning and Scheduling to Maintenance

- Defines and measures workload
- Permits advance determination of staffing by area, work unit or by type of craft skill
- Establishes realistic priorities
- Identifies best methods and procedures

- Identifies risks and all HSSE factors
- Anticipates bottlenecks and interruptions

Figure 11.19 Benefits to the maintenance organization #1.

Selling the Benefits of Planning and Scheduling to Maintenance

- Coordinates manpower, material and equipment to include;
  - Craft labor
  - Parts and materials
  - Special tools and equipment
  - Shop and other support
  - Off-site job preparation to minimize downtime
  - Equipment access; internal & external rentals

Figure 11.20 Benefits to the maintenance organization #2.

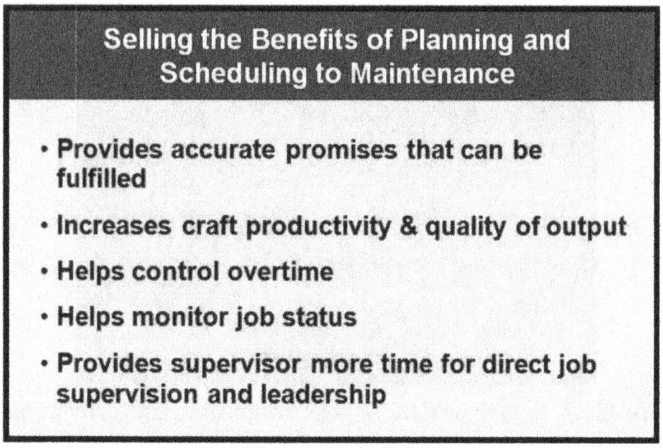

Selling the Benefits of Planning and Scheduling to Maintenance

- **Provides accurate promises that can be fulfilled**
- **Increases craft productivity & quality of output**
- **Helps control overtime**
- **Helps monitor job status**
- **Provides supervisor more time for direct job supervision and leadership**

Figure 11.21 Benefits to the maintenance organization #3.

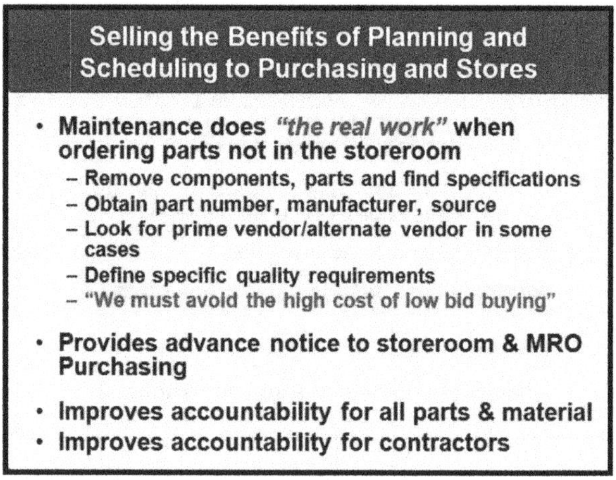

Selling the Benefits of Planning and Scheduling to Purchasing and Stores

- **Maintenance does *"the real work"* when ordering parts not in the storeroom**
  - Remove components, parts and find specifications
  - Obtain part number, manufacturer, source
  - Look for prime vendor/alternate vendor in some cases
  - Define specific quality requirements
  - "We must avoid the high cost of low bid buying"
- **Provides advance notice to storeroom & MRO Purchasing**
- **Improves accountability for all parts & material**
- **Improves accountability for contractors**

Figure 11.22 Benefits to the storeroom and purchasing organization #1.

Selling the Benefits of Planning and Scheduling to Purchasing and Stores

- **Helps insure parts ordering with more lead time to**
  - Reduce number of emergency purchase$
  - Reduce cost of expre$$ freight
  - Reduce Fedex, UPS, DHL emergency shipping cost
- **Helps optimize maintenance inventory**
  - Note: *Modernizing Your Maintenance Storeroom*
  - *Your storeroom: One Cornerstone for Maintenance Excellence*
- **Improves information for equipment specifications to include;**
  - "Part Where Used"
  - Parts list by asset/equipment

Figure 11.23 Benefits to the storeroom and purchasing organization #2.

Figure 11.24 Benefits to the bottom line.

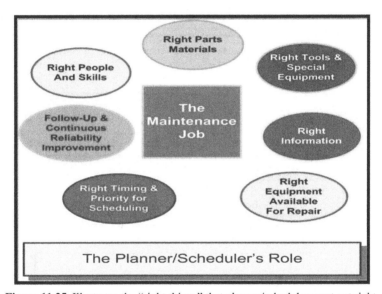

Figure 11.25 Illustrates the "right things" that planner/scheduler must get right.

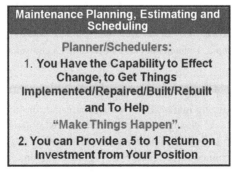

Figure 11.26 Planner/scheduler must make things happen.

Maintenance Planning, Estimating and Scheduling Provides Gained Value		
Typical Maintenance Worker's Day	Reactive No Planning	Proactive Planned
Receiving Job Instructions	5%	3%
Obtaining Tools and materials	12%	5%
Travel To and From Job (Both with and without tools & materials)	15%	10%
Coordination Delays	8%	3%
Idle at Job Site	5%	2%
Late Starts and Early Quits	5%	1%
Authorized Breaks and Relief	10%	10%
Excess Personal Time (Extra breaks, phone calls, smoke breaks, slow return from breaks/lunch)	5%	1%
Subtotal Non-Productive Time	**65%**	**35%**
Direct Wrench Time Available for Work	**35%**	**65%**

Figure 11.27 Typical workday: reactive versus proactive planned work.

Productivity Improvement for a Central Maintenance Operation	
Central Maintenance Without Planning	**Central Maintenance With Planning**
0 Planner 30 Total Crafts 35% Wrench Time 11 Equivalent Full Time	1 Planner 29 Total Crafts 65% Wrench Time 19 Equivalent Full Time Workers
Net Gain of 8 Equivalent Craft Positions	
Net Gain in Craft Capacity = 73%	
Gained Value = 8 crafts x $25/Hr Avg. x 40 Hrs/Wk x 52 Wks/Yr = $416,000 Gained Value	
A Real Case Study	Wrench time improvement of 30% from an estimated baseline of 35% to 65% is very realistic with an effective planning and scheduling process in place.

Figure 11.28 Example: 1 planner equals an additional 8 equivalent crafts positions.

least one backup planner, designated and trained to support an existing planner during time off and possibly during major shutdown, turnaround, or outage (STO) periods.

CATEGORY 9: Work Management and Control: Maintenance and Repair and CATEGORY 11: Shop Level Reliable Planning, Estimating and Scheduling from the Scoreboard for Maintenance Excellence™ in Figure 11.58 and Figure 11.59.

What is the Possible Gained Value of Increased Wrench Time Across GRIDCO?			
Practical Exercise: Gained Value			
Without Effective Planning, & Scheduling		**With Effective Planning & Scheduling**	
Total Crafts	600	600	
Wrench Time Estimate	X ___.35___ %	X ___.55___ %	
Equivalent Crafts	210	330	**NET GAIN CRAFTS** 120
Potential Gained Value	Net Gain Crafts X ___120___	Annual Salary $54,000 c/yr	**TOTAL GAINED VALUE** 6,480,000 cedi = $3,000,000 USD

Figure 11.29 Example: gained value of 120 equivalent crafts positions.

What is the Possible Gained Value of Increased Wrench Time in Your Operation?			
Practical Exercise: Gained Value			
Without Effective Planning, & Scheduling		**With Effective Planning & Scheduling**	
Total Crafts	___	___	
Wrench Time Estimate	X _____ %	X _____ %	
Equivalent Crafts	___	___	**NET GAIN CRAFTS**
Potential Gained Value	Net Gain Crafts X_____	Annual Salary $_____	**TOTAL GAINED VALUE** = _____

Figure 11.30 What is possible gained value of increased wrench time in your operation?

Category 10: Work Management and Control: Shutdowns, Turnarounds and Outages (STO), Category 12: STO and Major Maintenance Planning/Scheduling with Project Management, and Category 13: Contractor Management from The Scoreboard for Maintenance Excellence™ are included as Figure 11.60, Figure 11.61, and Figure 11.62. Contractor work planning during a major shutdown, turnaround, or outage (STO) will be an important part of the total work package of an STO.

Figure 11.31 Ensure your storeroom promotes the planning process.

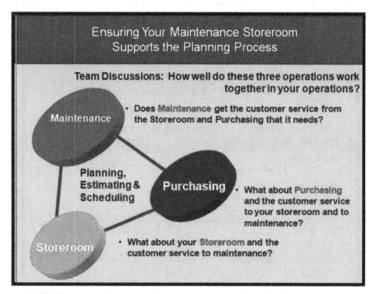

Figure 11.32 Ensure maintenance, the storeroom, and purchasing work well together.

Types of Maintenance & Storeroom Organizations

The Three Major Options

1. Storerooms Reporting to Maintenance

2. Storerooms Reporting to Finance or Purchasing

3. Storerooms are Contracted Services

Figure 11.33 Three major options.

Storerooms Reporting to Finance/Purchasing

Key Points

1. Finance may not understand true maintenance needs

2. Also requires best practices, leadership and collaboration

3. Also requires total accountability by maintenance for all costs

4. Financial side must balance inventory $'s with cost of downtime

5. May create a "We versus They" culture.

Figure 11.34 Reporting to finance or purchasing.

Storerooms Reporting to Maintenance

Key Points

1. Recommended structure: Maintenance directly responsible

2. Requires best practices and strong leadership

3. Requires total accountability by maintenance for all costs

4. Financial side must monitor accountability

5. Allows maintenance to operate as a "business"

Figure 11.35 Reporting to maintenance (recommended).

Being Successful Regardless of Organization

Key Points

1. Contracted Stores has "Pros" and "Cons"

2. Finance must understand true maintenance needs

3. Requires best practices, leadership and collaboration

4. Requires total accountability by maintenance for all costs

5. Financial side must balance inventory $'s with cost of downtime

6. Must be "Team Effort" and not a "We versus They" culture.

Figure 11.36 Keys to successful storeroom regardless of organizational position.

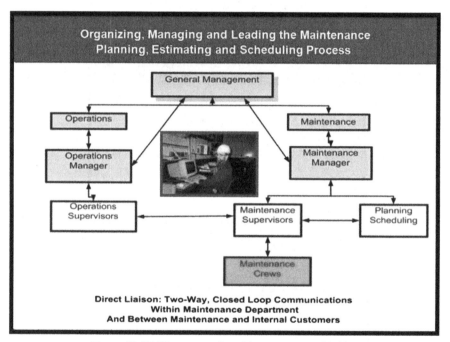

Figure 11.37 Illustrates a closed-loop-type organization.

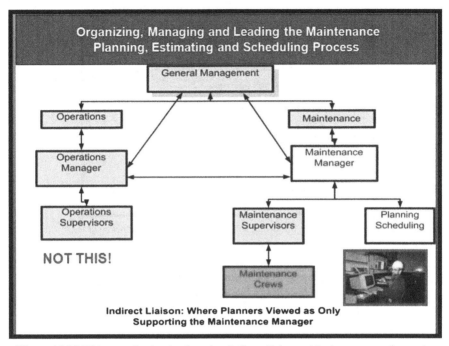

Figure 11.38 This organization is viewed as indirect liaison with planner reporting to the maintenance leader.

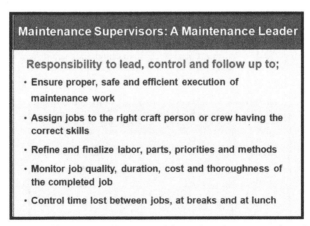

Figure 11.39 The important role of the maintenance supervisor #1.

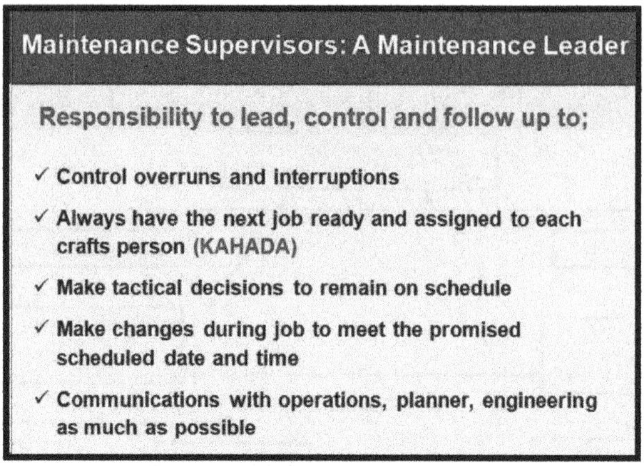

Maintenance Supervisors: A Maintenance Leader

Responsibility to lead, control and follow up to;

✓ Control overruns and interruptions

✓ Always have the next job ready and assigned to each crafts person (KAHADA)

✓ Make tactical decisions to remain on schedule

✓ Make changes during job to meet the promised scheduled date and time

✓ Communications with operations, planner, engineering as much as possible

Figure 11.40 The important role of the maintenance supervisor #2.

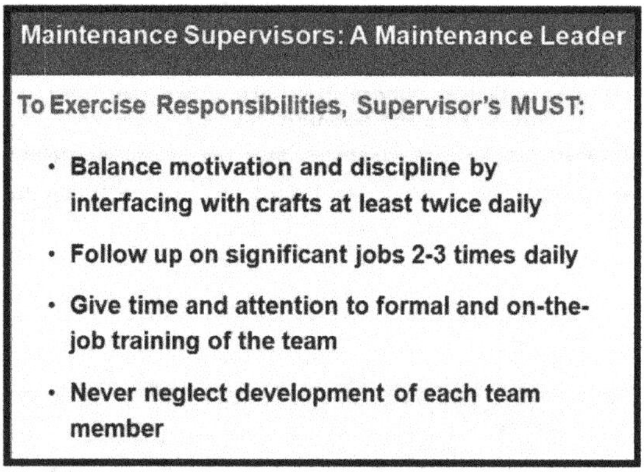

Maintenance Supervisors: A Maintenance Leader

To Exercise Responsibilities, Supervisor's MUST:

• Balance motivation and discipline by interfacing with crafts at least twice daily

• Follow up on significant jobs 2-3 times daily

• Give time and attention to formal and on-the-job training of the team

• Never neglect development of each team member

Figure 11.41 The important role of the maintenance supervisor #3.

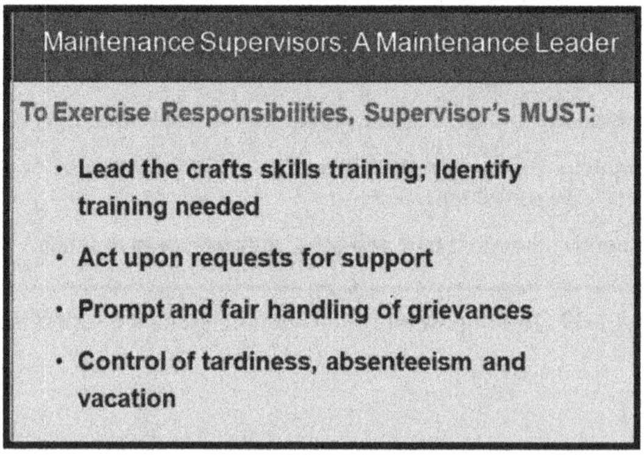

Maintenance Supervisors: A Maintenance Leader

To Exercise Responsibilities, Supervisor's MUST:

• Lead the crafts skills training; Identify training needed

• Act upon requests for support

• Prompt and fair handling of grievances

• Control of tardiness, absenteeism and vacation

Figure 11.42 The important role of the maintenance supervisor #4.

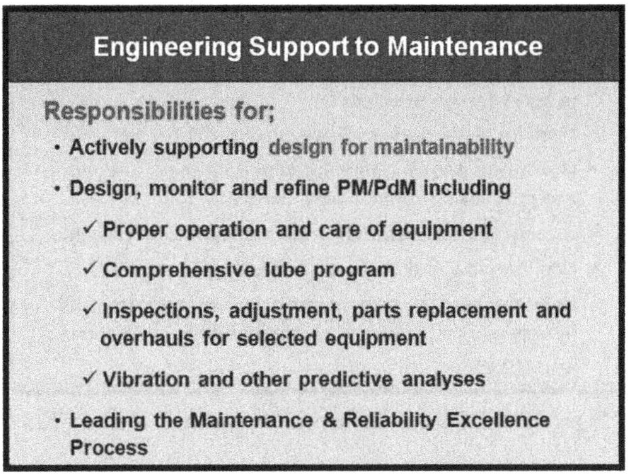

Engineering Support to Maintenance

Responsibilities for;

- Actively supporting design for maintainability
- Design, monitor and refine PM/PdM including
  - ✓ Proper operation and care of equipment
  - ✓ Comprehensive lube program
  - ✓ Inspections, adjustment, parts replacement and overhauls for selected equipment
  - ✓ Vibration and other predictive analyses
- Leading the Maintenance & Reliability Excellence Process

Figure 11.43 Engineering support to maintenance #1.

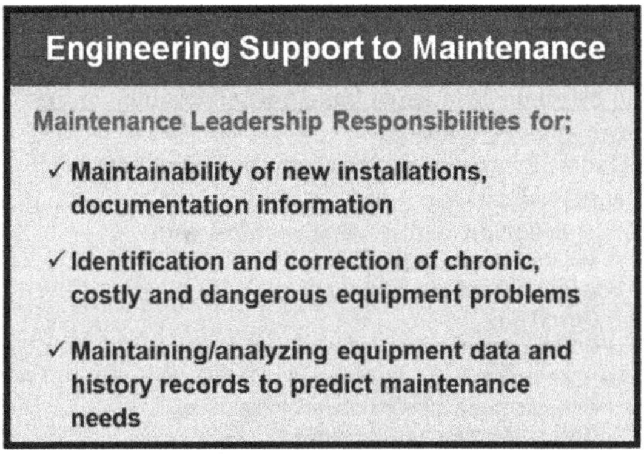

Engineering Support to Maintenance

Maintenance Leadership Responsibilities for;

- ✓ Maintainability of new installations, documentation information
- ✓ Identification and correction of chronic, costly and dangerous equipment problems
- ✓ Maintaining/analyzing equipment data and history records to predict maintenance needs

Figure 11.44 Engineering support to maintenance #2.

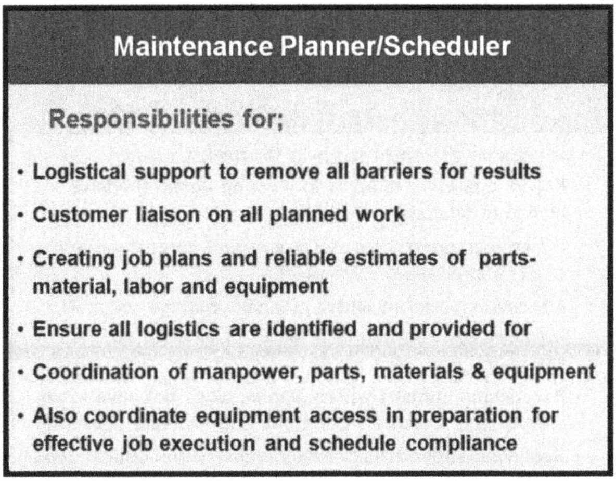

Maintenance Planner/Scheduler

Responsibilities for;

- Logistical support to remove all barriers for results
- Customer liaison on all planned work
- Creating job plans and reliable estimates of parts-material, labor and equipment
- Ensure all logistics are identified and provided for
- Coordination of manpower, parts, materials & equipment
- Also coordinate equipment access in preparation for effective job execution and schedule compliance

Figure 11.45 Responsibilities of the planner/scheduler #1.

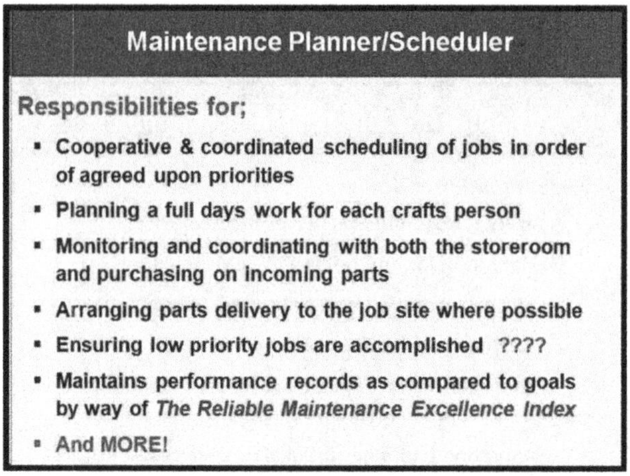

Maintenance Planner/Scheduler

Responsibilities for;

- Cooperative & coordinated scheduling of jobs in order of agreed upon priorities
- Planning a full days work for each crafts person
- Monitoring and coordinating with both the storeroom and purchasing on incoming parts
- Arranging parts delivery to the job site where possible
- Ensuring low priority jobs are accomplished ????
- Maintains performance records as compared to goals by way of *The Reliable Maintenance Excellence Index*
- And MORE!

Figure 11.46 Responsibilities of the planner/scheduler #2.

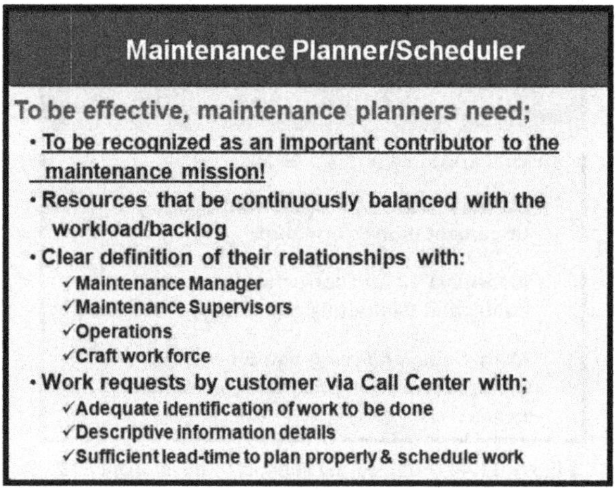

Maintenance Planner/Scheduler

To be effective, maintenance planners need;
- <u>To be recognized as an important contributor to the maintenance mission!</u>
- Resources that be continuously balanced with the workload/backlog
- Clear definition of their relationships with:
  - ✓Maintenance Manager
  - ✓Maintenance Supervisors
  - ✓Operations
  - ✓Craft work force
- Work requests by customer via Call Center with;
  - ✓Adequate identification of work to be done
  - ✓Descriptive information details
  - ✓Sufficient lead-time to plan properly & schedule work

Figure 11.47 Responsibilities of the planner/scheduler #3.

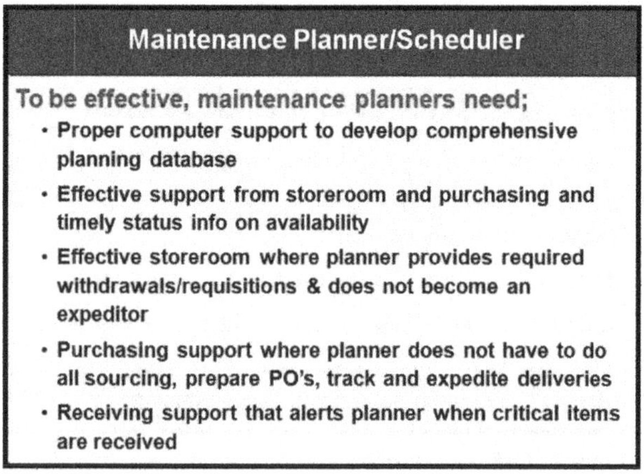

Maintenance Planner/Scheduler

To be effective, maintenance planners need;
- Proper computer support to develop comprehensive planning database
- Effective support from storeroom and purchasing and timely status info on availability
- Effective storeroom where planner provides required withdrawals/requisitions & does not become an expeditor
- Purchasing support where planner does not have to do all sourcing, prepare PO's, track and expedite deliveries
- Receiving support that alerts planner when critical items are received

Figure 11.48 Planner/scheduler needs for effectiveness #1.

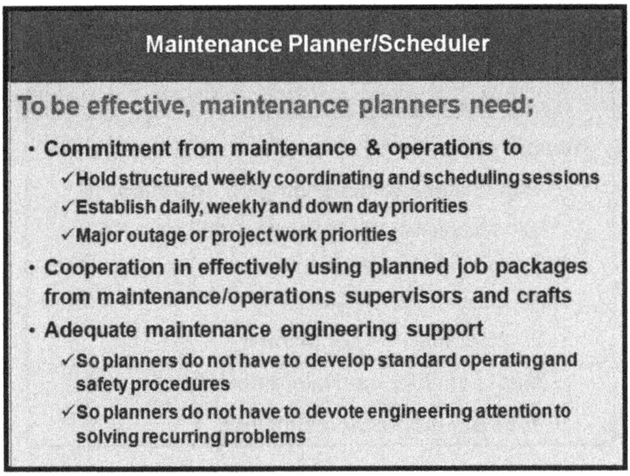

Maintenance Planner/Scheduler

To be effective, maintenance planners need;

- Commitment from maintenance & operations to
  - ✓ Hold structured weekly coordinating and scheduling sessions
  - ✓ Establish daily, weekly and down day priorities
  - ✓ Major outage or project work priorities
- Cooperation in effectively using planned job packages from maintenance/operations supervisors and crafts
- Adequate maintenance engineering support
  - ✓ So planners do not have to develop standard operating and safety procedures
  - ✓ So planners do not have to devote engineering attention to solving recurring problems

Figure 11.49 Planner/scheduler needs for effectiveness #2.

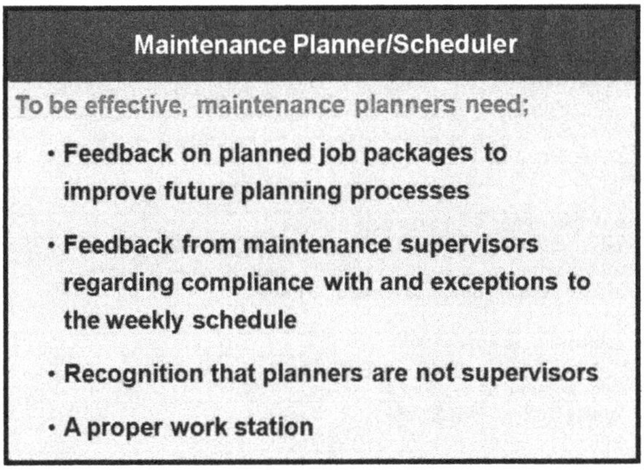

Maintenance Planner/Scheduler

To be effective, maintenance planners need;

- Feedback on planned job packages to improve future planning processes

- Feedback from maintenance supervisors regarding compliance with and exceptions to the weekly schedule

- Recognition that planners are not supervisors

- A proper work station

Figure 11.50 Planner/scheduler needs for effectiveness #3.

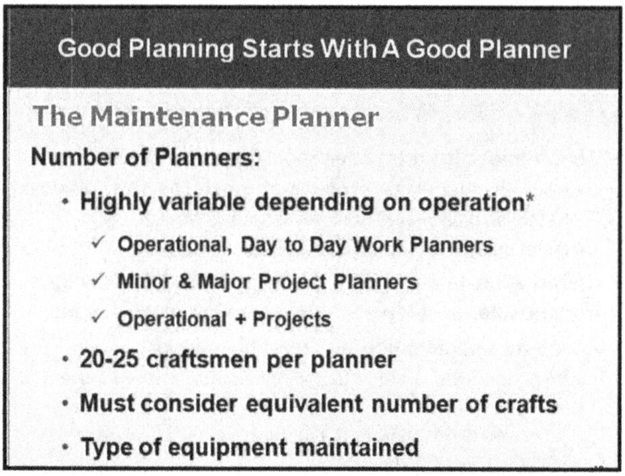

Figure 11.51 Defining number of planner/schedulers required.

Worksheet for Ratio of Crafts to Planners	
	Points Assigned
Planning and Scheduling Structure	
• Separate from material coordination (vertical structure)-1 Point	
• Combined (Horizontal structure)-2 points	
Number of Crafts Coordinated	
• One-1 Point	
• Two- 2 Points	
• Three-3 Points	
• Four-4 points	
Level of Planning	
• Craft and general description with schedule -1 Point	
• Graft, general description, special tools and major materials with schedule -3 Points	
• Craft, general instructions, tools, materials, prints and schedule - 5 Points	
• All of the above plus work methods described - 7 Points	
Level of Estimating	
• Estimates or historical averages 1 Point	
• Slotting against past jobs in labor library -3 Points	
• ACE Team-analytical estimating- 5 Points	
Inappropriate Responsibilities (Additive)	
• Sourcing -1 Point	
• Procuring -1 Point	
• Expediting -1 Point	
• Receiving -1 Point	
• Stocking -1 Point	
• Picking and kiting -1 Point	
• Staging and securing- 1 Point	
• Delivery to scheduled job site -1 Point	
Total Points	

Figure 11.52 Worksheet for ratio of crafts to planners.
From Maintenance Planning, Scheduling and Coordination by Don Nyman and Joel Levitt.

Conversion Table	
Total Points	Crafts to Planner Ratio
4-7	30:1
8-12	25:1
13-17	20:1
18-22	15:1
23-26	12:1
27-30	10:1

Figure 11.53 Conversion table for worksheet: total points on worksheet for ratio of crafts to planners.

Factors Influencing Number of Planners (Planner to Craft Ratio)

• Number of crafts performing planned/scheduled work

• Current state of maintenance management organization

• Complexity of craft structure in place

• Level of planning/scheduling needed

• Method of estimating used

• Current state of planner support system (labor and material libraries, CMMS etc

Figure 11.54 Factors influencing number of planners #1.

Factors Influencing Number of Planners (Planner to Craft Ratio)

• Complexity of the operations supported by maintenance

• Level of liaison and coordination required by planners

• Structure of the planning and scheduling organization

• Other support available:
  ✓ Maint Engineering
  ✓ Clerical support
  ✓ PM Coordinator
  ✓ Material coordinators
  ✓ Maintenance Planning Coordinator etc.

Figure 11.55 Factors influencing number of planners #2.

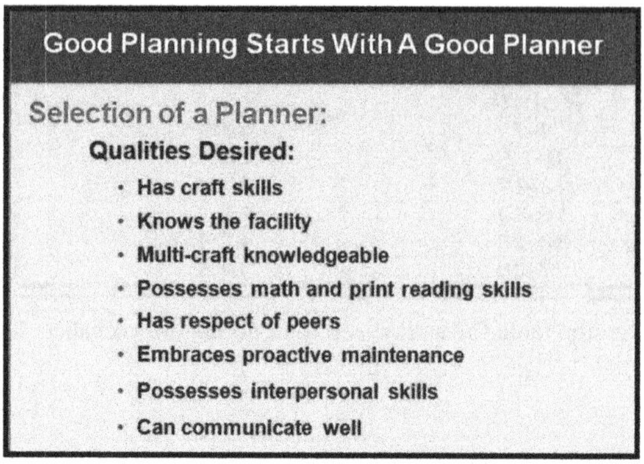

Good Planning Starts With A Good Planner

Selection of a Planner:

Qualities Desired:

- Has craft skills
- Knows the facility
- Multi-craft knowledgeable
- Possesses math and print reading skills
- Has respect of peers
- Embraces proactive maintenance
- Possesses interpersonal skills
- Can communicate well

Figure 11.56 Qualities desired in a planner/scheduler #1.

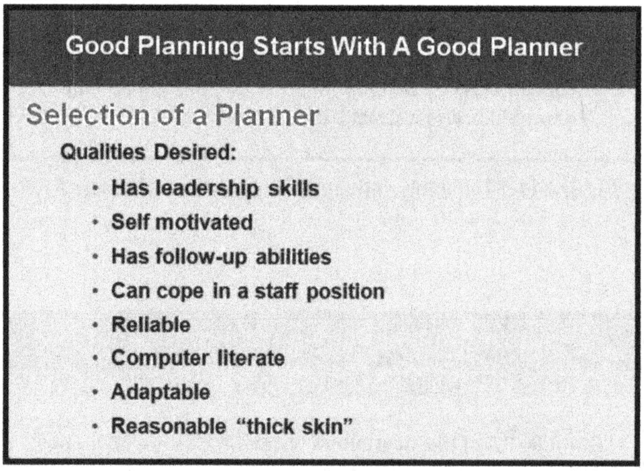

Good Planning Starts With A Good Planner

Selection of a Planner

Qualities Desired:

- Has leadership skills
- Self motivated
- Has follow-up abilities
- Can cope in a staff position
- Reliable
- Computer literate
- Adaptable
- Reasonable "thick skin"

Figure 11.57 Qualities desired in a planner/scheduler #2.

CATEGORY 9. WORK MANAGEMENT AND CONTROL: MAINTENANCE AND REPAIR (M/R)

**Rating: Excellent – 10, Very Good – 9, Good – 8, Average – 7,
Below Average – 6, Poor – 5**

ITEM #		RATING
1.	A work management function is established within the maintenance operation generally crafted along functionality of the CMMS.	
2.	Written work management procedures which governs work management and control per the current CMMS is available.	
3.	A printed or electronic work order form is used to capture key planning, cost, performance, and job priority information. 10=Bar coded assets, parts and work order.	
4.	A written procedure which governs the origination, authorization, and processing of all work orders is available and understood by all in maintenance and operations.	
5.	The responsibility for screening and processing of work orders is assigned and clearly defined.	
6.	Work orders are classified by type, e.g. emergency, planned equipment repairs, building systems, PM/PdM, project type work, planned work created from PM/PdM's.	
7.	Reasonable "date-required" is included on each work order with restrictions against "ASAP", etc.	
8.	The originating departments are required to indicate equipment location and number, work center number, and other applicable information on the work orders.	
9.	A well-defined procedure for determining the priority of repair work is established based on the criticality of the work and the criticality of equipment, safety factors, cost of downtime, etc.	
10.	Work orders are given a priority classification based on an established priority system.	
11.	Work orders provide complete description of repairs performed, type labor and parts used and coding to track causes of failure.	
12.	Work management system provides info back to customer; backlogs, work orders in progress, work completed, work schedules and actual cost charge backs to customer.10= real time system	
	9. Work Management and Control: Maintenance and Repair SUBTOTAL SCORE POSSIBLE:	**120**

Figure 11.58 CATEGORY 9: Work management and control: maintenance and repair.

CATEGORY 11. SHOP LEVEL RELIABLE PLANNING, ESTIMATING AND SCHEDULING (M/R)

ITEM #	Rating: Excellent – 10, Very Good – 9, Good – 8, Average – 7, Below Average – 6, Poor – 5 or less	RATING
1.	A formal maintenance planning function has been established and staffed with qualified planners in an approximate ratio of 1 planner to 20-25 crafts people.	
2.	The screening of work orders, reliable estimating of repair times, coordinating of repair parts and planning of repair work is performed as a support service to the supervisor.	
	Planner/Schedulers realize their primary scope and role of planning and scheduling is to improve craft labor productivity and quality through the elimination of unforeseen obstacles such as potential delays coronation parts machine time and available resources.	
	Planner/Schedulers clearly understand the scope of their defined roles and responsibilities within your organization and are in an organization structure that promotes close coordination, cooperation and communications with their customer in operations.	
3.	The planner uses the priority system in combination with parts and craft labor availability to develop a start date for each planned job to be scheduled	
4.	A daily or weekly maintenance work schedule is available to the supervisor who schedules and assigns work to crafts personnel with multiple week "look a heads" if required.	
5.	The maintenance planner develops reliable and well accepted estimated times for planned repair work and includes on work order for each craft to allow performance reporting, backload levels and even documentation of work competency for selected jobs.	
6	A day's planned work is available for each crafts person with at least a keeping a half a day ahead (KAHADA) during the working day known in advance.	
7.	A master plan for all repairs is available indicating planned start date, duration, completion date, and type crafts required to define "total maintenance requirements".	
8.	The master plan is reviewed and updated by maintenance, operations, and engineering as required with project type work expected from maintenance. Care is taken not to overload maintenance with project work that causes PM/PdM and other work to be neglected.	
9.	Total maintenance requirements are a total of Total Backlog + Ready Backlog that has all resources (except labor or equipment availability) available to be scheduled.	
10.	A firm rule of thumb is never to put anything on the schedule without parts in house. But have contingency plan if needed parts arrive for critical equipment.	

Figure 11.59   Cont'd

#		
11.	When parts arrive for critical equipment and can be inserted to the current schedule this is very proactive maintenance cooperation with operations.	
12.	Scheduling/progress meetings are held periodically with operations to ensure understanding, agreement and coordination of planned work, backlogs, and problem areas.	
13.	Operations cooperate with and support maintenance to accomplish repair and PM schedules.	
14.	Operations staff signs off the agreed upon schedule and are responsible to approve change in schedules and are accountable to TOP Leaders for adverse results.	
15.	Set-ups and changeovers are coordinated with maintenance to allow scheduling of selected maintenance repairs, PM inspections, and lubrication services during scheduled downtime or unexpected "windows of opportunity" to insert Ready Backlog jobs into the schedule.	
16.	Planned repairs are scheduled by a valid priority system, completed on time and in line with completion dates promised to operations and measured accordingly.	
17.	Deferred maintenance is clearly defined on the master plan and increased costs are identified to management as too the impact of deferring critical repairs, overhauls, etc.	
18.	Maintenance planners and production planners work closely to support planned repairs, to adjust schedules and to ensure schedule compliance in a mutual goal.	
19.	The planning process directly supports the supervisor and provides means for effective scheduling of work, direct assignment of crafts and monitoring of work in progress by the supervisor.	
20.	Planners training has included formal training in planning/scheduling techniques, super user training on the CMMS, report generating software or via Excel and on the job training to include developing realistic planning times for craft work being planned. Understand use of MS Project or the company's larger project management system such as Primavera 6.	

Figure 11.59 Cont'd

21.	Benefits of planning/scheduling investments are being validated by various metrics that document areas such as reduced emergency work, improved craft productivity, improved schedule compliance, reduced cost and improved customer service.	
22.	Planning and scheduling procedures have been established defining work management and control procedures, the planning/scheduling process, the priority system, etc.	
23.	A reasonable number of backup planner/schedulers are selected and properly trained and used to cover for the full time staff. The number is based on the size and type work being planned. Ideally just like the full time planner should have good shop experience and sound technical experience.	
24	If a maintenance coordinator is assigned within a unit of a large refinery or any production the planner will coordinate with that person about the job request, location within the unit, the problems to be repaired and related risks. In many cases this is an engineer or experienced operator who should be able to define complete requirements or a work request and in some case prepare a risk assessment for the planner to use for the job.	
25.	If issues and of any nature arise and are not readily resolved by the planner and operations, the maintenance leader should be the next step that the planner/scheduler takes for resolution.	
26.	If planned work orders involve participation by several shops or functional crews they are crossed over to a planner in that area. However a single planner/scheduler must plan and then coordinate various functional crews with the respective supervisors during the scheduling process.	
27	Planners are in an excellent position to ensure critical spares by asset are accounted for as well as to recommend items to consider for including within the storeroom as critical spares.	
29.	Planners see what is repetitively coming up for non-stock item purchasing as well as what is being repaired over and over again, Are your planners active in this area can support improving reliability and uptime.	
30.	Planners help ensure that warranted parts or equipment is denoted in the equipment file and that work orders for warranted parts or equipment are flagged during the planning process to document supplier reimbursements.	
	11. Shop Level Reliable Planning, Estimating and Scheduling M/R SUBTOTAL SCORE POSSIBLE:	300

Figure 11.59 CATEGORY 11: Shop level reliable planning, estimating, and scheduling.

ITEM #	Rating: Excellent – 10, Very Good – 9, Good – 8, Average – 7, Below Average – 6, Poor – 5 or less	RATING
1.	Work management and control is established for major overhaul repairs, shutdowns, turnarounds and outages (STO) and includes effective work management and control by in house staff and contracted resources.	
2.	Work management and control of major projects provide means for monitoring project costs, schedule compliance and performance of both in house and contracted resources with a robust project management system.	
3.	Work orders are used to provide key planning info, labor/material costs and performance info for major all STO and overhaul work.	
4.	Equipment history is updated with info from work orders generated from major overhaul repairs, and SATO work.	
5.	The responsibility for screening and processing of work orders for major repairs is assigned to one person or unit.	
6.	Change order procedures and control are clear to all and approved at the appropriate level based on company requirements.	
7.	Change orders are reviewed by planners just as they review all jobs; scope of work, key job steps, equipment required and total additional cost and impact on total STO duration and appropriate approvals received before work execution.	
8.	Work orders for major repairs, shutdown, and overhauls are monitored for schedule compliance, overall costs and performance info including both in house staff and contracted services.	
9.	Cost variances are measured at key milestone with cost variance info so extreme variance can be investigated sooner than later when it is too late. A 5%-10% variance is set as maximum with clear reason for increased scope of work.	
10.	Has the current level of plant maintenance/asset management achieved the desired reliability to make to make an STO a) needed at longer timeframe than normal b)needed at a shorter timeframe or c) needed the appropriate period based on age and state of asset capabilities in their operation context.? a)= 10,9,8; b)=5; c=7,6	
11.	The organization has the capability to manage the turnaround program and be cost effective as compared to the best in the sector, has a strategy for reducing costs in the face of an ageing plant and rising manpower and material costs and where can we get high level technical advice?	
12.	The organization knows what manpower is available in-house, the competence levels where to get additional resources, who will lead the site team, who will do the work plus the cross functional team to design, monitor and control the event organization?	
13.	Have STO's received significant level of attention companies, have a history of tolerating higher than necessary downtime have older age of plants, see STO as a "necessary evil" are striving to lengthen the STO intervals from 12 – 24 months and sometimes 8 years	
14.	The organization's history of planning & preparation for STO's has been: carried out more carefully, alignment of capital programs, has been scrutinizing and challenging scope of work	
15.	A process of assessing plant equipment deterioration is in place, the likely impact on reliability is known, the planning stocks and safeguarded and has partnering with major plant overhaul engineering contractors and have a learning organization from past history to manage STO's effectively	
16.	The plant beginning an STO has personnel available when required and capable of performing design specifications economically and a) safely for life of plant ,b) knows sum of activities performed to protect reliability of the plant , c) helps provide consistent means of production, d) help generate profit all with e) reducing the Total Cost of Ownership (TOC)	
17.	Top Leaders clearly understand that STO is a significant maintenance and engineering event during which new plant is installed, existing plant overhauled, and redundant plant removed which has a direct connection between successful accomplishment and the company's profitability	
18.	The company includes profit lost during period of STO is considered part of turnaround cost because they know the total true cost of event and the real impact can be assessed	
19.	All involved with STO's realize the potential hazard to plant reliability or can diminish or destroy reliability if not :properly planned, prepared, executed, poor decisions by managers and engineers, bad workmanship, use of incorrect materials and damage done while plant is being shut down, overhauled, restarted	
20.	Technical uncertainty due to occurrence of unforeseen problems can be accurately reported, knowing when cost estimates are being exceeded, event's duration must be extended, how both cost and duration increases be justified, Are reasonable cost and time contingencies built into an STO plan with accurate loss of revenue/profit considered.	
21.	Have Top Leaders created their business strategy to manage the STO basic objectives to eliminate STO's all together unless proven it is absolutely necessary,	
22.	If an STO is proven to be necessary, the Top Leader ensures that it will align with maintenance objectives, production requirements, business goals.	
23.	When beginning as STO the Top Leader has formed a chartered leadership driven, self-managed (not a committee) forming a cross functional staff to help a committed company get the best STO value.	
24.	The STO team has senior managers , responsible for long-term strategy and meet at regular intervals throughout year to review current performance and formulate high-level strategies for management of events such as a long-term STO program	
25.	Is an STO truly aligned to overall business strategy which include an evolution of asset management's driven search for change to preventive/predictive maintenance, being driven by technical considerations and a philosophy of maintenance prevention and continuous reliability improvement.	
26.	STO"s are driven by business needs and question every maintenance practice to determine if it can be eliminated by addressing cause that generated the need and examines the largest maintenance initiatives first during an STO.	
	10. Work Management and Control: Shutdowns, Turnarounds and Outages (STO) SUBTOTAL SCORE POSSIBLE:	**260**

Figure 11.60 Category 10: Work management and control: shutdowns, turnarounds, and outages (STO).

ITEM #	Rating: Excellent – 10, Very Good – 9, Good – 8, Average – 7, Below Average – 6, Poor – 5 or less	RATING
1.	The planning and scheduling function includes major repairs, overhauls and shutdown, turnaround and outage (STO) type work not considered as part of normal maintenance work and any work requiring an STO event,	
2.	The planner team is a resource (or member) for the STO team of senior managers and the Maintenance Leader. Planners should meet at regular intervals to review current jobs awaiting a planned STO event.	
3,	Are your Total Backlog jobs coded and planned effectively to await an STO event. In large plants and refineries planners support the plant schedulers with normally detailed job packages for estimates of all required resources for an STO job	
4.	Schedulers from Item #3 coordinate parts/materials develop daily or weekly schedules, monitor status of work along with onsite observations, from the supervisor input and from a planner's job package which could include several crews, defined job steps and estimated time for each step. With real time reporting to a project management system or CMMS status including costs can be readily determined from progress reporting.	
5,	Current planning/scheduling manpower is available with the competency levels needed to support the site team during an STO.	
6.	All planners and schedulers involved with STO's must realize the potential hazard to plant reliability or can diminish or destroy reliability if not: properly planned, prepared and execute.	
7.	There may be even poor decisions by managers and engineers, bad workmanship, use of incorrect materials and damage done while plant is being shut down, overhauled or restarted. Planners realize that properly planned, well prepared work and work executed to all HSSE requirements is essential.	
8.	The use of work orders, estimating of repair times, coordinating and staging of repair parts/materials and planning/scheduling of internal resources and contractor support is also included for major work and STO work not considered day to day maintenance and repair…	
9.	A project work schedule or formal project management system is used to manage status and cost variance for STO work.	
10.	The current CMMS is integrated and linked to the project management system in real time when STO work orders or a change order work is approved.	
11.	Estimated labor and materials are established prior to project start using work orders with effective labor and material reporting to track overall cost, work progress, schedule compliance, etc.	
12.	The master plan for all major STO repairs, overhauls and new installation is available indicating planned start date, duration, completion date, and type crafts required.	
13.	Resources required for day to day maintenance work are not compromised by having to perform major repair type work, installation, modifications etc, consuming in house resources required for PM's and other day to day type work.	
14.	Scheduling/progress meetings are held periodically with operations to ensure understanding, agreement and coordination of major work and problem areas such as asset being ready for scheduled work.	
15.	Major work performed by contractors is preplanned, scheduled and includes measuring performance of contracted services.	
16.	Planning and scheduling procedures have been established for project type work.	
	12. STO and Major Maintenance Planning/Scheduling with Project Management SUBTOTAL SCORE POSSIBLE:	160

Figure 11.61 Category 12: STO and major maintenance planning/scheduling with project management.

ITEM #	Rating: Excellent – 10, Very Good – 9, Good – 8, Average – 7, Below Average – 6, Poor – 5	RATING
1.	Contracted work is clearly defined because the better the definition at the early stages the better the job will go.	
2.	Loose specifications for both materials and work to be done are avoided	
3.	Communication of your ideas to a contractor is included in the Scope of Work and make sure they understand with meeting of the minds at kick off and status update meetings.	
4.	Ensure that the contractor understands the quality of materials needed from clear specifications.	
5.	Negotiation and award of the contract has had key due diligence by the key owner's representative.	
6.	For larger jobs, owners may check-out finances, credit, insurance, and staff.	
7.	Owners may visit other jobs to see contractor quality of work and call references.	
8.	Maintain at shop level a copy of the contract documents and keep a fair and complete set of contract info including requests for changes of scope.	
9.	Be aware of and avoid, if possible, low ball bids and negotiate a schedule of extras if applicable (i.e. "the high cost of low bid buying")	
10.	Avoid a common ploy where low balling the bid to get the job and floods the company with small extras.	
11.	Always add in for clauses like "all extras not included in the original price have to be agreed to in writing prior to the commencement of the work."	
12.	Are deduction clauses in the contract that spell out what you will charge back and when you will charge it? Examples would be debris removal, clean-up, missing firm completion dates.	
13.	Negotiate cancellation clauses and spell out how and why you can cancel the contract. Otherwise you could find yourself with a mechanic's lien against you over an inadequate job after you did not pay the final payment.	
14.	For ongoing service bids avoid both too short of a contract term and too long of a contract term for two reasons: 1. If the term is too short then the contractor will charge excessively for mobilization costs. 2. If the term is too long you might be stuck with a barely adequate vendor with no easy way to improve the situation.	
15.	Is the contract as clear as possible about responsibilities on, who supplies what, where to unload, site rules (safety, owner contact, clean-up, security, keys, etc.)?	
16.	Are there statements about how the site is to be left at the end of each work day?	
17.	Ensure who is responsible for locking up, barricades, traffic management, cleaning, and debris removal.	
18.	Does the agreement also include who is responsible for municipal permits, job plans, and all health, safety, security and environmental (HSSE) issues?	
19.	Are contractor's insurance policies reviewed with an agreement about what happens when (if) the contractor damages your property?	
20.	Damage to a neighbor's property that then might sue you or might spoil a good relationship is included as required.	
21.	Is all Insurance certificates up to date covering: General liability, Casualty (property damage), Workmen's compensation, Auto liability	
22.	If the contractor did a design build then Malpractice and Errors & Omissions is included.	
23.	Define performance as to what would a good job look like.	
24.	Add clause like "all work is expected to be done in a professional and workmanship manner and all work will be in compliance with applicable codes".	
25.	Owners should prepare the area to be worked on and remove as much as possible to avoid breakage/theft and isolate area so contractors have no reason to wander around the plant or a large multi-operational site such as refinery.	
26.	Does the owner manage the contractor and keep a record of the job/project as it unfolds & provides feedback?	
27.	Does owner perform frequent inspections and document results with a functional planned schedule and compare progress to projections with problems being identified as early as possible for resolution?	
28.	Clear agreements have been made about when and amounts of progress or final payments are to be made, etc.	
29.	To avoid sloppy record keeping all contractor work is documented on the owner on their CMMS/EAM.	
30.	Owner requires copy of paid receipts to prove subcontractors and material vendors have been paid.	
31.	Owner should get a "release of all liens form" signed before last payment because: a) You could have paid off the general contractor and still be hit with liens from unpaid jobs b) Consult with your legal department about lien laws in your state or country and be sure you are covered	
	13. Contractor Management SUBTOTAL SCORE POSSIBLE:	**310**

Figure 11.62 Category 13: Contractor management.

Why the Work Order Is a Prime Source for Reliability Information

12

The work order is the single most important document within maintenance. Beginning as a work request, it becomes a description of work to be accomplished. The work order would be viewed as the main document if maintenance was operating as a business; in essence, it would be the invoice given to the customer with total costs of labor, parts, and materials used. We strive to have our craftspeople understand this analogy so that the work order is accurate and correctly documents the work that was performed.

Work orders may come in many formats, as shown by the examples in Figures 12.1–12.4.

A beneficial job plan requires that a logical step-by-step process be followed. Job planning encompasses verification of all aspects of the job, as well as resources such as material manpower and equipment required to complete the work in an orderly manner and at optimum costs.

The job objectives and scope must be defined and described by the planner while listing the steps to be performed, thus defining *what* is to be done. There may have been a similar job performed in the past with a job plan already developed, which would be in the planner's database for job plans. In a case where the planner is not familiar with the job, a visit to the jobsite may be necessary for discussion with the requester as a possible next step. At this point, the planner may not have experience in the trade that will be required for the repair. Nonetheless, the planner must try to identify the cause of the failure and possible consequences of the failure. In some cases, plan troubleshooting may be necessary to get the full details of the repair job and the extent of repair parts required.

On the occasions where troubleshooting is required and disassembly is needed, we are in a good position to define the cause of the problem. The planner and the craftsperson involved with this event should be concerned with why the failure occurred. At times like this, a member of maintenance engineering, plant engineering, or reliability engineering may be called to the scene to give their analysis of the condition. Planners should never allow these types of events to pass without documenting the cause of failure, whether it is the operator or wear-and-tear of a system component.

To complete the work order correctly, a number of important criteria should be established. These criteria are included in Figure 12.5.

Two key questions are: (1) What work orders should be planned? (2) How much planning is enough? Here, we are talking about the approximate degree of detail to which jobs should be planned. During the early installation phases of planning and scheduling, there may be insufficient planning capacity to plan all jobs. Therefore, as proactive maintenance replaces firefighting, the organization plant will settle down into more capability for planning more jobs as planning and scheduling matures. So, the level of detail is less during the early phases of implementation than in the later phases.

Reliable Maintenance Planning, Estimating, and Scheduling. http://dx.doi.org/10.1016/B978-0-12-397042-8.00012-7

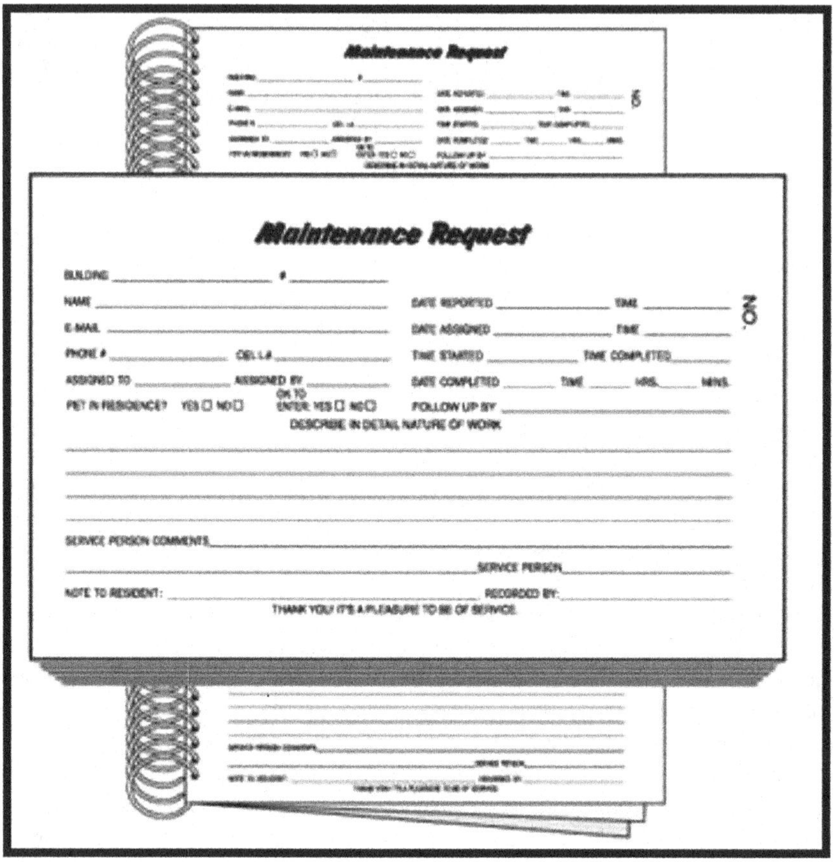

Maintenance Request

BUILDING #
NAME DATE REPORTED TIME NO.
E-MAIL DATE ASSIGNED TIME
PHONE # CELL # TIME STARTED TIME COMPLETED
ASSIGNED TO ASSIGNED BY DATE COMPLETED TIME HRS. MINS.
PET IN RESIDENCE? YES ☐ NO ☐ OK TO ENTER: YES ☐ NO ☐ FOLLOW UP BY
DESCRIBE IN DETAIL NATURE OF WORK
SERVICE PERSON COMMENTS
SERVICE PERSON
NOTE TO RESIDENT: RECORDED BY:
THANK YOU! IT'S A PLEASURE TO BE OF SERVICE

Figure 12.1 Work request example #1.

- In general, large jobs are planned first. Larger jobs are usually accompanied by delays and conflicts, therefore giving greater opportunity for benefits from planning.
- Cutoffs of 4–8 h may initially be established for the magnitude of jobs to be planned.
- The selected cutoff point should be progressively reduced as the planning process matures.
- Delays encountered in smaller jobs have a more dramatic percentage impact, but planning coverage should ultimately include all jobs that can benefit.
- Detailed planning of large jobs requires much more effort than what is justified on simple jobs.
- The usual tendency is to underplan large jobs and overplan small jobs.
  - Smaller jobs require less planning.
  - However, a 1-h job that is missing essential materials can cause much wasted time due to unnecessary travel.
  - Also, a small job will have a greater unproductive time as a percentage of total time than a larger job.
  - Focusing on repetitive jobs is important during early stages of planning and scheduling.
  - As the planner's library/database of plan job packages increases, this leads to reduced work.

MAINTENANCE WORK REQUEST FORM

Date of Request:

Location:

Please be specific Building, Room #, etc.

Description of Work required:

Requested by: Extension No:

Completed by:

Leave maintenance request in our mailbox, fax: 768-8112, inner-office mail, or email to MAINTENANCE via outlook.... If the problem is an **Emergency**, call extension 171, but don't forget to turn in a maintenance request form after the phone call. Hours are fr. 8am-3:30pm.

MAINTENANCE WORK REQUEST FORM

Date of Request:

Location:

Please be specific Building, Room #, etc.

Description of Work required:

Requested by: Extension No:

Completed by:

Leave maintenance request in our mailbox, fax: 768-8112, inner-office mail, or email to MAINTENANCE via outlook.... If the problem is an **Emergency**, call extension 171, but don't forget to turn in a maintenance request form after the phone call. Hours are fr. 8am-3:30pm.

Figure 12.2 Work request example #2.

- These benefits are a prime reason to provide ample planning capacities (such as a backup planner) to have sufficient planning capacity during the early phases of program installation.
- We must remember the Pareto principle during early phases of installation, whereby 80% of the benefits are derived from 20% of the effort.

The planning process should cover 80% or more of the available man-hours. The remaining 20%, which constitutes emergencies and unplanned work for one reason

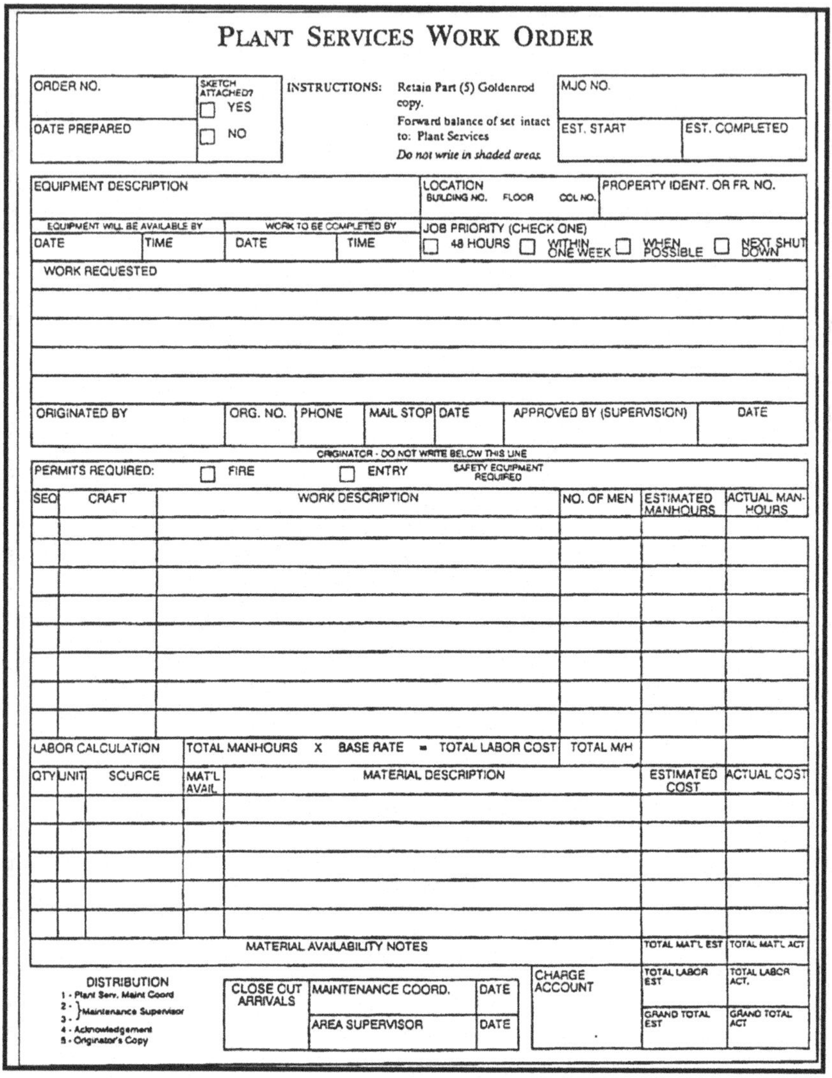

PLANT SERVICES WORK ORDER

ORDER NO. | SKETCH ATTACHED? ☐ YES ☐ NO | INSTRUCTIONS: Retain Part (5) Goldenrod copy. Forward balance of set intact to: Plant Services. *Do not write in shaded areas.* | MJO NO.
DATE PREPARED | EST. START | EST. COMPLETED

EQUIPMENT DESCRIPTION | LOCATION BUILDING NO. FLOOR COL NO. | PROPERTY IDENT. OR FR. NO.

EQUIPMENT WILL BE AVAILABLE BY: DATE | TIME
WORK TO BE COMPLETED BY: DATE | TIME
JOB PRIORITY (CHECK ONE) ☐ 48 HOURS ☐ WITHIN ONE WEEK ☐ WHEN POSSIBLE ☐ NEXT SHUT DOWN

WORK REQUESTED

ORIGINATED BY | ORG. NO. | PHONE | MAIL STOP | DATE | APPROVED BY (SUPERVISION) | DATE

ORIGINATOR - DO NOT WRITE BELOW THIS LINE

PERMITS REQUIRED: ☐ FIRE ☐ ENTRY SAFETY EQUIPMENT REQUIRED

SEQ | CRAFT | WORK DESCRIPTION | NO. OF MEN | ESTIMATED MANHOURS | ACTUAL MAN-HOURS

LABOR CALCULATION | TOTAL MANHOURS X BASE RATE = TOTAL LABOR COST | TOTAL M/H

QTY | UNIT | SOURCE | MAT'L AVAIL | MATERIAL DESCRIPTION | ESTIMATED COST | ACTUAL COST

MATERIAL AVAILABILITY NOTES | TOTAL MAT'L EST | TOTAL MAT'L ACT

DISTRIBUTION
1 - Plant Serv. Maint Coord
2 - } Maintenance Supervisor
3 -
4 - Acknowledgement
5 - Originator's Copy

CLOSE OUT ARRIVALS | MAINTENANCE COORD. | DATE | AREA SUPERVISOR | DATE
CHARGE ACCOUNT | TOTAL LABOR EST | TOTAL LABOR ACT. | GRAND TOTAL EST | GRAND TOTAL ACT

Figure 12.3 Work order example #1.

or another, falls upon the supervisor and his team to execute effectively with as much productivity as possible. Chapter 16 focused on scheduling, we will talk about when the planner does not plan for emergency work. However, the planner **can plan for emergency work**. In this case, selected craftspeople are designated to lower priority planned jobs, but they can leave that job and do it now (DIN, as part of a DIN squad) as emergencies emerge. In this case, the planner knows the percent of emergency work versus planned work and can have the supervisor's scheduled staff loaded accordingly.

When working with the Facilities Management Division for Raleigh, North Carolina state government facilities, we received numerous hot/cold calls and calls to

```
---------------------------------------------------------------
  WORK ORDER              W O R K   O R D E R            DATE
     192           ---------------------------------------  05/03/86
---------------------------------------------------------------
  PRIORITY: C          SCHEDULED DATE: 05/03/86  COMPLETION DATE:   / /
---------------------------------------------------------------
  REQUESTER: D. BLANN
---------------------------------------------------------------
  DESCRIPTION: REPLACED BEARINGS
---------------------------------------------------------------
  EQUIPMENT #:  2010-RM          EQUIPMENT NAME: REC. MOTOR, FIN TUN
                                     LOCATION: PD        DEPT.200
---------------------------------------------------------------
                         L A B O R
---------------------------------------------------------------
    Craft/man                                        Reg    OT
    ---------   ----------------------------------   -----  -----
  1     101     REPLACE MAIN MOTOR BEARINGS           1.0
  2                       AND BUSHES
  3
  4
  5
  6
  7
  8
  ---------
  REMARKS:                 TOTAL CRAFT/MAN 1    101    1.0
                           TOTAL CRAFT/MAN 2
                           TOTAL CRAFT/MAN 3
                                    TOTALS:      HRS    1.0
---------------------------------------------------------------
          M A T E R I A L S                      C O D E S
---------------------------------------------------------------
  QTY     STOCK #        DESCRIPTION      COST, $     REPAIR TYPE
  ---     --------   -----------------   ----------   -----------
 1   1  SET        BEARINGS             $   30.00  ----------------
 2   2  B-93       BRUSHES              $    2.95    FAILURE CODE
 3                                      $            ----------------
 4                                      $            ACCOUNT NUMBERS
 5                                      $            LAB:        .
                        TOTAL MATERIALS  $   35.90   MAT:
                                                     SUB:
---------------------------------------------------------------
```

Figure 12.4 Work order example #2.

unstop drains, sinks, etc. We solved this challenge by having one repair van plus typically used parts and one plumber assigned to take all so-called emergency calls that were dispatched by our central call center. This was back in the "good old days" of papers work orders and fat files of completed work orders. We did have an estimator for jobs with other Raleigh state government agencies, and they paid our department for the work.

- There is a need shown by the work order, work request or other documentation defining the content and scope of the job
- An inquiry has asked the questions a) should the job be done b) what priority should be given c) have expenditures been approved?
- Conduct a thorough analysis which breaks down the job into smaller subtasks. From this the planner can;
  - ✓ identify required skills
  - ✓ make a reliable estimate of repairtime(total labor hours and Job duration)
  - ✓ determine parts and material needs
  - ✓ ensure parts and materials are available before the job to scheduled
  - ✓ determine any special tools that may be needed, rented or reserved
  - ✓ determine if selected specifications, drawings repair manuals and other documents are provided for
  - ✓ ensure that all safety and legal permissions are provided for
- identify all processes that may have to be rerouted shutdown or back to and notify the affected owner/customer
- list and coordinate all preparatory andrestart activitiesthat are needed regardless of whether they are theresponsibilities of maintenance or operations

Figure 12.5 Important criteria to be established for selected work order.

So, **Why is the Work Order the Prime Source for Reliability Info?** The work order provides the most comprehensive description of work accomplished, repairs parts used, reasons for failure, and causes of failure. Over the life of the equipment, the work order tracks total cost of ownership and life cycle cost as they accrue over time, allowing replacements or major repair decisions to be made at the most economical point in time. As work orders are completed and signed off by the supervisor, the planner makes one final review before the work order is closed in maintenance. This process allows the planner to have a continuing picture of equipment problems of all types and magnitude. From this pure observation of work orders plus the analysis of work orders, the planner can play a very important role within a strategy of continuous reliability improvement. As we will see in Chapter 15, the planner/scheduler is a key member of an A Consensus of Experts (ACE) team that reviews basic job plans, large and small, for key steps and wrench time for each step and the total job. Teams established by maintenance engineers or reliability engineers have their ideal team member in the planner position.

Detailed Planning with a Reliable Scope of Work and a Complete Job Package

## 13

When there is sufficient planning capacity, "all jobs that can benefit" should be planned. We mentioned that planners should avoid getting involved with emergency work, but nonetheless they should know the percentage of man-hours normally required for emergency work. Then as I said before, "they can plan for emergency work," and factor this into the available hours for planned work. With repetitive jobs, the planner can invest more time because the details of planning these jobs will have a positive impact in reducing planning time in the future.

The planner can also use the "building block of time" approach where previously planned work is often applicable as a portion of a new job package example. Figure 13.1 illustrates the building block approach.

In Figure 13.1, tasks 1, 2, 3, or 4 maybe repetitive-type jobs serving as templates for equivalent work content that can be used over and over again by the planner. There are many conditions that support the maturing stage s of s planning program. Some of these are listed in Figure 13.2.

One of the most important sources of information is from within the maintenance technical library. This is where information such as equipment history, equipment manuals, parts manuals, parts lists, and assembly drawings can be found along with site drawings of equipment that is in place. This should be a controlled area with Internet access, drawing files, large reference tables, and proper shelving for books and catalogs. Here there may be parts catalogs for specific equipment or suppliers of specific components. This is the one place that should have ease of accessibility yet be controlled consistently. It should not be like the example shown in Figure 13.3.

There are many benefits for a planner to have a well-organized library of the documents that are catalogs and the computer network established with Internet. These benefits include:

- the planner's job is simplified, accelerated and more productive
- the planner can maintain job plans that are more consistent and with good quality
- this library becomes the universal resource for other groups, such as maintenance engineering defining specifications and problem solving
- and this foundation can be good for further computer assistance via electronic document management systems.

Key records and the maintenance technical library are very important and should include:

Equipment records: These contain all the pertinent data for the equipment and subsystems such as installation data, make model, vendor capacity for service, and

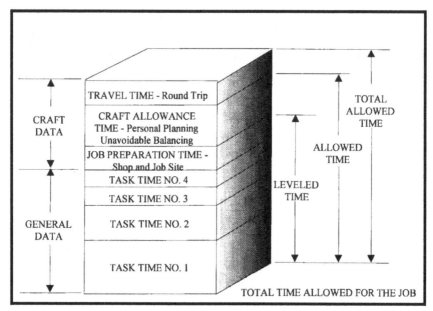

Figure 13.1 Building blocks of time.

parts support. One important item is the original asset capacity and specifications, not to be confused with equipment history of all repairs performed on the equipment.

Equipment history: This captures all the work order history of repairs performed on the equipment.

Library of planning aids: There are a number of means to simplify the planning process for machine repair and overhaul by classifying the debacle groups of machines and then building libraries of preplanned work element sequences along with the bill of materials. This concept establishes and documents the work sequence news for each type of equipment, class by clients. Here the documentation records the procedure needed to take the equipment completely apart and then put it back together with replacement parts as needed. In this case, when the ACE Team Process is used a number of individual benchmark jobs can be developed from the disassembly part of the job as well as the assembly part of the job.

Labor libraries: Labor libraries should be filed so that document retrieval is as convenient as possible. This might be by unit equipment specific type of skill required or by the job code. Having a labor library supports development of job steps sequences and labor resource requirements listings:

- Crew size by skill
- Job duration
- Man-hours by skill

Labor estimating: Having the term *estimating* within the actual title of this book strives to emphasize the importance that estimating plays in so many areas of planning and scheduling. There is an old saying that "what gets measured gets done."

- The RCM process the best maintenance strategy/plan has developed for the equipment. The maintenance plan may include PM., PdM, remove and rebuild, remove and replace, or even run to failure.
- Potential problems are reported by operations in a timely manner to provide lead time for planning
- There is a maintenance technical library, controlled and updated as new equipment is placed into the operation. In some cases this may be an electronic document management system with all site drawings and related documents.
- Over time the planner has built a good library of planned job packages, easily retrieved via an electronic system in some cases.
- For utilization of an existing CMMS system is present with functionalities needed by the planner for areas such as scheduling which often is a weak link.
- The planner receives good feedback on completed jobs. From the supervisors and craft employees
- Within the equipment master file there is good repair history to include failure codes and causes of failure
- PM/PdM task descriptions are such that anyone new can perform the inspection
- Within the organization and there is dedicated resources to failure analysis supported by the maintenance planner, supervisors and crafts people
- Equipment vendors and parts of vendors provide good support to plant staff when needed
- New technologies for repair are reviewed for site applications
- There is open dialogue on equipment problems from the operators
- Engineering is capable of providing qualified support to major problem areas
- Overhaul and rebuild capabilities are available either in-house or via contracted services
- Crafts personnel provide good workmanship and display PRIDE in Maintenance

Figure 13.2 Signs of a maturing planning process.

Figure 13.3 Invest in a good area for the maintenance technical library.

This ranges from maintenance strategy goals to personal goal setting down to work at the shop floor level. Even where a labor library is in place, some form of reliable estimating is required to extend it to as much work as possible. Later we will discuss some methodologies along with the ACE Team Benchmarking Process for the task level work content shown previously in Figure 13.1.

Material library: For critical units of equipment, this library should include parts lists and bills of materials. The material library supports identification of materials or parts requirements for jobs to be planned, including:

• the parts involved
• stockroom number of each item
• manufacturing ID for each part
• storeroom location
• unit of issue

Purchasing/Stores: These catalogs can be in hard copy or electronic format and are essential for all parties. They contain much the same information that is in the material libraries used by the planner. So some form of storeroom catalog or vendor catalog is used to develop material libraries. We can sort these catalogs by:

• component description
• assets where used
• stock number
• vendor serial number

If searching a computerized parts inventory, we can search by keywords and perform many types of analysis such as doing an ABC analysis, looking at parts usage, or identifying obsolete items.

Planned job packages: This library of planned job packages is very important to the planner. When completed and reviewed, job plans become viable templates to save the planner much time when the same job is planned again. In addition, job plans within major job packages can become jobs that the ACE Team uses for benchmark job analysis.

Other references: The library would include services and any other information from experienced supervisors, crafts people, as well the planner. One of the things that can be done is to take a video of critical repair jobs as they are being performed by an SME. This could be an SME nearing retirement and he is the "go to guy" for a specific repair. Therefore, he becomes the star of his own personal training film.

Standard operating procedures: Within complex operations such as refineries, these SOPs are included for lockout tag out, safety, risk assessment, and management of change compliance etc. Figures 13.4–13.12 illustrate work flow within a refinery unit where (1) operators can do some repair or help crafts person, (2) the unit has an asset coordinator, and (3) work is carefully reviewed when job has been completed. This was a standard operating procedure for this refinery.

Job packages: When planning work within a refinery you are faced with large and complex equipment that is physically linked to form the process. When planning work within this environment, many factors are involved and plan job packages are often the best approach. Especially during shutdown turnaround and outage planning, all jobs must be detailed with a job package, which is all of the documentation from the planning efforts. In most cases, job packages are reviewed with the supervisor and the person making the requested work. They are also reviewed between a supervisor and the craftsperson who will be assigned the work.

During the verbal exchange between the maintenance supervisor and the requester clarification of work to be performed is confirmed between strategic planning phase and the tactical execution phase as a work is being completed. Figures 13.13 and 13.14 summarizes a list of items typically included in a job package:

Work ID to Post-Job Completion

Figure 13.4 Work ID to job completion–refinery.

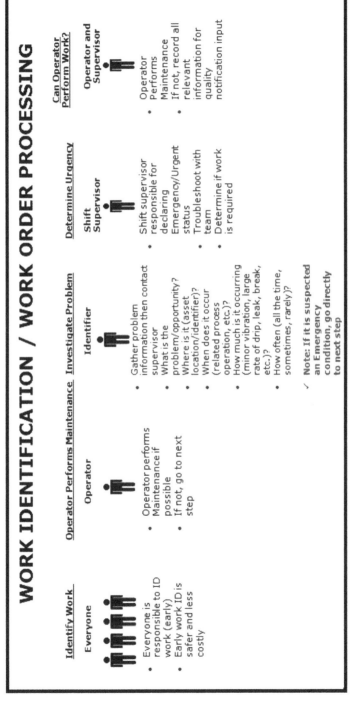

Figure 13.5 Work ID. Work order processing–refinery #1.

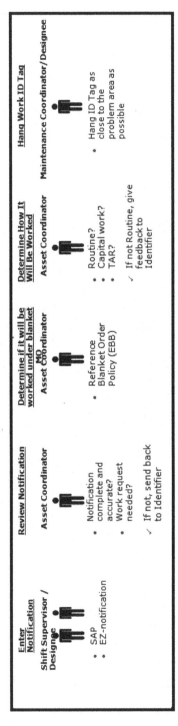

Figure 13.6 Work ID. Work order processing–refinery #2.

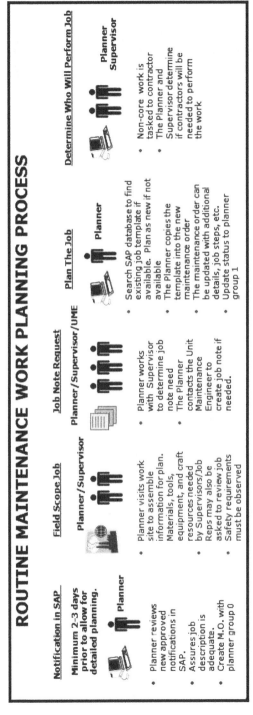

Figure 13.7 Routine maintenance work planning process–refinery #1.

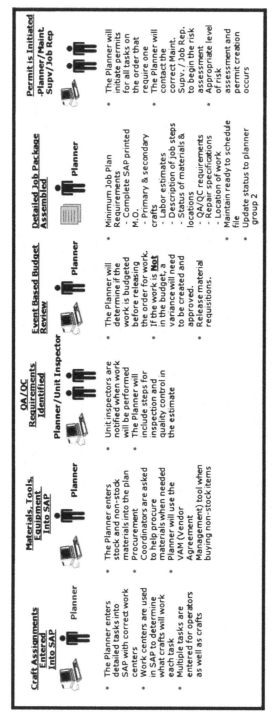

Figure 13.8 Routine maintenance work planning process–refinery #2.

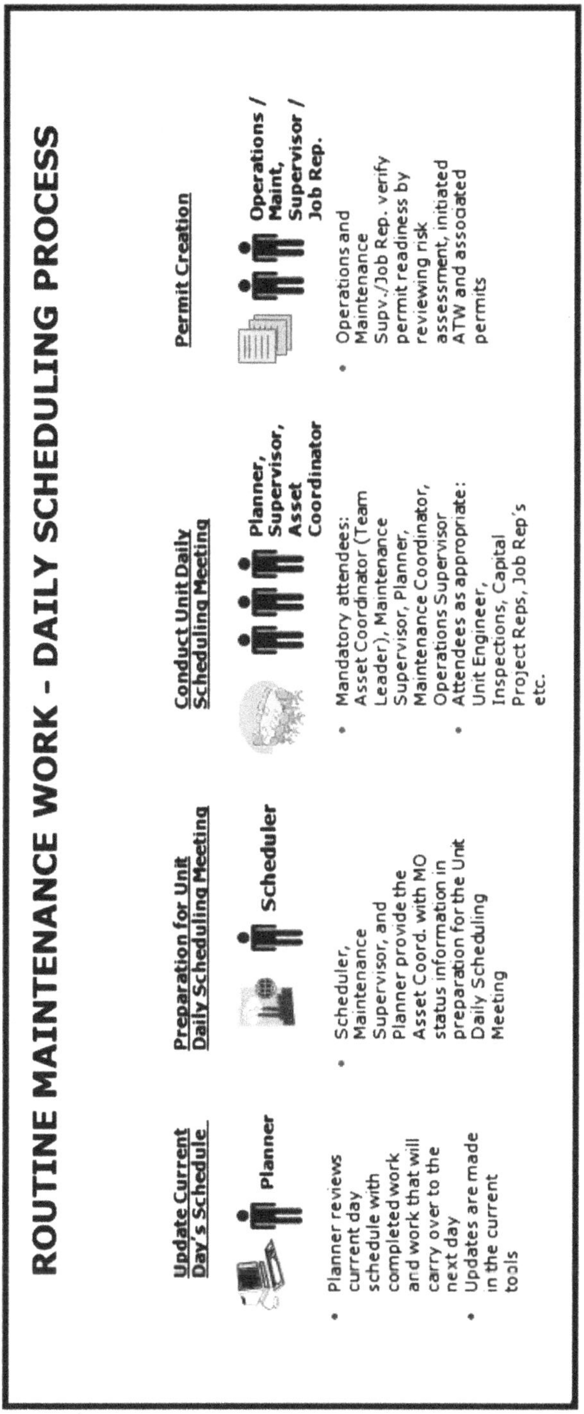

ROUTINE MAINTENANCE WORK – DAILY SCHEDULING PROCESS

Update Current Day's Schedule

Planner

- Planner reviews current day schedule with completed work and work that will carry over to the next day
- Updates are made in the current tools

Preparation for Unit Daily Scheduling Meeting

Scheduler

- Scheduler, Maintenance Supervisor, and Planner provide the Asset Coord. with MO status information in preparation for the Unit Daily Scheduling Meeting

Conduct Unit Daily Scheduling Meeting

Planner, Supervisor, Asset Coordinator

- Mandatory attendees: Asset Coordinator (Team Leader), Maintenance Supervisor, Planner, Maintenance Coordinator, Operations Supervisor
- Attendees as appropriate: Unit Engineer, Inspections, Capital Project Reps, Job Rep's etc.

Permit Creation

Operations / Maint, Supervisor / Job Rep.

- Operations and Maintenance Supv./Job Rep. verify permit readiness by reviewing risk assessment, initiated ATW and associated permits

Figure 13.9 Routine maintenance work daily scheduling process–refinery #1.

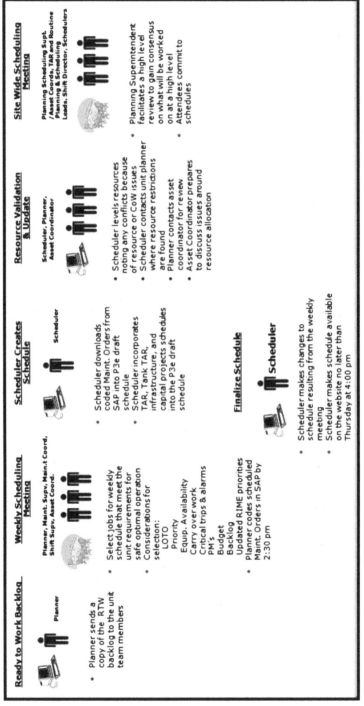

Figure 13.10 Routine maintenance work daily scheduling process--refinery #2.

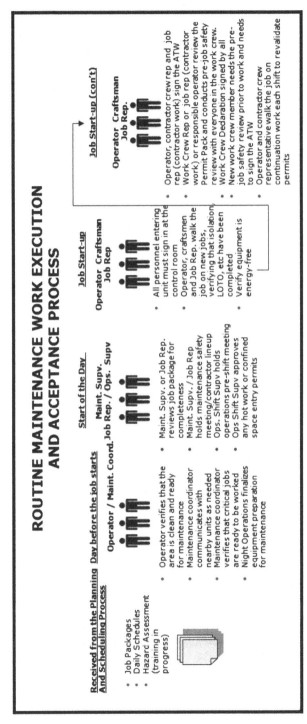

Figure 13.11 Routine maintenance work execution and acceptance process–refinery #1.

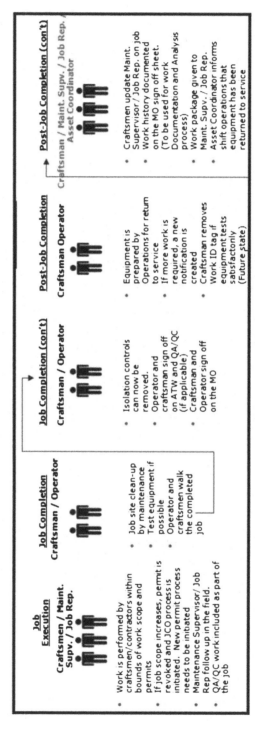

Job Execution
Craftsmen / Maint. Supv. / Job Rep.

- Work is performed by craftsmen/contractors within bounds of work scope and permits
- If job scope increases, permit is revoked and JCO process initiated. New permit process needs to be initiated
- Maintenance Supervisor / Job Rep follow up in the field.
- QA/QC work included as part of the job

Job Completion
Craftsman / Operator

- Job site clean-up by maintenance
- Test equipment if possible
- Operator and craftsmen walk the completed job

Job Completion (con't)
Craftsman / Operator

- Isolation controls can now be removed.
- Operator and craftsman sign off on ATW and QA/QC (if applicable)
- Craftsman and Operator sign off on the MO

Post-Job Completion
Craftsman Operator

- Equipment is prepared by Operations for return to service
- If more work is required, a new notification is created
- Craftsman removes Work ID tag if equipment tests satisfactorly (Future state)

Post-Job Completion (con't)
Craftsman / Maint. Supv. / Job Rep. / Asset Coordinator

- Craftsmen update Maint. Supervisor / Job Rep. on job
- Work history documented on the MO sign off sheet. (To be used for work Documentation and Analysis process)
- Work package given to Maint. Supv. / Job Rep.
- Asset Coordinator informs shift operations that equipment has been returned to service

Figure 13.12 Routine maintenance work execution and acceptance process–refinery #2.

✓ The work order that has initiated and been approved for the work to be accomplished

✓ A work planning sheet as a basic checklist (See Figure 13.13)

✓ The job plan with details for each task and the step-by-step sequence

✓ A labor deployment plan by craft and skill including labor hour estimates. This should also include duration and consider contract as well as in-house resources.

✓ If the job is complex and requires multiple crews is then the use of a GANNT bar chart might be needed to help plan task sequencing to assign crews

✓ For a pre-shutdown fabrication or other preparation this should be completed in advance so as to minimize equipment being out of service

✓ Bill of materials: here a list of all materials needed for the job is developed along with an acquisition plan for major items. We need to determine if the material is authorized inventory or a direct purchase item. The planning package should include spares reservation and stage location. However, wiping out stock levels of critical spares should be avoided.

✓ Develop a requisition and purchase order reference list

✓ Provide estimated time this step/task summarized by resource group and for the total job

✓ Provide a set down plan as to where everything used for major tear down will go

✓ A copy of all required permits clearances and tag outs

✓ ensure that contractors bring qualified people and have knowledge of all site safety requirements

✓ and for job packages are number of things that can be supplied; prints, sketches, digital pictures, specifications, special procedures and any references that the assigned crew is likely to need

Figure 13.13 List of items typically included in a job package.

PLANNING WORKSHEET

DATE _____   PLANT LOCATION _____
WORK ORDER# _____   COST CENTER _____
CRAFTSMAN _____   PLANNER _____
COMPLETION DATE _____ TOTAL EST HRS _____
DESCR OF WORK _____

_____

MATERIALS REQ'D _____

_____

_____

SPECIAL TOOLS _____

PROCEDURES
1. _____   10 _____
2. _____   11 _____
3. _____   12 _____
4. _____   13 _____
5. _____   14 _____
6. _____   15 _____
7. _____   16 _____
8. _____   17 _____
9. _____   18 _____

CRAFT					TOTAL
ESTIMATED HOURS					

SUGGESTIONS _____

_____

ADDITIONAL WORK PERFORMED _____

_____

DELAYS _____

_____

Figure 13.14 Planner worksheet example.

Understanding Risk-Based Maintenance by Using Risked-Based Planning with Risk-Based Inspections

14

Basic Overview[1]

In RBI analysis, *risk* is calculated as the product of the probability of failure and the consequence associated with a failure. Planning must always consider safety and all health, safety, security & environmental (HSSE) factors when developing small job plans or large job packages. In refineries as we saw in the work process flow process Chapter 14, there is extreme care in making sure that work is properly identified, accepted for quality, and all mitigation for risk related to the job has been considered. So risk-based maintenance (RBM) is something that a planner must understand within their special work environment. This chapter strives to introduce the key concepts of RBM and how selected software systems can help define risk-based inspections (RBI). First, let us define what risk is and how it is calculated. And, as I always say, it is like gambling with maintenance costs and equipment conditions that are deferred to a later date.

Risk is calculated as the product of the probability of failure and the consequence associated with a failure:

Risk = Probability of Failure × Consequence of Failure

Risk is usually considered a better measure for prioritization than either the probability of failure alone or the consequence of failure alone, because it is more descriptive of the actual damage/loss caused. As an example, if you need to prioritize two assets where one asset has a high probability of failure but low consequence of failure, and the other asset has a low probability of failure but a high consequence of failure, the analysis would yield completely opposite results if you considered only one factor or the other. The use of *risk* eliminates this ambiguity.

The probability of failure (*POF*) is determined using applicable damage factors (mechanisms), a generic failure frequency, and a *management system* factor:

$$POF(t) = 1 - e^{-gff \times FMS \times Df(t)}$$

where:

- *gff* is the generic failure frequency
- *FMS* is the management system factor
- *Df(t)* is the overall damage factor

---

[1] This overview of basic RBI concepts is reprinted with permission from ReliaSoft Corporation based on "Introduction to Risk Based Inspection (RBI)," *Reliability HotWire*, Issue 151 (September 2013). http://www.weibull.com/hotwire/issue151/hottopics151.htm

The generic failure frequency is based on industry averages of equipment failure. The management system factor is a measure of how well the management and labor force of the plant are trained to handle both the day-to-day activities of the plant and any emergencies that may arise due to an accident. The overall damage factor is the combination of the various damage factors that are applicable to the particular piece of equipment being analyzed.

The consequence of failure is calculated as the combined values of the consequences for damage to the failed equipment, damage to the surrounding equipment, loss of production, the cost due to personnel injury, and the damage to the environment. The consequence of failure can include both a financial consequence (FC) and an area (safety) consequence (CA):

$$FC = FC_{cmd} + FC_{affa} + FC_{prod} + FC_{inj} + FC_{environ}$$

$$CA = max\left(CA_{equip}, CA_{personnel}\right)$$

- FC_{cmd} is the financial consequence to failed equipment
- FC_{affa} is the financial consequence to surrounding equipment
- FC_{prod} is the financial consequence due to production downtime
- FC_{inj} is the financial consequence due to personnel injury
- $FC_{environ}$ is the financial consequence due to environmental damage/cleanup
- CA_{equip} is the area consequence to surrounding equipment
- $CA_{personnel}$ is the area consequence to nearby personnel

For further detail on calculating probability of failure and/or consequence of failure, please consult API RP 581.

Evolution of Maintenance Strategies to Create Transition Between ReliaSoft

Since the 1950s, maintenance strategies have evolved through an event-based response, time based, condition based, reliability based, and then risk based. Figure 14.1 illustrate development of these maintenance strategies. RBM is extremely important within large complex operations like refineries. In addition, as you will see, the analysis for risk leads to what we call RBI (Figures 14.2–14.10).

There are a number of software systems that can support the analysis of reliability data and in turn support RBM with defining the best RBI plan. The following will describe such a system that is available from ReliaSoft. This material is reprinted with permission from ReliaSoft Corporation based on "Introduction to Risk Based Inspection (RBI)," *Reliability HotWire*, Issue 151 (September 2013). For more information about the RBI software tool, visit http://www.ReliaSoft.com/rbi/.

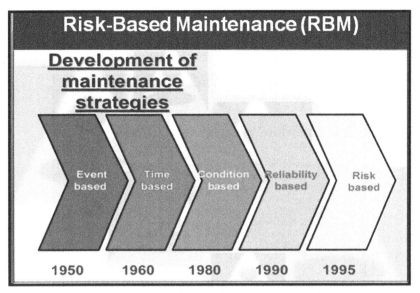

Figure 14.1 Development of maintenance strategies.

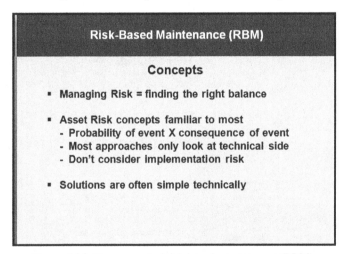

Figure 14.2 The concept of risk-based management (RBM).

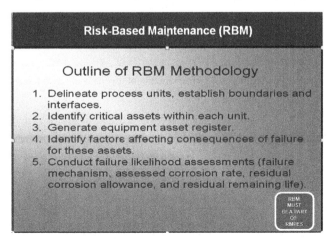

Figure 14.3 Outline of risk-based maintenence (RBM) methodology (Part 1).

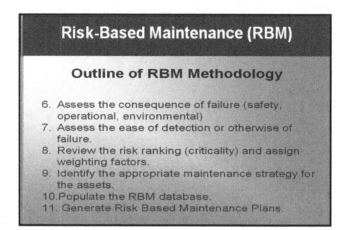

Figure 14.4 Outline of risk-based maintenence (RBM) methodology (Part 2).

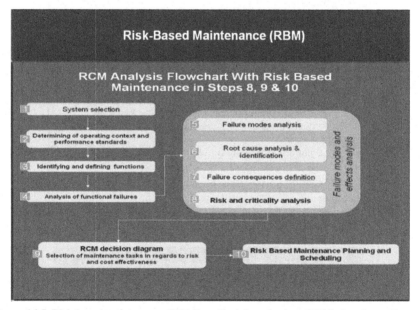

Figure 14.5 Risk-based maintenence (RBM) methodology leads to RBM planning and scheduling.

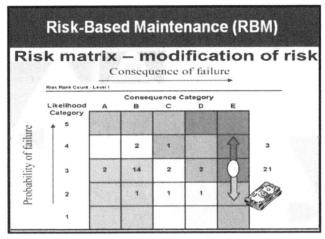

Figure 14.6 Risk-based maintenence (RBM) risk matrix.

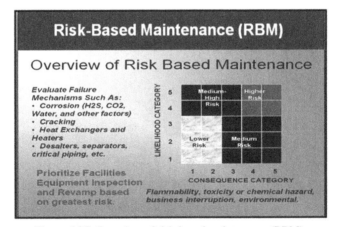

Figure 14.7 Overview of risk-based maintenance (RBM).

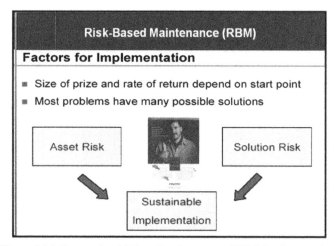

Figure 14.8 Factors for risk-based maintenance (RBM) implementation.

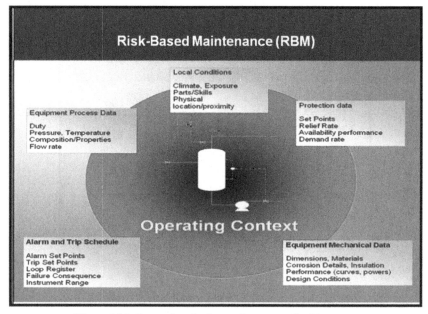

Figure 14.9 Remember the "operating context" of the asset.

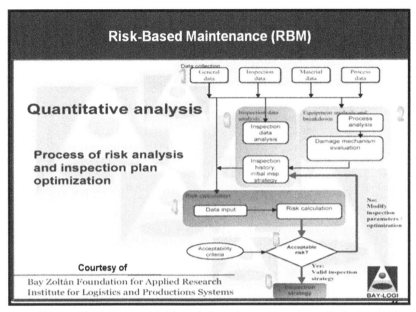

Figure 14.10 Process of risk analysis and inspection plan optimization.

Introduction to the RBI Software from ReliaSoft Corporation

ReliaSoft's new RBI software tool facilitates RBI analysis for oil and gas, chemical, and power plants in adherence to the principles and guidelines presented in the American Petroleum Institute's recommendations in the API RP 580 and RP 581 publications, as well as the American Society of *Mechanical Engineers'* recommendations in the ASME PCC-3-2007 publication. RBI also includes all of the standard features available in RCM++, a software tool that is widely used for Failure Modes and Effects Analysis (FMEA) and Reliability Centered Maintenance (RCM).

The RBI interface is identical to RCM++, with the addition of RBI specific options on the System Hierarchy tab.

The equipment and component types that are available for RBI analysis are currently limited to those addressed API RP 581. The equipment and component types that are available for RBI analysis are currently limited to those addressed in API RP 581.

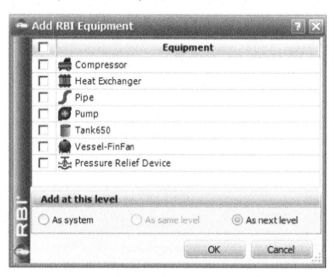

Each equipment type also has associated components. For example, the Vessel-FinFan equipment has the specific components shown below available for analysis.

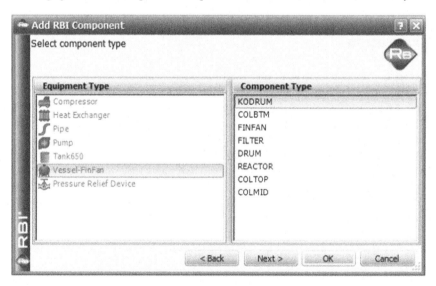

While the system hierarchy can contain items that are not RBI related, those items will be ignored when performing an RBI analysis. For example, in the following picture the system is the hydrogen generation unit, which has several items that receive RBI analysis and several that do not.

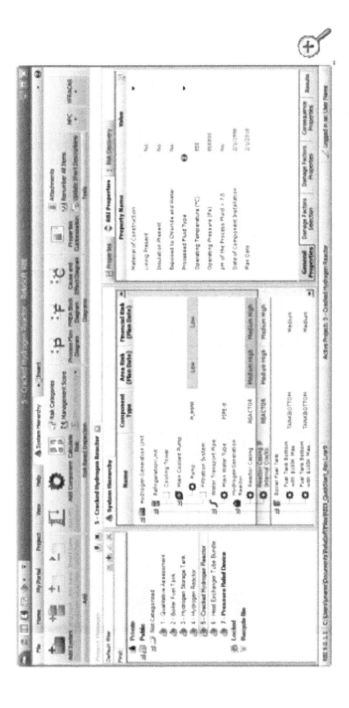

Once you create the assets, you must answer questions and fill out the relevant properties for the asset to be analyzed. All of these inputs are used to create a failure model that determines the probability of failure, and calculates the consequences of failure. The results also include the recommended inspections, if any, that should be performed to keep the asset under the maximum allowable risk.

Case Study Example

A small town wants to do a risk analysis on a proposed high-pressure 24-inch pipe carrying crude oil for which the oil transportation company is willing to pay a rent of $500,000 in advance for the next 20 years. The company will also perform an inspection on the pipe halfway through the rental period. Since the city is self-insured and is not willing to take any financial chances, the city council would like to know if the $500,000 offsets the possible risk associated with the pipe. The city has requested the required information from the oil transportation company to conduct its own RBI analysis of the pipe. The pipe is composed of carbon steel. The heavy crude, which contains 100 ppm H_2S, is being pumped through at an operating temperature of 25 °C with a pressure of 4.5 MPa.

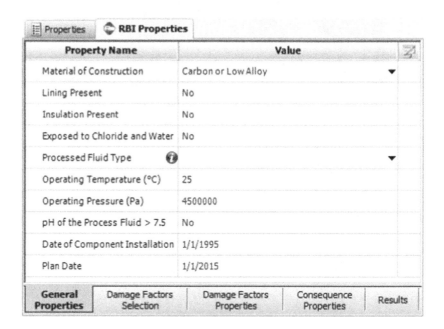

The only two damage mechanisms expected are general thinning and external corrosion.

Damage Factor	Description	Applies?	Comments
Thinning Damage Factor	This is a required factor that applies to all components.	• Yes	This is a required damage factor.
SCC Damage Factor-Caustic Cracking	The component is composed of a carbon or low alloy steel and there are any concentrations of caustic elements in the process environment.	No	
SCC Damage Factor-Amine Cracking	The component is composed of a carbon or low alloy steel and the process environment has any concentration present of acid gas treating amines (e.g., DIPA, MEA, etc.).	No	
SCC Damage Factor-Sulfide Stress Cracking	The component is composed of carbon or low alloy steel and it operates in an environment that contains any concentration of H2S and water.	No	
SCC Damage Factor-HIC/SOHIC-H2S	The component is composed of a carbon or low alloy steel and there is H2S and water in any concentration in the process environment.	No	
SCC Damage Factor- HSC-HF	The component is composed of a carbon or low alloy steel and it is exposed to any concentration of hydrofluoric acid.	No	
SCC Damage Factor-HIC/SOHIC-HF	The component is composed of a carbon or low alloy steel and it is exposed to any concentration of hydrofluoric acid.	No	
External Corrosion Damage Factor- Ferritic Component	Select this factor if the component is un-insulated and subject to any of the following conditions: • Some areas are exposed to steam vents, deluge systems or cooling tower mist overlays. • Some areas are subject to acid vapors, process spills or the ingress of moisture. • The component is composed of carbon steel and the operating temperature is between 23°C and 121°C. • The component does not normally operate between 12°C and 177°C, but does periodically heat or cool in this range. • The component is subject to frequent outages. • The component has deteriorated wrapping or coatings. • The component consistently operates below the atmospheric dew point. • The component has un-insulated protrusions or nozzles in cold conditions.	Yes	

Tabs: General Properties | **Damage Factors Selection** | Damage Factors Properties | Consequence Properties | Results

For their analysis, the city assumed that the effectiveness of the performed inspections was average. All other property values were estimated based on other similar pipes used elsewhere.

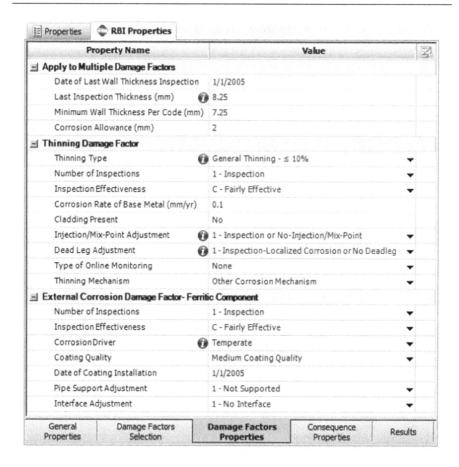

The city based the financial portion of the consequences on the current population density and the property values around the proposed pipe area. For the initial estimate, the city did not take into account inflation or any possible losses associated with property value changes, or possible resident dislike of the pipeline across city land.

Properties	⟳ RBI Properties	
Property Name	**Value**	⇄
⊟ Flow Rates / Flammability		
Representative Process Fluid	C25+	▼
Storage Phase	Liquid	▼
Atmospheric Temperature (°C)	20	
Detection Classification	A - Good	▼
Isolation Classification	B - Average	▼
Component Mass (kg)	150000	
Inventory Mass (kg)	1500000	
Flammability Mitigation System	No Mitigation System	▼
⊟ Toxicity		
Toxicity Mitigation Reduction (%)	ⓘ 0	
Mass Fraction Hydrofluoric Acid (HF)	0	
Mass Fraction Hydrogen Sulfide (H2S)	0.0001	
Mass Fraction Ammonia (NH3)	0	
Mass Fraction Chlorine (Cl2)	0	
Mass Fraction Aluminum Chloride (AlCl3)	0	
Mass Fraction Carbon Monoxide (CO)	0	
Mass Fraction Hydrogen Chloride (HCl)	0	
Mass Fraction Nitric Acid (HNO3)	0	
Mass Fraction Nitrogen Dioxide (NO2)	0	
Mass Fraction Phosgene (COCl2)	0	
Mass Fraction Toluene Diisocyanate (TDI)	0	
Mass Fraction Ethylene Glycol Monoethyl Ether (EE)	0	
Mass Fraction Ethylene Oxide (EO)	0	
Mass Fraction Propylene Oxide (PO)	0	
⊟ Financial		
Component Material	Carbon Steel	▼
Equipment Outage Multiplier	1	
Personnel Density Within Unit (personnel/m^2)	0.01	
Equipment Cost ($/m^2)	2500	
Loss of Production Due to Component Downtime ($/day)	1E-10	
Cost Associated with Injury or Fatality of Personnel ($)	1000000	
Environmental Cleanup Cost ($/bbl)	1000	
Risk Target for Consequence Area (m^2)	1000	
Risk Target for Financial Consequence ($)	500000	

General Properties	Damage Factors Selection	Damage Factors Properties	**Consequence Properties**	Results

Results

The analyzed results show that in case of a containment failure of the pipe, the expected cost associated with a failure would be almost $4.9 million. This far exceeds the $500,000 payment to be received. However, the probability that a failure will occur within 20 years under normal circumstances is estimated to be only 0.7%. Therefore, the expected *financial risk* is the product of the two, or a little over $36,000. This is well under the $500,000 payment to be received.

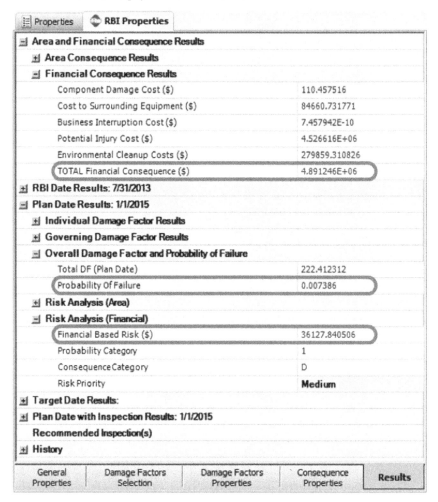

In terms of a cost analysis of the risk, and not including the potential loss in property value by having the pipe run across city property, the city council recommended that the community accept the proposed pipeline as an additional revenue source with minimal risk.

Category 22 on Risk Based Maintenance from The Scoreboard for Maintenance Excellence

The following pages include category 22 for risk-based maintenance (RBM).

22. Risk-Based Maintenance (RBM)

Item#	Rating: excellent—10, very good—9, good—8, average—7, below average—6, poor—5	Rating
1.	Maintenance in this company has been carried out by integrating analysis, measurement, and periodic test activities to standard preventive maintenance. The gathered information is viewed in the context of the environmental, has been related to process conditions of the equipment in the system in its operating context. This company has a process and resources to perform *the asset condition and risk assessment* and define the appropriate maintenance program. All equipment displaying abnormal values is *refurbished* or *replaced*. In this way it is possible to extend the useful life and guarantee over time high levels of reliability, safety, and efficiency of the plant.	
2.	A RBM strategy is in place that prioritizes maintenance resources toward assets that carry the most risk if they were to fail. It is a methodology for determining the most economical use of maintenance resources. This is done so that the maintenance effort across a facility is optimized to minimize the total risk of failure. A RBM strategy is based on two main phases: • Risk assessment • Maintenance planning based on the risk	
3.	This company realizes that the objective of maintenance over the last 20 years has steadily shifted from a "prevention" approach to "risk"-based approach. This thinking has been spawned within the oil, gas, and petro chemical sectors by the following factors: • Increasing complexity of systems with relatively high reliability at the component level, but failures have become less predictable at the system level • Equipment chained together into continuous processes, bigger pieces of equipment with higher capacities. Bigger losses when failures occur • Lean processes with less in-process storage, lower inventory levels. Short duration downtime affecting output • Rising expectations of customers wanting higher quality at lower prices in real terms. Organizations unable to adapt to these rising demands go out of business • Workers, unions, passengers, consumers, and the general public demand higher safety and environmental compliance. Industrial action, lawsuits, and consumer resistance more prevalent	

Continued

22. Risk-Based Maintenance (RBM)—Cont'd

4.	Physical asset operators and owners within the oil, gas, and petrochemical sectors are increasingly being held to a higher "standard of care" with regard to their physical assets.	
5.	Preventing all failures is generally not feasible from either a technical or an economical point of view. The airline industry attempted failure prevention until the mid-1970s and found with the increase in technology of modern equipment that their traditional maintenance strategies did not have the desired result. Is your organization still applying these ineffective strategies or has not developed a physical asset management strategy discussed in Benchmark Area 2? Maintenance strategies must evolve to support the technological requirements of modern equipment and the challenges of a competitive and legislated environment. To the same degree that financial auditors apply "due diligence" to the management of an organization's financial assets, maintainers must apply due diligence in the management of physical assets. To the same degree that the justice system requires "due process" in the implementation of the law, so should physical asset managers apply due process in the management of physical assets? Unfortunately, many physical asset managers are still managing their assets using gut feelings. It is no wonder that:	
6.	Your maintenance strategies have evolved to support the technological requirements of modern equipment and the challenges of a competitive and legislated environment, to the same degree that financial auditors apply due diligence to the management of an organization's financial assets. Maintainers are applying due diligence in the management of physical assets to the same degree that the justice system requires due process in the implementation of the law, so should physical asset managers apply due process in the management of physical assets?	
7.	When looking closely at your company, most people will not find a disorientating mismatch between the long-term nature of your liabilities and then see an increasingly short-term nature of your assets management strategy.	
8.	RBM: Your maintenance strategy considers RBM as an evolution of reliability-centered maintenance (RCM) since RCM is based in the equipment condition and the importance of equipment to the overall process/system, but RCM is limited because it does not solve the quantification of failures.	
9.	Your company has the capability to quantify problems with probability as well as level of consequences of failure and RBM has been successfully applied and has achieved significant direct savings or cost avoidance in plants such as oil and gas, petrochemical, and power generation and power distribution networks, etc. And, it achieves important savings.	
10.	API RP 580 standard defines risk as the combination of the probability of an event occurring during a time period and the consequences associated with the event. In mathematical terms: Where risk factor = %probability of failure × consequence of that failure. Does your company define risk factors to obtain an economic value (if the consequence is valued) or a classification by a risk matrix?	

22. Risk-Based Maintenance (RBM)—Cont'd

11.	API considers the RBI as the next generation of inspection interval settings, focusing attention specifically on the equipment and associated deterioration mechanisms representing the most risk to the facility. It recognizes the ultimate goal of inspection is the safety and reliability of facilities. Has your company integrated risk factors into PM/PdM inspections or stand-alone RBI tasks for critical processes?	
12.	F. I. Khan and M. M. Haddara propose an RBM methodology broken down into three modules: • Module I: Risk estimation, including a failure scenario development, a consequence assessment, and a probability failure analysis; it can be conducted using fault tree analysis • Module II: Risk evaluation, setting up acceptance criteria, and applying these criteria to the estimated risk for each unit in the system • Module III: Maintenance planning, optimizing the maintenance plan to reduce the probability of failure, reducing the total risk level of the system	
13.	Some see RBM as a simpler methodology than RCM; it also requires an initial reliability study but includes an economic risk assessment, so it allows doing financial analysis and makes it easier to choose timed-based and on-condition tasks as well as complex actions such as spare parts quantity and location, redesign of equipment, or changes in the process. Does you approach go beyond traditional RCM?	
14.	Your strategy includes maintenance based on equipment performance monitoring and the control of the corrective actions taken as a result. *The real actual equipment condition is continuously assessed* by the online detection of significant working device parameters and their automatic comparison with average values and performance. Maintenance is carried out when certain indicators give the signaling that the equipment is deteriorating and the failure probability is increasing. This strategy, in the long term, allows drastic reduction of the costs associated with maintenance, thereby minimizing the occurrence of serious faults and optimizing the available economic resources management.	
15.	The maintenance frequency and type are prioritized based on the risk of failure. Assets that have a greater risk and consequence of failure are maintained and monitored more frequently. Assets that carry a lower risk are subjected to a less stringent maintenance program. By this process, the total risk of failure is minimized across the facility in the most economical way. The monitoring and maintenance programs for high-risk assets are typically condition-based maintenance (CBM) programs. Some use signals from wireless input to a central data collector from numerous PdM technologies within control loops or on stand-alone assets.	

Continued

22. Risk-Based Maintenance (RBM)—Cont'd

16.	RBM is a suitable strategy for all maintenance plants as well as your site. As a methodology, it provides a systematic approach to determine the most appropriate asset maintenance plans. Upon implementation of these maintenance plans, the risk of asset failure will be acceptably low.	
17.	A framework for determining risk is in place. Here the RBM framework is applied to each system in a facility. A system, for example, may be a high-pressure vessel. That system will have neighboring systems that pass fluid to and from the vessel. The likely failure modes of the system are first determined. Then, the typical RBM framework is applied to each risk. Is this framework present at your plant?	
18.	**Collect data**: For each risk that is being identified. Accurate data about the risk needs to be collected. This will include information about the risk, its general consequences, and the general methods used to mitigate and predict the risk.	
19.	**Risk Evaluation**: At the risk-evaluation stage, your operation considers both the probability of the risk and the consequence of the risk, quantified in the context of the facility/process under consideration.	
20.	**Rank risks**: With the risk evaluation complete, the probability and consequence are then combined to determine the total risk. This total risk is ranked against predetermined levels of risk. As a result, your decision on risk is either acceptable or unacceptable.	
21.	**Create an inspection plan**: If the risk is unacceptable, you evaluate a plan to inspect the system using a *condition monitoring* approach where possible. Or, if it is more cost appropriate, and technically feasible, a *preventative maintenance* program might be selected.	
22.	**Propose mitigation measures**: At this stage, you have a proposal for mitigating the risk, using the condition monitoring and maintenance approach.	
23.	**Reassessment**: Finally, your proposal is evaluated against other factors, such as legal and regulatory requirements. If the proposal needs are deficient, then the process starts again. Otherwise, the maintenance proposal is put into place.	
24.	RBM decision methods can be categorized into qualitative, quantitative, and semiquantitative method. However, RBM is not widely applied, and most of the applications focus on the area of process plant [3] and petroleum transport system.	
	22. Risk-based maintenance **Subtotal score possible**: 240	

Further Reading

American Petroleum Institute, 2008. API RP 581 Risk-based Inspection Technology, second ed. American Petroleum Institute, Washington, D.C.

Developing Improved Repair Methods and Reliable Maintenance Planning Times with the ACE Team Process

Work measurement within maintenance has never been nor will it be an exact science. However, we must have a reliable method to establish maintenance repair time that is acceptable, accurate within limits, and reasonable from the point of view of the crafts-people. There have been several ways that time standards for maintenance work have been developed. These include

1. **Construction Trade Estimates (Commercially published)**: These types of standards are used by contractors when bidding on construction jobs. Therefore, they are not recommended for in-house maintenance work because construction is not recommended for skilled craft technicians. These standards include engineering safety standards in the interest of better safety, and most relate to construction rather than maintenance.
2. **Educated Guesses**: Many planning and scheduling processes start, they normally began with pure estimates or educated guesses. Of course, they are the least costly and least time-consuming. Disadvantages include that they reflect personal judgment, whether by a supervisor or planner. In addition, the time is typically the time in which the person feels they could accomplish the repair. One key point here is that as a planner gets started and the time estimated is needed for the backlog, there must be a starting point and educated guesses can be useful. It is very important to quantify backlog not in jobs but rather in man hours. Remember that the common denominator is man hours. As time and experience continues, other methods may be considered.
3. **Historical Averages**: This method considers work orders of the same type and taking out extreme highs and lows and then taken an average of a specific job. Some computerized maintenance management systems (CMMS) have this functionality, but results are just averages with all of the current nonrelated maintenance activities included. The craft performance (CP) factor of overall craft effectiveness (OCE) is standard labor hours divided by actual labor hours. Using historical averages, CP is normally 100%. Therefore, this makes performance and schedule compliance looked favorable when using historical estimates. Thus, we might ask "how demanding are these expectations." If work order descriptions are not complete, then the averages will not be reliable. In essence, they reflect the current work environment, methods, and tooling rather than standard procedures and methods.
4. **Adjusted Averages**: Adjusted averages provide the first step toward true expectancy or a standard. These averages require a base, perhaps 6 months, during which time averages are collected for repetitive jobs and activity work sampling is concurrently performed. This work sampling looks to establish the average CP for various crews. If a given job averages 10 labor hours and the crew averages 70% performance during the base, then 7 labor hours becomes the adjusted average. This method is not recommended in any situation.

Reliable Maintenance Planning, Estimating, and Scheduling. http://dx.doi.org/10.1016/B978-0-12-397042-8.00015-2

5. **Slotting**: This is what the ACE Team Process is all about. However, it goes way beyond just an estimate of time. It includes finding ways to improve repair methods and identify how the failure occurred and how future failures can be detected before actual failure. It is a method that uses experienced craftspeople, planners and supervisors to provide benchmark job estimates on jobs they understand and have had a hand in developing "work content time for benchmark jobs."

6. **Universal Maintenance Standards (UMS)**: UMS look at method analysis and use "standard data" for every detailed motion of using a maintenance tool. It was first used by the U.S Navy and then by consulting firms such as H. B Maynard. I take back what I said earlier about the *Harvard Business Review*. They did publish a great White Paper on UMS, covering it very well.

Why the ACE Team Process: The ACE Team exists to provide a well-qualified team of experienced craftspeople, technicians, and supervisors to establish benchmark repair jobs and work content time for these jobs. The ACE Team is chartered to help develop the ACE System for establishing maintenance performance standards at each site.

Process: *What are the steps to be followed, and what are the questions to be answered by this team?*

1. Orientation, charter review, and charter acceptance or modification.
2. Ensure that all team members understand team objectives and agree on what needs to be achieved and the criticality of this initiative to the planning and scheduling process.
3. Understand the current concepts of the ACE System as defined in your organization's maintenance planning and scheduling standard operating procedure (SOP).
4. Understand the basics of the new XYZ system for CMMS/enterprise asset management (EAM), the characteristics, functionality, and performance.
5. Determine critical repair jobs that should be used as representative benchmark jobs; define key steps and elements for each benchmark job; and define any special tools, safety requirements, and other special requirements for the job.
6. Determine ways to improve doing the jobs being analyzed as benchmark jobs, considering better tools, equipment, skills, and even better preventive maintenance (PM)/predictive maintenance techniques to avoid this type of failure problem.
7. Conduct the ACE Team Process as outlined in the 10-step approach from the SOP.
8. Develop a team consensus on work content times for all of the benchmark jobs selected.
9. Continuously improve the ACE Team Process within the XYZ organization as an element of our continuous reliability improvement efforts.

Evidence of Success: (*What results are expected, in what periods, for this team to be successful?*)

1. A sufficient number of benchmark jobs will be developed as to individual tasks and steps along with estimated work content times to complete the site's ACE Team spreadsheets.
2. The actual period to complete the initial spreadsheets will depend on the time allocated by the ACE Team at each site.
3. ACE Teams from one site are expected to share their results with the other sites. Because of the similar nature of equipment, the sharing of benchmark job write-ups and even work content times that ACE Teams have developed can be shared throughout the operation.

4. Overall success will be determined by each planner having adequate spreadsheets that cover all construct areas as well as types of crafts work (mechanical, electrical, etc.) so that planning times can eventually be established for 80% or more of the available craft hours.

Resources: (*Who are the team members, team leader, and team facilitator that will support the team if needed? How much time should be spent in meetings and outside of meetings?*). The ACE Team should consist of the following representatives:

• One maintenance planner/team leader
• One maintenance supervisor
• Two to three crafts representatives from area 1
• Two to three crafts representatives from area 2
• Two to three craft representative from area 3

Note: Crafts representatives should rotate periodically and sufficient numbers should be designated so as to have at least two to three representatives from each craft area when benchmark jobs from these areas are being reviewed for job steps and estimated for work content time.

• An initial ACE Team meeting will be for 1 h or more. The team shall meet initially for at least 3 h each week. This team's activities and success will be considered as part of each team member's job.

Constraints: (*What authority does the team have, what items are outside of the scope of the team, and what budget does the team have?*)

1. No changes to organization structure are anticipated.
2. Benchmark job plans are to be reviewed and approved by the maintenance manager.
3. Each team must obtain buy in and overcome concerns from the other crafts on their estimates for benchmark jobs and repair methods recommended for each benchmark job.
4. Team presents implementation status reports as required and any additional recommendations to the XYZ maintenance excellence strategy team.
5. The ACE Team has the authority to recommend new and improved repair methods, new tools to help craft productivity and safety, and other improvements to improve asset reliability as developed during the ACE Team Process.

Expectations: (*What are the outputs, when are they expected, and to whom should they be given?*)

1. Spreadsheets for the site that cover all crafts areas and construct types are completed by _____.
2. Reliable planning times are provided for benchmark jobs so that effective planning, performance measurement, backlog control, and level of PM work can be established with a high level of confidence.
3. ACE Team provides a steady source of continuous improvement ideas to make repair jobs safer and easier.
4. Minutes are to be completed for all team meetings and sent to the maintenance manager and the XYZ maintenance excellence strategy team.

The Methodology for Applying the ACE Team Benchmarking Process

This section will outline the methodology for applying the ACE Team Benchmarking process. This very easy-to-use procedure will allow a planner/scheduler to be the central organizer of the ACE Team within an organization desiring to use this methodology for developing benchmark jobs, the use of slotting, and allowances to develop reliable planning times for scheduling. A graphical illustration of this process is included in Figure 15.1.

ACE Team Benchmarking Process

A New Maintenance Work Measurement Tool from the Maintenance Excellence Institute International

A True Team-Based Approach: Here, we will outline a new and highly recommended methodology for establishing team-based maintenance performance standards that we call **reliable planning times**. The ACE Team Benchmarking process (ACE System) was developed by The Maintenance Excellence Institute International founder back in the 1980s. It is a true team-based process that uses skilled craftspeople, technicians, supervisors, planners, and other knowledgeable people to do two things:

1. Improve current repair methods, safety, and quality
2. Establish work content time for selected *benchmark jobs* for planners and others to use in developing **reliable planning times**

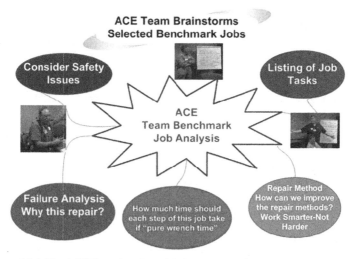

Figure 15.1 The ACE Team benchmark job analysis. ACE, a consensus of experts.

Benchmark Jobs: This is a proven process that uses "**a** consensus of experts" (ACE) who have performed these jobs and can also help improve them. In turn, relatively few representative "benchmark jobs" are developed for the major work areas/ types within the operation. Benchmark jobs are then arranged into time categories ("time slots") on spreadsheets for the various craft work areas.

Spreadsheets: By using spreadsheets to do what is termed "work content comparison" or "slotting", a planner is then able to establish planning times for many jobs using a relative small sample of "benchmark jobs". This publication also provides the step-by-step process on using the ACE System. Most importantly, it will illustrate how this method supports continuous reliability improvement and quality repair procedures for all types of maintenance repair operations.

Nearly every CMMS allows a user to enter "planned" or "standard" hours on a work order and then report on actual versus planned hours (the CP element of OCE) when the job is complete. This holds true for PM and corrective maintenance work orders as well as project-type work for renovation, major overhauls, and capitalized repairs. Most do not use this for one main reason: They do not have reliable planning times or standard hours available.

Determining the standard hours an average maintenance technician will require to complete a task under standard operating conditions provides everyone involved a sense of what is expected. The standards provide management with valuable input for backlog determination, manpower planning, scheduling, budgeting, and costing. Labor standards also form the baseline for determining craft productivity and labor savings for improved methods.

The ACE System Supports Continuous Reliability Improvement: Maintenance work by its very nature seldom follows an exact pattern for each occurrence of the same job. Therefore, exact methods and exact times for doing most maintenance jobs cannot be established as they can for production-type work. However, the need for having reliable performance measures for maintenance planning becomes increasingly important as the cost of maintenance labor rises and the complexity of production equipment increases. To work smarter, not harder, maintenance work must be planned, have a reasonable time for completion, use effective and safe methods that are performed with the best personal tools and special equipment possible and have the right craft skill using the right parts and materials for the job at hand.

Investment for Planners: With an investment in maintenance planners, there must be a method to establish reliable planning times for as many repair jobs as possible. The ACE System provides that method as well as a team-based process to improve the quality of repair procedures. Various methods for establishing maintenance performance standards have been used, including reasonable estimates, scientific wild average guess (SWAGs), historical data, and engineered standards such as UMS using predetermined standard data. These techniques generally require that an outside party establish the standards, which are then imposed upon the maintenance force. This approach often brings about undue concern and conflict between management and the maintenance workforce over the reliability of the standards.

The ACE System: A Team-Based Approach: Rather than progressing forward together in a spirit of continuous improvement, the maintenance workforce in this type of environment often works against management's program for maintenance improvement. The ACE System overcomes this problem with a team-based approach involving craft people who will actually do the work that will be planned later as the planning and estimating process matures. As shown later, the ACE System is truly a team-based process that looks first at improving maintenance repair methods, the reliability of those repairs to improve asset uptime, and then secondly to establish a benchmark time for the job.

Gaining Acceptance for Performance Standards: To overcome many of the inherent difficulties associated with developing maintenance performance standards, the ACE System is recommended and should be established as the standard process for modern maintenance management. Other methods such as the use of standard data can supplement the ACE System. The ACE System methodology primarily relies on the combined experience and estimating ability of a group of skilled craftspeople, planners, and other with technical knowledge of the repairs being made within the operation.

The objective of the ACE Team Benchmarking process is to determine reliable planning times for several selected benchmark jobs and to gain a consensus and overall agreement on the established work content time. This system places a very high emphasis on improving current repair methods, continuous maintenance improvement, and the changing of planning times to reflect improvements in performance and methods as they occur. The ACE System is a very progressive method to developing maintenance performance standards, a very hard area in itself to develop reliable and well-accepted planning times for maintenance. The complete 10-step approach to implementing the ACE Team Benchmark process within your current planning, estimating, and scheduling can be found at http://www.pride-in-maintenance.com.

Application Guide for the ACE Team Benchmarking Process

Generally, the ACE System parallels the concepts of the UMS approach. For the UMS and the ACE System, the "range of time concept" and "slotting" are used once the work content times for a representative number of "benchmark jobs" have been established. The ACE System focuses primarily on the development of work content times for representative "benchmark jobs" that are typical of the craft work performed by the group. An example of an actual UMS benchmark job that has been analyzed with standard data to establish work content time is included in Figure 15.1.

For the example illustrated in Figure 15.2, we see that through the use of UMS standard data, the eight elements of the job, including oiling of the parts, have been analyzed and assigned time values that total 1.07 h. Because the time value for this benchmark job falls within the time range of 0.9–1.5 h (see Figure 15.3), it is assigned a standard work content time of 1.2 h.

What this implies is that the actual work content for this benchmark job will generally be performed within the time range for the work group G (0.9–1.5 h)

Benchmark Analysis Sheet

Decription: Remove and reinstall 3 oil wiper rings	B.M. No:			
split type of air compressor 1950 cfm at 100 psi	Craft: Mech			
	Dwns: N/A			
	No. of Men: 1		Sh. 1 of 1	
	Analyst: JEB		Date:	

Line	Men	Operation Description	Reference Symbol	Unit Time	Freq.	Total Time
1	1	Remove and reinstall 1 crankcase cover	PWN-10-10	.030	2	.060
2		Remove and reinstall 2 bolts	PWN-10-1	.011	4	.044
3		Slide gland off and on	PWN-10-7	.012	2	.024
4		Unfasten 3 garter springs and refasten	PWN-10-8	.023	6	.138
5		Remove and reinstall 9 wiper ring	PWN-10-9	.012	18	.216
		segments				
6		Fit 9 ring segments to piston rod	PWN-5-2	.040	9	.360
7		Clean 12 springs and rings	PWN-8-1	.016	12	.192
8		Oil 12 parts	PWN-3-9			
		2 squirts per pint				
		.0023 + .0012(N2)		.0023	1	.002
		N2 = number of application		.0012	24	.029

Notes:		
	Benchmark time	1.07
	Standard work group	E

Figure 15.2 An example of a benchmark job analysis using UMS standard data. UMS, Universal Maintenance Standards.

with a confidence level of 95%. When we refer to "work content," the following applies: The work content of the benchmark job *excludes* things such as travel time, securing tools and parts, prints, delays, and personal allowances, etc. The benchmark time that is estimated does not include the typical "make-ready" and "put-away" activities that are associated with the job. Therefore, several allowances must be added to the work content time by a planner to get the actual planning time for the job.

ACE TEAM Benchmark Job Analysis Sheet						
Benchmark Job Description			Benchmark Job No.: MECH-AC-5			
Remove and reinstall 3 oil wiper rings, split type air compressor 1950 cfm @100psi			Craft: Mechanical			
			Ref. Drawing: AC-9999			
			No. of Crafts: 1			
			Analyst: JEB/ACE Team			
Line No.	No. of Crafts	**Operation Description**	Ref. Code	Unit Time	Freq	Total Time
1	1	Remove 2 bolts to remove crankcase cover				.10
2		Slide gland off and unfasten 3 garter springs				.10
3		Remove 9 wiper ring segments				.10
4		Clean 12 springs and rings and properly oil all 12 parts per Lube Spec #AC-2000	CLN-1			.25
5		Reinstall 9 wiper ring segments and fit to piston rod				.50
6		Slide glide on and fasten 3 garter springs				.10
7		Replace. crankcase cover and fasten with 2 bolts				.10
Notes: CLN-1 = Average benchmark time established to clean and oil this # of small parts per lube specification noted above.			**Benchmark Time for Work Content**			1.25
			Standard Work Group			E

Figure 15.3 Example of an ACE Team benchmark analysis. ACE, a consensus of experts.

Important Note: The estimated time for a benchmark job is for pure work content time and is made under these conditions:

1. The right craft skills and level of competency is available to do the job.
2. An average skilled crafts person, two-person team, or crew is doing the job giving 100% effort (i.e., a "fair day's work for a fair day's pay").

CRAFT: _____   CODE:_____

Task Area: Task areas would be mechanical, electrical, hydraulic, etc or by major areas such as fork lift, conveyor systems or building systems types of repairs.			
Group E 1.2 Hour	Group F 2.0 Hours	Group G 3.0 Hours	Group H 4.0 Hours
>.9 <1.5	>1.5 <2.5	>2.5.........................<3.5	>3.5.........................<4.5
Job Description A	Job Description E	Job Description I	Job Description M
Job Description B	Job Description F	Job Description J	Job Description N
Job Description C	Job Description G	Job Description K	Job Description O
Job Description D	Job Description H	Job Description L	Job Description P
NOTE:			

Figure 15.4 Example spreadsheets for workgroups E, F, G, and H.

3. The correct tools are available at the job site or with the craftsperson or crew.
4. The correct parts are available at the job site or with the craftsperson or crew.
5. The machine/process/asset is available and ready to be repaired.
6. The craftsperson or crew is at the job site with all of the above and proceeds to complete the job from start to finish without major interruption.

Once a sufficient number of benchmark job times have been established for craft areas and work types, these jobs are categorized onto spreadsheets. They are established on spreadsheets by craft and task area and according to the standard workgroups (Figure 15.4), which represent various ranges of time. This is exactly the concept behind the UMS approach. Figure 15.5 provides an example of a spreadsheet for workgroups E, F, G, and H with jobs that have benchmark times of 1.2, 2.0, 3.0, and 4.0 h, respectively. A complete set of ACE Team Spreadsheets for Workgroup A (0.1 h) up to Workgroup T (30.0 h) is available in the forms section of Appendix G.

ACE System Work Groupings and Time Ranges: Figure 15.6 includes a listing of the ACE System work groupings and the respective time ranges for each workgroup from A to T. Likewise, spreadsheets for Workgroups A, B, C, and D would be developed with benchmark jobs having a work content time below 1.2 h. Spreadsheets for Workgroups I, J, K, and L for benchmark jobs having work content times from 5.0 to 9.0 h, respectively, would also be developed, as they were needed. Figure 15.7 shows the ACE System standard work groupings and time ranges for Workgroup U (from 32 h) to Workgroup CC (68 h).

Spreadsheets Provide Means for Work Content Comparison: After sufficient spreadsheets have been prepared based on the representative benchmark jobs from various craft/task areas, a planner/analyst now has the means to establish planning

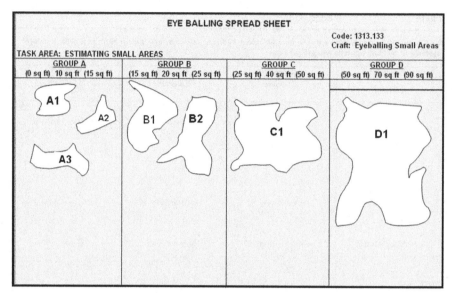

Figure 15 5 Example spreadsheets for comparing areas for Workgroups A, B, C, and D.

The ACE System Time Ranges

	ACE SYSTEM TIME RANGES		
WORK GROUP	FROM	STANDARD TIME (Slot time)	TO
A	0.0	.1	.15
B	.15	.2	.25
C	.25	.4	.5
D	.5	.7	.9
E	.9	1.2	1.5
F	1.5	2.0	2.5
G	2.5	3.0	3.5
H	3.5	4.0	4.5
I	4.5	5.0	5.5
J	5.5	6.0	6.5
K	6.5	7.3	8.0
L	8.0	9.0	10.0
M	10.0	11.0	12.0
N	12.0	13.0	14.0
O	14.0	15.0	16.0
P	16.0	17.0	18.0
Q	18.0	19.0	20.0
R	20.0	22.0	24.0
S	24.0	26.0	28.0
T	28.0	30.0	32.0

Figure 15.6 ACE System standard work groupings and time ranges: A to T.

ACE TEAM BENCHMARKING SYSTEM: WORK GROUPS & TIME RANGES Up to 69 Hours			
WORK GROUPS	RANGES FROM (Hrs)	BENCHMARK TIME (Slot Time)	UP TO (Hrs)
U	32.0	34.0	36.0
V	36.0	38.0	40.0
W	40.0+	42.0	44.0
X	44.0	46.0	48.0
Y	48.0	50.0	52.0
Z	52.0	54.0	56.0
AA	56.0	58.0	60.0
BB	60.0	62.0	64.0
CC	64.0+	66.0	68.0

Figure 15.7 ACE System standard work groupings and time ranges: 32–68 h.

times for many different maintenance jobs using a relatively few benchmark jobs as a guide for work content comparison. By using work content comparison (or slotting as it is called) combined with a good background in craft work and knowledge of the benchmark jobs, a planner now has the tools to establish reliable performance standards consistently, quickly, and with confidence for many different jobs.

Because the actual times assigned to the benchmark jobs are so critical, it is very important to use a technique that is readily acceptable. The ACE System provides such a technique because it is based on the combined experience of a team of skilled craftspeople and others. It is their consensus agreement on the range of time for the benchmark jobs. A consensus of experts who know the mission-essential maintenance work that is to be done.

The 11-Step Procedure for Using the ACE System

1. **Select "benchmark jobs"**: Review past historical data from work orders and select representative jobs that are normally performed by the craft groups. Special attention should be paid to determine the 20% of total jobs (or types of work) that represent 80% of the available craft manpower. Focus on determining repetitive jobs where possible in all craft areas.

2. **Select, train, and establish a team of experts (ACEs)**: It is important to select craftspeople, supervisors, and planners who, as a group, have had experience in the wide range

of jobs selected as benchmark jobs. All craft areas should be represented in the group. To ensure that this group understands the overall objectives of the maintenance planning effort, special training sessions should be conducted to cover the procedures to be used, reasons for establishing performance measures, etc. A total of 6–10 knowledgeable team members is the recommended size for the team.

3. **Develop an ACE Team Charter**: At this point, it is highly recommended that a formal ACE Team charter is established. Section 1.0 of the Appendix provides a sample charter format that can easily be tailored for each site. The ACE Team has an important task that will take time to accomplish. However, the task of the ACE Team will be important and, in turn, their success as a team can significantly contribute to continuous reliability improvement and increased asset uptime.

4. **Develop major elemental breakdown for benchmark jobs**:

   a. For each benchmark job that is selected, a brief element analysis should be made to determine the major elements or steps for completing the total job. Another example is shown in Figure 15.3. Here, the elements of the same job that we illustrated in Figure 15.2 are used but arranged in a more logical sequence of the actual repair method. In this example, a standard time allowance for cleaning and oiling a small group of parts had already been established; therefore, reference to CLN-1 task was made. This task referenced back to a standardized lube specification (#AC-2000).

   b. This listing of the major steps of the job should provide a clear, concise description of the work content for the job under normal conditions. It is important that the work content for a benchmark job be described and viewed in terms of what is a normal repair and not what may occur as a rare exception. All exceptions along with make-ready and put-away time are accounted for by the planner when the actual planned time is completed.

   c. An excellent resource to consider for doing the basic element analysis for each benchmark job is the craftspeople (ACEs) that are selected for doing the estimating or even other craftspeople within the operation. Brief training on methods/operations analysis can be included in the initial training for the ACEs. Very significant methods improvements and methods to improve reliability can be discovered and implemented as a result of this important step.

   d. The ACE Team Process must include and also lead to getting answers to the following questions
      – Are we using the best method, equipment, or tools for the job?
      – Are we using the safest method for doing this job?
      – Are we using the best-quality repair parts and materials, or is this a part of our problem?
      – What type of preventive task and/or predictive task would help identify or eliminate the root cause of the problem?
      – Where can we work even smarter, not necessarily harder?

   e. Major exceptions to a routine job should be noted if they are significant; an exception will generally be analyzed as a separate benchmark job along with an estimate of time required for such repair.

   f. This portion of the ACE Team Process ensures that the work content of each benchmark job is clearly defined so that each person/planner doing the estimating has the same understanding about the nature and scope of the job. When the benchmark jobs are finally categorized onto "spreadsheets", the "benchmark job description" information developed in this step is then used as key information about the benchmark job on the ACE Team spreadsheets.

5. Conduct first independent evaluation of benchmark jobs:

   a. Each member of the group is now asked to review the work content of the benchmark jobs and to assign each job to one of the UMS time ranges or slots. Each member of the group provides an independent estimate, which represents an unbiased personal estimate of the *pure work content* time for the benchmark job. It is essential that each team member do an independent evaluation of each benchmark job and not be influenced by others on the team with their first evaluation.

   b. Focus on work content time: It is important here for each member of the team to remember that only the work content of the benchmark job description is to be estimated and not the make-ready and put-away activities associated with the job. This part of the procedure is concerned only with estimating the pure work content excluding things such as travel time, securing tools and parts, prints, delays, and personal allowances, etc.

   c. The estimate should be made for each job under the following conditions:
   - An average skilled craftsman is doing the job giving 100% effort, i.e., a "fair day's work for a fair day's pay".
   - The correct tools are available at the job site or with the craftsperson.
   - The correct parts are available at the job site or with the craftsperson.
   - The machine is available and ready to be repaired.
   - The craftsperson is at the job site with all of the above and proceeds to complete the job from start to finish without major interruption.

   d. Therefore, the work accomplished under these conditions represents the "pure work content" of the job to be performed. Establishing the "range of time" estimate for this pure work content is the prime objective of the first evaluation.

   e. It is important for each ACE Team member to remember that to develop a planning time requires the pure work content time plus additional time allowances to cover "make-ready" and "put-away" type activities associated with each job as illustrated in Figure 15.8. The "make-ready time" and "put-away time" will be accounted for as the planner adds time and allowances for these elements as the actual planning time is completed. Make-ready and put-away times are established specifically for each operation and added to the work content time to get the total planned time for the job being estimated.

6. Summarize first independent evaluation:

   a. Results of the first evaluation are then summarized to check the agreement among the group as to the time range for each benchmark job. A coefficient of concordance can be computed from the results if required, but normally this level of detail is not needed. A coefficient of concordance value of 0.0 denotes no agreement whereas a value of 1.0 denotes complete agreement or consensus among the ACEs. A consensus can generally be reached by the ACE Team within one, two, and at the most three rounds of evaluations.

   b. Define high and/or low estimates. Team members who are significantly higher or lower than the rest of the group for a particular benchmark job are then asked to explain their reasons for their respective high or low estimates. They explain their reason for their estimate to the group, discussing the method, condition, or situation for their initial time estimate. This information will then be used during the second evaluation to refine the next round of time estimates from the entire group.

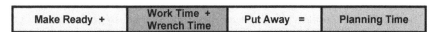

Figure 15.8 Planning time elements.

7. Conduct second independent evaluation of benchmark jobs:

    a. A second evaluation is conducted using the overall results from the first evaluation as a guide for the entire team. Various reasons for high or low estimates from the first evaluation are provided to the group before the second evaluation. This can normally be done in an open team discussion with team members making personal notes to use in their second independent evaluation.

    b. The second round allows for adjustment to the first estimates if the other ACE Team members' reasons for a higher or lower estimated time are considered to be valid. In other words, results from the first evaluation plus reasons for highs and lows will allow each team member to reconsider their first estimate. In many cases, a review of the repair method or scope of work will be more clearly defined, causing a change to the time estimate for the second evaluation.

8. Summarize second evaluation:

    a. Results of the second evaluation are then summarized to evaluate changes or improvements in the level of agreement. The goal is a consensus among the ACEs as to the time range (workgroup) for each benchmark job. The second round should bring an agreement as to the time range.

    b. The second independent evaluation should produce improved agreement among the group. If an extreme variance in time range estimates still exists, then further information regarding the work content, scope, and repair method for the job may be needed. Here, those with high/low estimates should review their reasons for their estimates again, with the team describing the scope of work that they see is causing differences from the rest of the team.

9. Conduct third independent evaluation if required: This evaluation is required only if there remains a wide variance in the estimates among the group members.

10. Conduct a review session to establish final results: This session serves to finalize the results achieved and to discuss any of the high or low estimates that have not been completely resolved. A final team consensus on all time ranges is the objective of this session.

11. Develop spreadsheets:

    a. The benchmark jobs with good work content descriptions and agreed-upon time ranges can now be categorized onto spreadsheets. From these "spreadsheets," which give work content examples for a wide range of typical maintenance jobs, a multitude of individual maintenance performance standards can be established by the planner through the use of work content comparison.

    b. The basic foundation for the maintenance planning system is now available for generating consistent planning times that will be readily acceptable by the maintenance workforce that developed them. Attachment B to these appendices provides a graphical illustration of the ACE System.

The ACE Team approach combines the DELPHI technique for estimating along with a proven team process plus the inherent and inevitable ability of most people to establish a high level of performance measures for themselves. As used in this application, the objective for the ACE Team Process is to obtain the most reliable, reasonable estimate of maintenance-related work content time from a group of experienced craftspeople, supervisors, and planners. This process provides an excellent means to evaluate repair methods and safety practices and even to do a risk analysis on jobs that leads to improved safety practices. The ACE Team Process can significantly contribute to continuous reliability.

The ACE Team approach allows for independent estimates by each member of the group, which in turn builds into a consensus of expert opinion for a final estimate.

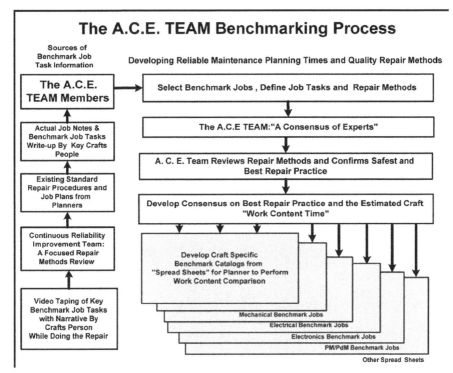

Figure 15.9 The complete ACE Team Process. ACE, a consensus of experts.

Therefore, the final results are more readily acceptable because they were developed by skilled and well-respected craftspeople from within the work unit. Application of the ACE System promotes a commitment to quality repair procedures and provides the foundation for developing reliable planning times for a wide range of maintenance activities. Figure 15.9 illustrates the complete ACE Team Process.

Successful Scheduling by Keeping the Promise and Completing the Schedule

## 16

In this chapter, we look at scheduling, beginning with how the planner can use the ACE team process for determining reliable planning times for the schedule. First of all, the planner must have an accurate backlog that he or she has calculated based upon current craft performance. Now, we will look at defining a schedule with standard reliable estimates for an expectancy of how long the job will take. Nyman and Levitt described this as "analytical estimating," which is very much like the slotting concept of the ACE team process. Figure 16.1 illustrates how a planner can define the work content for a job being estimated for the schedule. Appendix H provides an excellent baseline for a scheduling standard operating procedure.

At this point, the planner is considering the work content of the job at hand to be estimated. So, in Figure 16.1, the planner looks at jobs in Group D (1.5 h) and sees nothing that compares directly, but the job at hand is certainly greater than 1.5 h. Now, looking at jobs in Work Group E (2.0 h), the planner also sees nothing that matches. Then, when the planner looks at Work Group F (3.9 h), he or she knows for sure that the job at hand is less than 3 h. Because the job is greater than 1.5 h and less than 3 h,

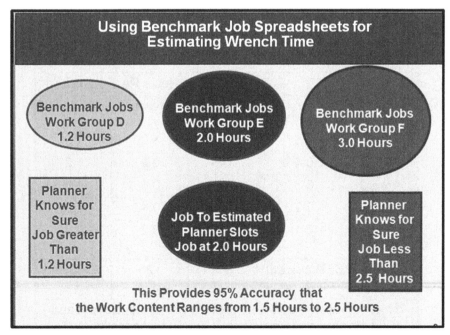

Figure 16.1 Using benchmark spreadsheets for estimating wrench time.

Reliable Maintenance Planning, Estimating, and Scheduling. http://dx.doi.org/10.1016/B978-0-12-397042-8.00016-4

the planner slots the job in Work Group E at 2.0 h. This method has proven to be 95% accurate that the job will range from 1.5 h to 2.5 h.

This wrench time must have three types of allowances added for it to become a reliable estimate for the schedules. The three types of allowances are shown in Figures 16.2 and Figures 16.3, Figure 16.4 and Figure 16.5.

Factors in Determining Total Planned Time for the Schedule

The Planner now must consider adjusting the work content time <u>adding allowances</u> for computing the <u>planned time for scheduling.</u> These include;

- **Travel Time Table of Allowances (TT)**
- **Miscellaneous Allowances (MA)**
- **Personal Fatigue and Delay Allowances (PF&D)**

Figure 16.2 Allowances for determining the total planned time for the schedule.

Example: Travel Time Table of Allowances

From Shop To:	Round Trip Hours	Allowed Hours Per Person		
		Simple	Average	Complex
Area A	0.5	0.5	1.0	1.5
Area B	0.4	0.4	0.8	1.2
Area C	0.3	0.3	0.6	0.9
Area D	0.2	0.2	0.4	0.6
All Other Areas	0.1	0.1	0.2	0.3
Round Trips Provided		1	2	3

Note: This table would be based on site or plant layout and distances and walking pace of 250 feet per minute

Figure 16.3 Example of travel time.

Example		Miscellaneous Allowances Table		
		Simple	Average	Complex
1	Feedback (paperwork)	WO	Plus PM Checklist	Detailed Feedback
2	Receive Instructions	Up to 3 minutes	Approx. 5 minutes	Approx. 10 minutes
3	Gather tools	Tool Belt	From personal tool box	Special shop tools needed
4	Follow Safety Procedures	Normal	Lockout	Rope area, special clothing
5	Obtain Parts and Materials	None	Bin Stock	Storeroom requisition
6	References, Job Plan, Drawings, etc	None	Job Plan	Multiple References
	Allowed Hours Per Person	0.1	0.3	0.5
Note: Appropriate estimates must be developed for each plant/operation				

Figure 16.4 Example miscellaneous allowance table.

Personal, Fatigue and Delay Allowances (PF&D)

Personal, Fatigue and Unavoidable Delay Allowance Table

Nature of Allowance	%	Light	Average	Heavy
Personal Time, Breaks & Clean Up	5	5	5	5
Fatigue	5-10	5	7	10
Unavoidable Delay	5	5	5	5
Sub-Total	15-20	15	17	20
Crew Balance: Multi-Person	3	3	3	3
Crew Balance: Multi-Craft	2	2	2	2
Total	15-25	15-20	17-22	20-25

Notes:
1. Use 15% on Travel Time and Job Preparation
2. Use Appropriate Fatigue % as applies to Direct Work Only
3. Crew Balance% applies to Direct Work only
4. Most jobs need only 15% PF&D Allowance

Figure 16.5 Example personal fatigue and delay (PF&D) allowance table.

How Allowances Apply to Typical Job

A Typical Maintenance Job Sequence of Steps and Example Where Allowances Are Applied

Source	Task
MA	Get ready and receive instructions for doing the job. Miscellaneous Allowances (MA) includes receiving job instructions from supervisor, collecting personal tools, obtaining parts from storeroom and gathering special tools and equipment
TT	Travel to job site (Outbound)
MA	Listen to operations input regarding the problem
MA	Make preliminary diagnosis and trouble shoot prior to shutdown
MA	Shutdown and Lockout. This may be done jointly with line supervisor, control tech and/or line operator
WT	Partial or total disassembly to reach the problem area
MA	Determine full extent of the problem
MA	Identify any additional parts or material , obtain from storeroom or initiate direct purchase
WT	Reassembly of equipment using the replace parts needed
MA	Check proper job completion, test operability of equipment, clean up job site and put away tools
TT	Travel back to shop (inbound)
MA	Report on job and return unused parts, special tools and equipment
PF&D	Personal Fatigue and Unavoidable Allowances for entire job

Legend:

Most Jobs 15% to 17%	MA: Miscellaneous Allowances
	TT: Travel Time
	WT: Wrench Time from Spreadsheets
	PF & D: Personal Fatigue and Unavoidable Delay Allowances

Coordination Required by Planners for Successful Scheduling

➢ **Downtime will schedule itself at worst possible time**

➢ **Planner liaison with operations must be permanent**

➢ **Planners must learn and take interest in customer problems**

➢ **Planners can help operations to think in advance**

➢ **Planner can facilitate planned work to maximize operational plans**

Preparation for the Schedule Coordination Meeting: Daily-Weekly-Monthly??

☐ Upkeep of backlog; current, cleansed and accurate

☐ Issuance of backlog report by job status
  ✓ Send electronically the day before weekly schedule meeting to;
  ✓ Maintenance leaders
  ✓ Operations staff
  ✓ Storeroom & Purchasing (denoting jobs awaiting parts)
  ✓ Engineering (as Required)
  ✓ Top Leaders

☐ Planner looks to resolve possible conflicts:

☐ Planner must have good answers to better chair and coordinate any scheduling meetings

☐ Determine resources available for backlog relief
  ✓ Ensure realistic capacities available (in house or contractors)
  ✓ Ensure schedule commitments can be met

Preparation for the Weekly Schedule Coordination Meeting

☐ Planner groups jobs for optimization
  ✓ Link multiple jobs on same equipment or area
  ✓ Link multiple jobs requiring same equipment, etc

☐ Have list of all jobs "Ready to be Scheduled"
  ✓ Set up by "Required Start Date"
  ✓ Do not include jobs where material might arrive during the "schedule week"
  ✓ Typically plenty of work is in "Ready Backlog"

Preparation for the Weekly Schedule Coordination Meeting

❑ Planner develops preliminary (first cut) of next week's schedule considering;
   ✓ Resources needed for next week's PM/PdM
   ✓ Resources needed for any carry-over work
   ✓ Preliminary schedule that has been reviewed by maintenance
      ▪ Based upon "Required Start Dates" and "Ready Backlog"
      ▪ Current craft capacities and priorities
      ▪ Equipment availability
      ▪ Contractor support

The Weekly Schedule Coordination Meeting

❑ The best case; Operations approves the schedule!
❑ Operations is the approver of "schedule-breaks" for "urgent" jobs not on the current weekly schedule

Key Procedures for Effective Scheduling

**These three must be performed concurrently
for developing an effective schedule;**

✓ Job Loading: setting up the right jobs, balancing
immediate needs of operations with long term
maintenance needs for reliability and asset care

✓ Job Scheduling: Sequencing the loaded jobs through
the schedule based on meaningful estimates of
duration and agreed upon access to equipment

✓ Manpower/Equipment Commitment: ensure optimal
utilization of resources via a labor and equipment
deployment plan to achieve schedule

Weekly Scheduling and Available Resources

10 Crafts x 40 Hrs. =   400.0 Hrs.

    Less Carryover   _____

    Less Late & Early   _____

    Less Vacation   _____

    Less PM   _____

    Less Standing WO   _____

    Less Emergency   _____

Net Hours. Available   _____

Key Guidelines for Completing the Scheduling Process

✓Prepare a schedule form for each supervisor; week beginning date, foreman or crew leader etc.

✓Planner must determine from supervisors, vacation time, known absences etc.

✓Define net available hours for next week's schedule

✓Review all jobs in backlog, start with incomplete jobs from current or previous schedules

✓Review Planned job Packages

   ✓Ensure all are complete for scheduling and assignment

   ✓Final confirmation of parts, equipment & special tools

   ✓Permits and safety instructions

Key Guidelines for Completing the Scheduling Process

✓ Plan strategy on a weekly basis; rigid enforce rule that schedules complete for each supervisor by Friday...

✓ If coordination meeting on previous Thursday

✓ Work schedules must be balanced against available craft hours and equipment

✓ Schedule to consume all available craft resources

✓ Schedule what can be done, not necessarily what needs to be done

Key Guidelines for Completing the Scheduling Process

✓ Schedule a full day of productive work for each person

✓ Schedule majority of crews for important work that needs to start and be completed without interruption

✓ But to a "few good people" who are flexible

   ✓ Assign to lower priority jobs

   ✓ 10% to 15% of scheduled hours

   ✓ Jobs that can be sacrificed for emergencies

   ✓ So they can break away for urgent schedule-breaks

   ✓ Come back to scheduled job with minimal loss of efficiency

Do Not Schedule a Job Until All of These Things are in Place

Key Guidelines for Completing the Scheduling Process

✓ List jobs in descending order of important starting with PM/PdM first based upon agreements reached in weekly schedule meeting

✓ Determine most logical time to schedule PM/PdM's
  ✓ Early morning most likely for breakdown
  ✓ End of day bad because they might not get done
  ✓ Mid morning or mid afternoon is best

✓ Add jobs totaling about 10% to 15% of schedule hours as provisional jobs.
  ✓ If scheduled job is delayed
  ✓ If scheduled job completed in less than estimated time

Key Guidelines for Completing the Scheduling Process

✓ Establish contingency section of the schedule

  ▪ Jobs where equipment was not expected to be available when the schedule was firmed up

  ▪ But if equipment is available they are more important than some jobs on primary schedule

  ▪ Provisional jobs must also be properly planned

  ▪ Moving these into the schedule is proactive and counts toward schedule compliance

Key Guidelines for Completing the Scheduling Process

✓ Avoid duplicate shutdown by scheduling all work requiring common equipment when appropriate

✓ Save minor indoor jobs for severe temperatures and increment weather

✓ Eliminate unnecessary trips

   ✓ Jobs in the same location

   ✓ Jobs with same equipment, special tools & material

   ✓ Delivery additional parts & materials needed to site

Key Guidelines for Completing the Scheduling Process

✓ Schedule multi-person jobs as first job in morning, all there to start at same time

✓ Consider previous assignments when scheduling multi-person jobs later in the day

   • One craft a one hour job

   • Helper to a two hour job

✓ Think about crew balancing on multi-person jobs

   • Think about a 4 person job-Seldom will all be working

   • Often a small job in same area can be worked on concurrently

Key Guidelines for Completing the Scheduling Process

✓ Planner may have to allocate people to specific jobs with supervisor approval
  • Pick based on skill & aptitude
  • Experience over time helps here
  • Balance equipment specialization (Generators, Pumps)

✓ Planned job packages delivered to supervisor
  • Ensure nothing falls through crack
  • Responsibility now transitions from planner to supervisor
  • Later supervisor-craft transition for job execution

Key Guidelines for Completing the Scheduling Process

✓ Operations provided copy of schedule
  • Confirm and document all agreed upon commitments
  • Everyone understands how to handle "schedule-breaks
✓ Vital that schedules be studied and approved by everyone concerned
✓ Weekly schedule now becomes document where all parties have mutual contributions and ownership

Key Guidelines for Completing the Scheduling Process

✓ When urgent work done at expense of scheduled jobs;
   ✓ Schedule overload occurs
   ✓ Carried over to next period
   ✓ Overtime authorized
   ✓ Displaced job is one scheduled for area that initiates

✓ Schedule breaks should be approved by Maintenance Manager and Operations manager

Key Guidelines for Completing the Scheduling Process

✓ Finalize tactics on a daily basis when schedule is being executed
  • Updated each evening during week it is in force
  • If transitioning from reactive to proactive, the updating process can be a burden
  • Planner will have to do it until schedule compliance improves
  • Supervisor can and should do it later on

✓ Operations must advise ASAP when equipment can't be released

✓ Maintenance advises operations when schedule runs over. Could be the customer, the tenants etc

✓ Planner must ensure coordination takes place

Job Close Out and Follow Up

- Begins with good feedback from crafts and supervisor

- Job not complete without good feedback on work performed

- For well planned jobs this should be minimal
  - Approved out of scope work must be reported
  - Planner adds extra planned time in this case

- Most basic feedback is labor hours charged via labor reporting process

- Next is parts and materials crafts had to get from stores

- Equipment costs/rentals should be part of the plan

Schedule Compliance

✓Schedule compliance measures customer service
  - Planners plan for it
  - Supervisors make it happen
  - Crafts & crews really make it happen
  - All must keeps plan informed of both good & bad situations

✓Planner calculates schedule compliance each week

✓Not to place blame but to improve future performance

✓Schedule non-compliance is very serious when maintenance fails to meet the schedule (Keep the promise)
  - Customer concerns
  - Hinders future cooperation
  - Feeds distrust
  - Future reluctance to release equipment
  - Cost over runs "adds fuel to the fire"

Reasons for Schedule Non-Compliance

✓ As the Planning/Scheduling process begins, it reveals previously hidden problems

✓ Many where crafts can not do their job due to practices outside their control

✓ Reasons for non-compliance should be;
  ✓ Recorded
  ✓ Reported
  ✓ Studied for trends and ultimate improvement

✓ The following Figures provide typical reasons for schedule non-compliance and recommended codes for tracking each

Reasons for Schedule Non-Compliance

✓ FR: Operations fails to release equipment as promised

✓ EE: Excessive emergencies
  • Crafts are pulled off scheduled jobs for work on less important jobs
  • Common where operations is allowed to redirect maintenance
  • Failure to go through work management process

✓ PA: Poor assignment of crafts; crafts fall short of expectations,
  • Schedule must be feasible from the start
  • Not feasible unless detailed to level of individual or crew assignments

Reasons for Schedule Non-Compliance

✓ IC: Insufficient craft capacity; lack of cross training contributes to shortage of the right skills

✓ SO: Stock outs are frequent
  ✓ Inaccurate inventory control
  ✓ Excessive time finding what is in storeroom
  ✓ Parts not requisitioned

✓ PP: Planned jobs do not reflect reality
  ✓ Scope expands
  ✓ Plan for parts is incorrect
  ✓ Job steps incomplete
  ✓ Lockout, specifications and regulations not documented

Reasons for Schedule Non-Compliance

✓ DU: Failure to meet estimated job duration
  • Was the estimated time on the schedule wrong?
  • Did the crew perform poorly
  • Other problems; scope creep etc

✓ EA: Excessive absenteeism and simultaneous peak loads

Remember

If You Can't Measure IT,
You Can't Improve IT!

Calculation of Schedule Compliance: Two Methods

A. Schedule Compliance =
Scheduled Craft Hours Completed
Total Craft Hours Scheduled

800 Hrs Completed
1000 Hrs Scheduled = 80% Schedule Compliance

B. Schedule Compliance =
Scheduled Jobs Completed
Total Jobs Scheduled

12 Jobs Completed
15 Jobs Scheduled = 80% Schedule Compliance

Calculation of Schedule Performance

Schedule Performance =
Scheduled Craft Hours Completed
Total Craft Hours Available

800 Hrs Completed
1600 Hrs Available = 50% Schedule Performance

Measures % of craft hours available that were in fact scheduled and completed

Schedule Compliance and Schedule Performance

• Schedule Compliance can be high because only a small % of craft hours are scheduled

• But start formal scheduling measurement process knowing you have a Win-Win Situation
   ✓ Begin with lesser number of scheduled craft hours
   ✓ Work up to more hours as scheduling process matures
   ✓ Ensure operations understands this from Day One
   ✓ Operation Wins - Maintenance Wins – Customer Wins

Measuring Performance of the Planning and Scheduling Function

✓ Overall measurement of maintenance requires a number of metrics (10-15 possibly)
   • We will see this with The Reliable Maintenance Excellence Index
   • Planning/scheduling impacts most of these
   • Planner typically responsible for the overall process

✓ However only a few act as pure measures for planners

✓ Example; Schedule Compliance and Craft Performance depends on supervisor/crafts

✓ Planner job not to predict results

✓ Planner job is to establish expectations

Specific Measures for the Planning and Scheduling Function

✓ Per cent of Work Orders covered with Planned Job Packages

✓ Percent of Work Orders with reliable estimate of required craft hours

✓ Reliability of Backlog by status (Accuracy, completeness)

✓ Mean time from work request to "Ready to be Scheduled"

✓ Mean time between "Job Completion" & "Job Close Out"

✓ Steady and meaningful expansion of planner technical library

✓ Others as well

Specific Measures for the Planning and Scheduling Function

✓ Customer satisfaction with planner communication, coordination and feedback (periodic survey)

✓ Supervisor satisfaction with planner support on planned work orders (periodic survey)

✓ Crew satisfaction with thoroughness of planner's work (job plan survey)

✓ Timely posting of schedules and associated planned job packages

✓ Timely and accurate preparation, distribution or posting of control reports/trend charts (where responsible)

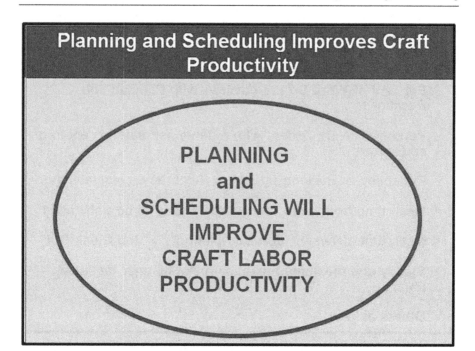

Maintenance, Repair, and Operations (MRO) Material Management: The Missing Link in Reliability

17

I am very pleased to have contributions from two authors, Phillip Slater and Art Posey, to kickoff this chapter on maintenance, repair, and operations (MRO) material management. As shown in this chapter, there must be close integration between maintenance, storeroom, purchasing, vendors, and even finance. Reliable spares and reliable vendors are part of a reliable MRO supply chain.

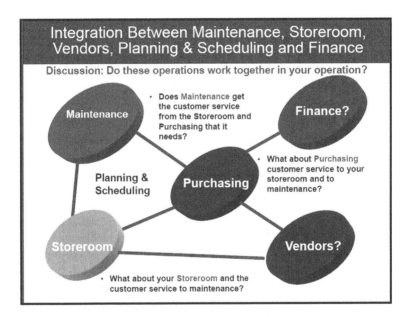

Introduction

"Oh no, not again!"—Is this the perennial cry of the maintenance and reliability professional? We all know this situation: The maintenance team rectified a failure and that repair has now failed again shortly after. This is both annoying and frustrating—after all, why can't we get these things right the first time? Usually, we are quick to blame the technician, the technique, or the design. But none of these may be the fault; the cause of the breakdown could be a poorly maintained spare part.

Reliable Maintenance Planning, Estimating, and Scheduling. http://dx.doi.org/10.1016/B978-0-12-397042-8.00017-6

Empirically, we can say that equipment failures result from a combination of poor design, incorrect operation, incorrect installation, inappropriate care activities, and poor cleanliness/environmental management. Therefore, it ought to be no surprise that we can also say that a vast majority of failures are self-induced, as we control four of those five factors.

Maintenance and reliability practitioners might also consider that reliable operations are achieved through a combination of equipment selection, maintenance program design, the use of tools to enable condition monitoring, and skills development of the team. However, many self-induced and premature failures are more likely the result of poor materials handling and storage methods for spare parts. For many maintenance and reliability practitioners, materials management is the missing link in achieving reliability.

Typically, most MRO professionals associate materials management with having, or not having, spare parts on hand. Despite their focus on optimizing plant performance, too little attention is paid to the maintenance of the spare parts that they hold in stock. The reason for this is that materials management is often thought of as a sideline or support activity rather than as a core function for achieving reliability, so the influence that materials management actually has on reliability is underestimated. The reality is that many companies could significantly improve their reliability outcomes by improving their materials management.

Why Your Data Does Not Tell You What You Think It Does

In order to explain this and determine what to do about it, let us initially try to understand why this problem is not obvious—that is, why your data does not tell you that you have a materials management problem.

Firstly, downtime is typically recorded as equipment failure rather than as a material or spares failure. So, your data is often at too high a level to recognize that the part failure is the root cause. In addition, root cause investigations are not performed on most failures, so assumptions are often made about the cause of the failure; usually, it is blamed on the installation, not a faulty part. This problem is further exacerbated by the time that may elapse between the installation of a spare part and its subsequent failure. If enough time has elapsed, we tend to disassociate the failure from the installation of the part, even though the spare part has exhibited an extremely shortened life span. Without some form of root cause analysis, we are often just too busy getting and keeping the plant going to consider this issue.

Other ways that we can mislead ourselves into not recognizing materials management issues as the cause of our problems include the following:

- Work orders were not closed off. If we are not disciplined in finalizing work orders, then the data is not available anyway. If we accept too long a delay in finalizing work orders, then not only is data timeliness an issue but we also end up relying on memory, which we all know can sometimes be selective!
- Configuration data are out of date. If our records do not reflect our equipment configuration, then how will we confidently track any problem, not just spares problems?

- Equipment hierarchies are not sufficiently granular. As mentioned above, if your equipment records are set at too high a level, then we will not be able to recognize problems.
- The team records their actions, not the equipment needs. That is, records reflect what was done (e.g., replaced fan belt) rather than observing equipment issues and requirements (e.g., broken fan belt seems relatively new, not sure why it was broken).
- Tracking of assets is inaccurate. Who really tracks their spares as assets? Is there an engineering spares register in the COMPUTERIZED MAINTENANCE MANAGEMENT SYSTEM (CMMS)?
- Multiple failure causes are rarely listed, even if they are known. If an equipment item fails due to a worn part and the repair is delayed due to a lack of availability of the part, is the delay time documented?
- Improper root cause conclusions can skew the data. If a motor is pulled from stores and prematurely fails, it may be attributed to a poor rewind, when the true root cause may have been poor storage techniques.

How Proper Spares Storage Can Significantly Improve Your Reliability

Let us define maintenance as the actions required to preserve equipment in a suitably operational state such that it operates as expected when required. In that case, perhaps we can extend that definition of equipment to our spare parts—after all, aren't they also part of our equipment and don't we want them to operate as expected when required? Many early-life failures can and do result from poor preservation of the spare parts, which results from improper storage practices.

The key issues for equipment operating in a plant are exposure to the environment and the effects of being in operation. For both mechanical and electrical elements of equipment, the effects of being in operation include the kinetic effects of wear, heat, and vibration. Our primary methods of preserving these parts are lubrication and observation.

For parts that we have in storage, the issues are similar. We still need to deal with the effects of exposure to the environment, but we also need to deal with the effects of *not* being in operation—that is, the effects of being stationary. Let us consider these issues.

Exposure to the Environment

The key things to consider are where the spare is kept and how is it stored. For example, is it the same environment as the operating equipment? Sometimes, we are very careful to protect the operating equipment with environmental controls, such as air conditioning or dust proofing, but then keep the spares in a storeroom without either of these.

Even items kept in suitable storage or under cover are exposed to the environment and exhibit failure modes from exposure, such as the following:

- Rust
- Oxidation of rubber components in seals, belts, and other parts
- Buildup of dust (especially important in electrical, rotating, and reciprocating equipment)

- Lubrication failure (through contamination, migration, and evaporation)
- Vibration from the operating facility (slight vibration over a period of time can degrade many spare parts, from bearings to electronic components)

Effects of Being Stationary

It is easy to assume that because a part is not in operation and is environmentally protected, then it will be acceptable to operate as expected when required. However, the main enemy of reliability for items that are stationary is gravity, and gravity is always with us!

The two main influences of gravity are

- Flat spots that result from a constant weight on one section in seals, shafts, and bearings.
- Lubricants flowing to the lower areas, leaving upper areas without lubrication.

It is really important to recognize that, although you can control exposure, you must manage gravity (i.e., the effects of gravity). Therefore, environmental control is likely to be passive (in that we provide infrastructure that provides the control) but managing the effects of gravity will need to be active (we need to perform tasks regularly to ensure the integrity of the part).

To demonstrate this visually, Figure 17.1 shows the relationship between parts in operation and parts in storage. Figure 17.2 shows the crossover between active and passive issues for parts in storage.

In Figure 17.1, you can see that whether a part is in storage or in operation, we must still manage the effects of rust, dust, lubrication (lack of), and other environmental impacts. This diagram alone indicates that, for reliable plant operation, we must maintain our parts because they are subject to many of the same issues as the items in operation.

Figure 17.1 The relationship between parts in storage and parts in operation.

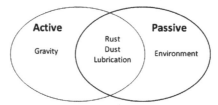

Figure 17.2 The crossover between active and passive issues for parts in storage.

In Figure 17.2, we see that those same core issues of rust, dust, and lubrication require active involvement if we are to ensure that a part will operate as required when needed. This is in addition to actions that must be taken in order to counter the effects of gravity and other environmental protection.

The Real Function of Your Storeroom

By viewing spare parts management in this way, we can see that the real function of your storeroom is not just to store and control access to parts but also to maintain them in a condition so that they are fit for use when required. This means ensuring an appropriate care and maintenance routine as well as providing environmental protection. Neatness and organization of spare parts only helps ensure storeroom efficiency; it does not guarantee spare parts integrity.

Some basic storeroom practices that affect reliability include the following:

* Proper storage—safe from the environmental effects of dust, water, vibration, and light
* Preservation of parts—appropriate lubrication and managing the effects of gravity
* Rigorous rotation of parts—using the oldest parts first (it seems that the newest parts are always pulled first by technicians)
* Proper labeling of parts so that the correct part is selected—for example, not selecting a 20 μm filter when a 5 μm filter is required
* Proper access and equipment for lifting devices—in order to avoid mechanical damage to parts when storing or retrieving them

Spare Parts Ownership Can Help Drive Reliability Outcomes

Two of the main problems with maintaining spare parts are identifying who is actually responsible for their care and whether they are qualified for the job. With parts in operation, it is usually clear who is responsible for their care and maintenance and for achieving reliability outcomes—the maintenance and reliability function. Spare parts in storage are often given over to the storeroom or warehouse, who is responsible for care and control. However, who sees the whole picture? Who really understands how those parts fit in with the overall reliability plans? Surely this is also the responsibility of the maintenance and reliability function.

Are warehouse clerks really clerks, or should they be warehouse technicians, where the care of the parts is as important as the oversight of receipts, issues, and counts? If so, a warehouse technician cannot perform a job to support plant reliability unless they have access to all manuals that reference storage requirements and have been trained in the following areas:

* Lubrication
* Effects of ozone on rubber components
* Requirements for storage of code materials (if applicable)

- Motor care and storage
- How hydraulic cylinders work and how proper storage can extend the life cycle
- How desiccants work and how they can enhance storage quality
- Bearing types, component storage, and preservation methods
- Valve types and lubrication requirements
- Temperature, humidity, cleanliness, and static electricity effects on electrical components

If you align your spare parts ownership with the need to deliver reliability outcomes, then you can start to see how you can achieve your outcomes with appropriate spare parts maintenance.

Establishing a Spares Maintenance Program

Establishing a spares maintenance program is really just like establishing the rest of your maintenance/reliability program. As we have discussed, the difference is that maintenance/reliability programs typically address kinetic effects (i.e., reliability issues that arise through the physical operation of the asset), whereas a spares maintenance program will address static effects (i.e., issues that arise through not operating). Both programs, of course, need to address environmental effects.

Therefore, establishing a spares maintenance program only really requires a change in mindset for your failure mode and effects analysis You now need to consider how the item will fail through *not* operating rather than how it will fail through operating. The rest of the program development is the same as with any other maintenance/reliability program: you need to identify the what, why, when, how, who, and where of your spares maintenance program.

This is an opportunity to expand your maintenance process and procedures to the personnel in the storeroom. A world-class spares maintenance program is not likely to require any more personnel than you currently have. It is an opportunity to transform the perception of your storeroom personnel from a simple receiving, issuing, and counting role to a role integral to the reliability of the facility. A written detailed program could add the structure and professionalism to transform your storeroom and your storeroom personnel.

Conclusion

Maintenance and reliability professionals spend an overwhelming amount of time and energy ensuring that they preserve the equipment in their care in a suitably operational state such that it operates as expected, when required. However, in doing this, they may overlook one of the single greatest causes of equipment failure and subsequent downtime: poor materials management. For many, this is the missing link in their reliability program. Poor materials management results from systemic issues with the way that reliability data is collected, inappropriate techniques for storage, a misalignment of responsibilities, poor training, and a lack of formal policy and procedures in this area.

Empirically, we know all this to be true; equally, we know from experience that correction of these issues can have a significant impact on the outcomes from any maintenance and reliability program. Materials maintenance is the missing link in most reliability programs and is vital for a reliability program to be a long-term success.

About the Authors

Art Posey is Senior Manager, Maintenance for Wheelabrator Technologies, Inc., a waste management company. Wheelabrator owns and/or operates energy-from-waste facilities, several independent power production facilities, and has operations in the United Kingdom and China. Art came to maintenance in 1997 after 16 years in the construction field. Art can be reached at artposey@gmail.com.

Phillip Slater is a materials and spare parts management specialist. He is the founder of the online training and best practice resource center at SparePartsKnowHow.com, and the author of eight books, including *Smart Inventory Solutions* and *The Optimization Trap*. For more information on Phillip's services, visit www.PhillipSlater.com.

How to Measure Total Operations Success with the Reliable Maintenance Excellence Index

(chapter number 18 shown in top right)

During this chapter we will look at the third area for benchmarking which is the reliable maintenance excellence index. As we saw previously we start with the scoreboard for maintenance excellence, then we can use the CMMS benchmarking system, the ace team system for measuring craft productivity at the shop for level and then compile the reliability maintenance excellence index for measuring overall total operations success. Figure 18.1 again illustrates the four levels of benchmarking.

THE SCOREBOARD for MAINTENANCE EXCELLENCE:
Benchmarking Current Operation Against Global Best Practices

THE CMMS BENCHMARKING SYSTEM:
Benchmarks Your Current CMMS to Achieve Maximum Utilization of Your IT Investment

The RELIABLE MAINTENANCE EXCELLENCE INDEX:
Defines Internal Benchmarks and KPI's to Measure and Validate Results of Maintenance and Reliability Excellence Actions

The ACE TEAM PROCESS:
Provides a Means for Improved Repair Methods and Benchmark Jobs for Establishing Reliable Planning Times for Scheduled Work

FOUR LEVELS OF MAINTENANCE BENCHMARKING

Figure 18.1 Illustrates the four levels of benchmarking.

Reliable Maintenance Planning, Estimating, and Scheduling. http://dx.doi.org/10.1016/B978-0-12-397042-8.00018-8

The Reliable Maintenance Excellence Index

- The RMEI should measure how all key resources that contribute
  - to profit & budget optimization,
  - greater customer service
  - and more effective physical asset management
- The RMEI should include measures for all maintenance resources
  - People resources; internal craft labor & outside contractors
  - Dollar resources; overall budget dollars of maintenance and the customer
  - MRO parts and material resources
  - Planning/scheduling resources, processes & customer service
  - Critical assets; uptime, availability or OEE and reliability
  - Information resources; how data becomes true information via effective CMMS

The Reliable Maintenance Excellence Index

The Reliable Maintenance Excellence Index provides:
- A very powerful, one page Excel spreadsheet
- 12-15 key metrics combined for a composite Total RMEI Performance Value
- A very "balanced scorecard" for the total facilities and maintenance process.
- An ideal method for measuring Continuous Reliability Improvement across a multiple operation & work units
- Support to applying and measuring the impact of standard best practices
- Maintenance planning & scheduling: Must define results!

The Reliable Maintenance Excellence Index

A-Performance Measures	B-Current Month Perf	C-Performance Goals	\ D-Baseline Performance Levels if Available / (10)	(9)	(8)	(7)	(6)	(5)	(4)	(3)	(2)	(1)	(0)	E-Performance Level Score	G-Weighted Value Metric	H-Performance Level Score (E) × Weight (G)	Date	J-Total RMEI Value
1. Actual Maintenance Cost Per Unit of Production	$1.30	$1.00	1.05	1.10	1.15	1.20	1.25	$1.30	1.35	1.40	1.45	1.50		4	10	40	Date	
2. % Major Work Completed within 5% of Cost Estimate	90	95	94	93	92	91	90	89	88	87	86	85		5	6	30	Jan	409
3. % Overall Maintenance Budget Compliance	94	98	96	94	92	90	88	86	84	82	80	78		8	6	48	Feb	435
4. % Overall Schedule Compliance	90	95	94	93	92	91	90	89	88	87	86	85		5	7	35	Mar	467
5. % Overall PM Compliance	94	100	98	96	94	92	90	88	86	84	82	80		7	11	77	Apr	
6. % Planned work	68	80	78	76	74	72	70	68	66	64	62	60		4	7	28	May	
7. % Craft Time for Customer Charge Back	75	85	83	81	78	75	72	69	66	63	60	57		6	6	36	June	
8. % Work Orders with Reliable Planned Time	50	60	58	56	54	52	50	48	46	44	42	40		5	7	35	July	
9. % Critical Asset Availability	90	95	94	93	92	91	90	89	88	87	86	85		5	13	45	Aug	
10. % Wrench Time (Craft Utilization)	34	50	48	46	44	42	40	38	36	34	32	30		2	8	16	Sept	
11. % Craft Performance	90	95	94	93	92	91	90	89	88	87	86	85		5	8		Oct	
12. % Inventory Accuracy	90	98	97	96	95	94	93	92	91	90	87			2	6	12	Dec	
13. Number of Stock Outs of Inventoried Stock Items	15	10	11	12	13	14	15	16	17	18	19	20		5	5	25	Jan	

F-Performance Level Scores — Perf Level: 10, 9, 8, 7, 6, 5, 4, 3, 2, 1, 0

E-Perf. Level × G-Weighted Value

I. Total RMEI Value: 467

Annotations: Current Month Perf · Baseline Perf. Level · Total RMEI Value · Weighted Value of Each Metric

The Reliable Maintenance Excellence Index

Example: 13 Performance Measures

I. Total RMEI Score: 481

A. Performance Measures	B. Current Month	C. Performance Goal	PL 10	PL 9	PL 8	PL 7	PL 6	PL 5	PL 4	PL 3	PL 2	PL 1	PL 0	E. Performance Level Score	G. Weighted Value of Metric	H. Performance Level Score (E) x Weight (G)	J. Date
1. Actual Maintenance Cost Per Unit of Production	1.30	1.00	1.00	1.05	1.10	1.15	1.20	1.25	1.30	1.35	1.40	1.45	1.50	4	10	40	7/08
2. % Major Work Completed Within 5% of Cost Estimate	90	95	95	94	93	92	91	90	89	88	87	86	85	5	6	30	8/08
3. % Overall Maintenance Budget Compliance	94	98	98	96	94	92	90	88	86	84	82	80	78	8	6	48	9/08
4. Overall Schedule Compliance	90	95	95	94	93	92	91	90	89	88	87	86	85	5	7	35	10/08
5. % Overall PM Compliance	94	100	100	98	96	94	92	90	88	86	84	82	80	7	11	77	11/08
6. % Planned Work	68	80	80	78	76	74	72	70	68	66	64	62	60	4	7	28	12/08
7. % Craft Time For Customer Charge Back	75	85	85	83	81	79	77	75	73	71	69	67	65	5	6	30	1/09
8. % Work Orders With Reliable Planned Time	50	60	60	58	56	54	52	50	48	46	44	42	40	5	7	35	2/09
9. % Critical Asset Availability	90	95	95	94	93	92	91	90	89	88	87	86	85	5	13	65	3/09
10. % Wrench Time (Craft Utilization)	34	50	50	48	46	44	42	40	38	36	34	32	30	2	8	16	4/09
11. % Craft Performance	90	95	95	94	93	92	91	90	89	88	87	86	85	5	8	40	5/09
12. % Inventory Accuracy	90	98	98	97	96	95	94	93	92	91	90	89	88	2	6	12	6/09
13. Number of Stock Outs of Inventoried Stock Items	15	10	10	11	12	13	14	15	16	17	18	19	20	5	5	25	6/09
F. Performance Level Scores / Perf Level			10	9	8	7	6	5	4	3	2	1	0				Score 481

D. Baseline Performance Levels (columns PL 10 through PL 0)

J. Total MEI Value Over Time

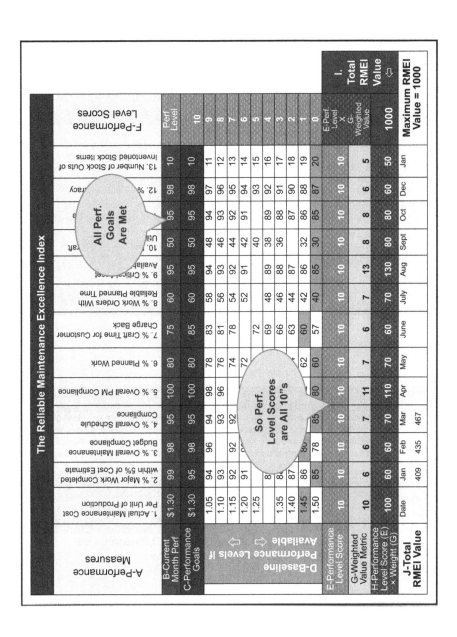

Step	Description	The 10 Step Process for RMEI Development
A	Performance Metrics	10 to 15 metrics are selected & agreed upon by the organization. Select metrics for total maintenance process + operational ones.
B	Current Month Performance	The actual monthly performance level for the metric. This value will also be noted in one of the incremental values blocks below the performance goal. This value will correspond to a value for F, the performance level scores which go from 10 down to 1.
C	Performance Goal	The pre-established performance goal for each of the RMEI metrics. For example, if the Current Month's Performance is at the Performance Goal level, the performance level score for that goal will be a 10, the maximum score.
D	Baseline Performance	The baseline performance level prior to start of RMEI performance measurement. Not always available. Use a 1–3 months performance average as a baseline after start of RMEI.
E	Current Performance Score	Depending on the current month's performance, a performance level score (F) will be obtained. This value then goes to the Current Performance Score row and serves as the multiplier for the (G) the Weighted Value of the Performance Metric.
F	Performance Level	Values from 10 down to one, which denotes the level of current performance, compared to the goal. If current performance achieves the predetermined goal, a performance value of 10 is given. Each metric is broken down into incremental values from the baseline to the goal. Each incremental value in the column corresponds to a performance level value. This value becomes the Current Performance Score.

Step	Description	The 10 Step Process for RMEI Development (cont)
G	Weighted Value of the Performance Metric	The values along this row are the weighted value or relative importance of each of the metrics. These values are obtained via a team process and a consensus on the relative importance of each metric that is selected for the RMEI. All of the weighted values sum to 100.
H	Performance Value Score	The Weighted Values (G) are multiplied by (E) the Current Performance Scores to get the Performance Value Score (H).
I	Total RMEI Performance Value	The sum of the Performance Value Scores for each of the metrics and the composite value of monthly maintenance performance on all RMEI metrics
J	Total RMEI Performance Values Over Time	Location for tracking Total RMEI Performance Values over a number of months

Important Notes

1. The RMEI is a composite of key metrics, each with relative importance
2. Each RMEI metric can be trended and linked to the monthly RMEI
3. The RMEI should be balanced across the total maintenance operation
4. Other metrics of lesser value can also be tracked.

#	Performance Metric and Purpose	Goal
1.	**% Overall Maintenance Budget Compliance:** To evaluate management of $ assets; Obtained from monthly financials	98%
2.	**Actual Maintenance Cost per Unit of Production or (Maintenance Cost Per Square footage Maintained:** To evaluate/benchmark actual costs against stated goals/baselines or against industry standards; Obtained from asset records and monthly CMMS WO file of completed WOs for the month. Obtained from production results and financial report. Provides ideal support to ABC Costing practices	TBD
3.	**% Customer or Capital Funded Jobs Completed as Scheduled and within +/- 5% of Cost Estimate:** To measure customer service & $ assets plus planning effectiveness; Obtained from funded WO types from the CMMS WO files, comparing date promised to date completed and estimated cost to actual cost	98%
4.	**% Other Planned Work Orders Completed as Scheduled:** To measure customer service and planning effectiveness; Obtained from a query of all planned WO types in CMMS WO files and comparing date promised to date completed. Could be expressed in % based on craft hours.	95%

#	Performance Metric and Purpose	Goal
5.	**Schedule Compliance:** To evaluate how effectiveness scheduling was in regards to executing to meet scheduled dates/time; Obtained from query of CMMS completed WO file where all scheduled jobs coded and their actual completion compared to actual planned completion date/time	95%
6.	**% Planned Work Orders versus % True Emergency Work Orders:** To evaluate positive impact of PM, planning processes and other proactive improvement initiatives (CRI,/RCM/etc); Obtained from a query of all true emergency WO types in CMMS WO files and comparing to total WOs completed. Could be expressed in % based on craft hours.	80% to 85% Planned
7.	**% Craft Time to Work** Order for Customer Charge Backs: To monitor craft resource Accountability for Internal Revenue Generation (or External); Obtained from a query of all WO types in CMMS WO files that are charged back comparing these craft hours to total craft hours paid	TBD
8.	**% Craft Time to Work Orders:** To monitor overall craft resource accountability and to support internal revenue generation ; Obtained from a query of all WO types in CMMS WO files and summation of actual craft hours	100%

#	Performance Metric and Purpose	Goal
9.	**% Craft Utilization (Actual Wrench Time):** To maximize craft resources for productive, value-adding work and to evaluate effectiveness of planning process; Obtained from a query of all craft hours reported to non craft work from CMMS time keeping WO files and summation of actual craft hours	60% to 70%
10.	**% Craft Performance (Against Reliable Estimates for PM and planned work):** To maximize craft resources, to evaluate planning effectiveness and also to determine training ROI; Obtained from completed WO file in CMMS	95%
11.	**Craft Quality and Service Level:** To evaluate quality and service level of repair work as defined by customer; Obtained from WO file in CMMS where all call backs are tracked and monitored via work control and planning processes	98% +

The Overall Craft Effectiveness Factor (%)

CU: Craft Utilization × CP: Craft Performance × CSQ: Craft Service Quality

OCE = %CU x CP x CSQ

#	Performance Metric and Purpose	Goal
12.	**Overall Craft Effectiveness (OCE):** To evaluate cumulative positive impact of overall improvements to Craft Utilization (CU), Craft Performance (CP) and Craft Quality and Service Excellence (CQSE) in combination; Obtained from using results of measuring all three OCE Factors: a) Craft Utilization, b) Craft Performance and c) Craft Quality and Service Excellence	65%
13.	**% Work Orders with Reliable Planned Times:** To measure planner's effectiveness at developing reliable planning times; Obtained from completed WO file in CMMS where panning times are being established for as many jobs as possible by planner/supervisor	70%
14.	**% Overall Preventive Maintenance Compliance:** (Could be by type asset, production department/location or by supervisory area): To evaluate compliance to actual PM requirements as established for assets under scope of responsibilities; Obtained from completed WO file in CMMS	100%

#	Performance Metric and Purpose	Goal
15.	**Gained $Value from Craft Utilization & Performance:** To determine actual gained $ value of craft productivity gains as compared to original estimate and/or the initial baseline; Obtained only from using results of measuring two of the OCE Factors: a) Craft Utilization, b) Craft Performance.	TBD
16.	**% Inventory Accuracy:** To evaluate one element of MRO material management and inventory control policies; Obtained from cycle count results and could be based on item count variances or on cost variance	98%
17.	**% or $ Value of Actual MRO Inventory Reduction:** To evaluate another element of MRO material management against original estimates and the initial baseline MRO inventory value; Obtained from inventory valuation summation at end of each reporting period	10%

#	Performance Metric and Purpose	Goal
18.	**Number of Stock Outs of Inventoried Stock Items:** To monitor actual stock item availability per demand plus to monitor any negative impact of MRO inventory reduction goals; Obtained from tracking stock item demand and recording stock outs manually or by coding requisition/purchase orders for the items not available per demand	????
19.	**$ Value of Direct Purchasing Cost Savings:** To track direct cost savings from progressive procurement practices as another element of MRO materials management. Could apply to contracted services, valid benefits received from performance contracting, contracted storerooms, vendor managed inventory; *Obtained via best method per a standard procedure that defines how direct purchasing savings are to be accounted for*	TBD
20.	**Overall Equipment Effectiveness (OEE):** World –class metric to evaluate cumulative positive impact of overall reliability improvements to Asset Availability A), Asset Performance (P) and Quality (Q) of output all in combination. (Similar to OCE above but for the most critical production assets); Obtained via downtime reporting process, operations performance on critical assets and the resulting quality of output	85%
21.	**% Asset Availability/Uptime:** To evaluate trends in downtime due to maintenance and the positive impact of actions to increase uptime; *Obtained via downtime reporting process*	TBD
	Many More Metrics Available	

Some Other Maintenance Metrics

Category	Benchmark
Yearly Maintenance Cost:	
Total Maintenance Cost/Total Manufacturing Cost	< 10-15%
Maintenance Cost/Replacement Asset Value of the Plant and Equipment	< 3%
Hourly Maintenance Workers as a % of Total	15%
Planned Maintenance:	
Planned Maintenance/Total Maintenance	> 85%
Planned & Scheduled Maintenance as a % of hours worked	~85-95%
Unplanned Down Time	~0%
Reactive Maintenance	< 15%
Run to Fail (Emergency + Non-Emergency)	< 10%
Maintenance Overtime:	
Maintenance Overtime/Total Company Overtime	< 5%
Monthly Maintenance Rework:	
Work Orders Reworked/Total Work Orders	~0%
Inventory Turns:	
Turns Ration of Spare Parts	> 2-3

Some Other Maintenance Metrics

Category	Benchmark
Training:	
For at least 90% of workers, hours/year	> 80 hours/year
Spending on Worker Training (% of payroll)	~4%
Safety Performance:	
OSHA Recordable Injuries per 200,000 labor hours	< 2
Housekeeping	~96%
Monthly Maintenance Strategies:	
✓**Preventive Maintenance**: Total Hours PM/Total Maintenance Hours Available	~20%
✓**Predictive Maintenance:** Total Hours PdM/Total Maintenance Hours Available	~50%
✓**Planned Reactive Maintenance**: Total Hours PRM/Total Maintenance Hours Available	~20%
✓**Reactive Emergency:** Total REM/Total Maintenance Hours Available	~2%
✓**Reactive Non-Emergency:** Total RNEM/Total Maintenance Hours Available	~8%
Plant Availability: Available Time/ Maximum Available Time	> 97%
Contractors: Contractors Cost/Total Maintenance Cost	35-64%

Making Maintenance Performance Based: Profit & Customer-Centered

Step 1: Planning

Understanding
Achievement Goals

Step 2: Making

Proactive Performance
a Part of Your
Maintenance Business
Plan

**Step 3: True
Maintenance
Leadership**

**Step 4:
Monitoring**

Your RMEI

---

3 R's of Performance Data Collection

- RELIABLE – Data Is Credible,
  Calculations Are Accurate and
  Consistent Over Time

- RELEVANT – Pertains to the Source It Is
  Intended to Measure

- REPRESENTATIVE – It Is Typical of the
  Service Being Measured

Planning & Scheduling of "Human Capital"

- Right People or Person
- Right Skills
- Right Place
- Right Time
- Right Parts
- Right Repair Method
- Right Tools

A Lot of "Rights" to Get Right!

How This Book Can Apply to the Very Small Work Unit in Oil and Gas or to Any Type of Maintenance Operation

19

This chapter gives a great case study on "Different Schools of Thought on Executing Plant Maintenance." It comes from Gary Royer, a highly experienced planner retired from a large, famous company known for maintenance excellence, who started his second career in a much smaller plant as more than "just a planner."

My experiences at the "King of Beer" (Anheuser-Busch) and "Queen of Chocolate" shaped up to be like the poles of the earth due to exact opposite support levels that the maintenance group received. The contrasting approach to maintenance appeared to be based on the perceived value that the corporate leadership had of the maintenance group's contributions.

Before I retired from Anheuser-Busch, The "King" recognized the value added back to the organization in the deliverance of a quality product to its consumers. Maintaining the plant equipment through its entire life cycle at peak efficiency and performance translated into higher throughput, less unplanned downtime, and ultimately added profits to the bottom line. Proactive maintenance was an integral part of the culture for the 90 plus machinists and 45 electricians at our 24/7 plant.

In contrast, the "Queen" acted as if maintenance was an unwelcomed cost to the business. The lack of support and commitment resulted in underperformance and missing its full potential for value-added cost savings to the organization. The bedrock of any maintenance organization begins with buy in by top management in any company.

At the "King," maintenance had a seat at the table with a senior vice president who was the "Champion for World Class Maintenance." He made sure the value-added contribution by the maintenance department was well known, respected, and appreciated. Funding and support for staffing, tools, and systems were provided. When it was time for the migration from the previous computerized maintenance management system (CMMS), the change from "Champs" to "SAP" was planned and executed. The mintenance champion made sure that adequate resources and timelines were provided to assure a successful rollout.

The SAP structure tree, equipment hierarchy, preventive maintenance (PM) tasks, and task instructions were preplanned. Workloads were proportionally divided into dedicated business units so that each assigned technical planner was able to do the best job possible in his or her area. Training throughout the SAP rollout was well planned and scheduled. Maintenance audits were later performed on an 18-month rotational schedule at the domestic breweries. Lessons learned, corporate benchmarks, and best practice PM enhancements were dividends earned from this process.

Adequate maintenance shop support staffing was available to assist in planned modular PMs as each business unit had planned scheduled outages per month. Annual

Reliable Maintenance Planning, Estimating, and Scheduling. http://dx.doi.org/10.1016/B978-0-12-397042-8.00019-X

weeklong shutdowns were scheduled for each production line for extended refresh overhauls. Corporate training resources were provided at the "Learning Center" to assist in developing technical skills and for professional development. True "world-class maintenance" requires and achieves maximum results when top management embraces it and demonstrates it by providing a "CHAMPION" for the maintenance team.

Little did I realize that the training and the years of experience with this world-class operation would later serve me well at my next employer, the "Queen." The supportive culture that I was accustomed was radically different. Everything I was taught and had experienced throughout my career with the "King" was put to the test with the "Queen."

Polar opposite support awaited my arrival. It became apparent that reactive, break-down maintenance was the norm after I completed an on-boarding assessment of the current state and performance evaluation of the maintenance group. The current CMMS was "MP2" and was not rolled out properly. It had very few PM tasks or instructions documented after 3 years in service. Prior to MP2, several years had trans-pired after the termination of the previous CMMS "Elke" system. During this time gap, production management tried what ultimately was a failed experiment and was actually directing the maintenance group via direct plant intercom calls to specific crafts people on shift. Proper data migration did not occur. Seamless communication between the accounting systems and MP2 was not established, and as a result, other useful data management tools did not exist. Double data input by maintenance was required for all purchase of parts by the accounting system.

The staff at the "Queen" consisted of fewer than 20 mechanics and electricians to cover three shifts of production that was steadily increasing to meet market demands. A centralized maintenance library for PDF and hard copy manuals and drawings was needed and was one of the earlier project priorities. A weekly maintenance metrics report was started to measure and track key performance indicator (KPI) progress.

One highlight, to my surprise, was that one experienced technician had been selected to get the maintenance storeroom in shape, which he was doing when I arrived. We decided to use barcodes, and he successfully laid out an effective storage arrangement with bins for some items and other appropriate storage and control methods for all items. We even used regular cycle counts to help achieve higher inventory accuracy. Now we were prepared to move forward with the next steps of bar-coded assets and work orders when that time might occur.

A major cultural change was needed. I soon realized that upper management did not support the required investment needed to turn this situation around in a timely fashion. Lip service was extended, but it lacked real meaning without the proper fund-ing and support. To expedite the positive change required with limited external back-ing, I scheduled communication meetings across the shifts to offer a "vision of the future" that could be done with a total internal commitment by all the skilled craft persons in the department.

As I have explained already, no one in maintenance likes to constantly be called for machine breakdowns and have to break away from a job in progress. Pressure experienced when production is halted, with operators not producing and management standing around watching and asking when will the machine or line be fixed, did not need to occur. The team was open to my proposed antidote to the reactive work with which they were so familiar.

Proactive maintenance was the solution, and they were willing to embrace the efforts required through this journey. This involved their help at times to assist in writing some of the earlier basic PMs that I later entered into the CMMS. It also required their willingness to identify PM repair work on the machinery before breakdowns occurred. The team knew that they had a vested interest in a successful outcome. I shared basic reminders of the importance for proper date-stamped work details documented in each work order and how accurate recorded parts usage would ultimately become a very good time-saving tool used by the team. Properly documented work histories, parts usage, bill of materials (BOM), tasks and task instructions by equipment were recognized as key tools in the turnaround from a reactive to a proactive maintenance culture.

The shift maintenance supervisors (from each of the three shifts) and I focused on the system tools, databases, and procedures that we needed to create, build on, and use to continue the turnaround momentum. Periodic training in the new systems were scheduled. Internal department efforts were steadily paying dividends, but further external work was needed to build and improve relationships with the service end users in production and process areas.

Maintenance planning meetings were arranged to bridge the differences between maintenance and operation groups that had developed and festered through the years. Regular meetings now occurred to continue working together to communicate and maximize uptime and minimize unplanned downtime. This partnership grew stronger over time as recognition of results of the maintenance team's proactive efforts had directly increased machine efficiencies throughout the plant.

The maintenance team was tasked with maintaining an aged facility and infrastructure along with installation of new technologies, new machinery, and new lines. This required another early initiative and led me to establish a maintenance contractor management program that included an standard operating procedures (SOP) that defines the procedures to follow for work scope identification with detail specifications, reviewing work scope to secure quotes, vendor selection, job scheduling, reviewing job progress, through completion.

It also included a final audit review before payment authorization. The program included a pool of approved contractors to safely perform specific work in a food manufacturing facility.

The attention to this process served us well as the business growth continued. Well-managed and -executed plant shutdowns using both internal and external maintenance resources became the norm.

A comprehensive multiyear preventative PM program was established to address and service all critically identified plant equipment and to provide for timely lubrication, calibrations, and other regulatory required work; utilities systems that provided electric power, boiler steam, condensate return systems, air compressors, dryers, and refrigeration equipment such as chillers, HVAC and exhaust fans, were also included. Proper PM task identification, schedule frequency, parts inventory, and parts kitting management are very important functions for any size manufacturing organization that requires constant review for frequencies and detailed task instruction improvements.

Then there are and use of root cause analyst tools to eliminate downtime occurrences to resolve issues before they become repetitive. A predictive PM program was begun with continual refinement for improved results. Recruitment, technical training, and development of qualified technicians along with succession planning are other challenges now faced with the baby boomer technicians currently exiting the workforce.

To those who accept the challenges encountered in a smaller or less-developed maintenance organization, be receptive to outside audits that can identify areas to improve and serve as a benchmark to compare how your group performance rates with your peers in industry. Use the maintenance metrics KPI report as a great communication tool with top management. This can demonstrate with facts the progressive positive results being achieved by a highly engaged and well-managed maintenance team.

Finally, with well-documented results showing successes, you should be able to go back and demonstrate the value added by such a well-run maintenance team. Of course our job is never done. Continue to educate others in your organization on the value derived from a well-supported maintenance team. We are not a cost but an investment in their success. In a globally competitive environment, maintenance planners and leaders must continue to evolve, grow, and meet the future challenges. Learn that smart delivery of maintenance service goes hand in hand with a more productive successful world-class organization.

Current and future planners must continually strive to grow in knowledge, experience, awareness of tools and systems available, and their use. You may never know what challenges await you. It is nice to work in a well-structured organization with resources and detailed processes provided for by other team members. However, opportunities do exist in smaller companies lacking all the bells and whistles for the maintenance group, which provides a chance to test your skills and personally grow and achieve greater personal success and satisfaction. The choice is yours, but be prepared. Best wishes to all pursuing and advancing their maintenance career.

By Gary Royer
Maintenance Planner
A Premier Chocolate Company

A Model for Success: Developing Your Next Steps for Sustainable and Reliable Maintenance Planning—Estimating and Scheduling

Ricky Smith, Jerry Wilson

A maintenance planning and scheduling process must be developed using known best practices and executed to the standards listed in this chapter. Alcoa Mt. Holly has been the standard for proactive planning and scheduling for over 30 years. I worked at this facility for a number of years and my time there helped me understand the advantage of proactive planning and scheduling. The numbers listed below have been validated by many external organizations over a period of many years. This process requires two things: a different way of thinking and discipline to follow the process.

Alumax Mt Holly (1997) vs. Alcoa Mt Holly (2012)

Category	Alumax-1997	Alcoa-2012
Maintenance Spending / RAV	3.4%	2.0%
Budget Compliance	-0.5%	+3.7%
Overtime / Straight Time	1.0	7.1%
Number of Crafts	4	3
Planners per Tradesperson	1:20	1:19
Absenteeism ·	1.6%	1.8%
Backlog in Crew Weeks (Per Tradesperson)	4.4	6.8 Total/6.25 Ready
Schedule Compliance	95%	85.7%
Percent of Urgent (Interruption) Work	10.5%	3%
Percent of PM / PdM to all Work Orders	32%	47.2%
PM Accomplishment	96%	85.7% (10% Rule)
Inventory Accuracy	96%	97.6%
Inventory Turns	3.31	2.86
Maintenance Training $'s as % Total Payroll $	4.2%;	1%
Wrench Time	62.3%	58.8%

February 15, 2013

Reliable Maintenance Planning, Estimating, and Scheduling. http://dx.doi.org/10.1016/B978-0-12-397042-8.00020-6

One of the major objectives of maintenance planning and scheduling is to optimize "wrench time."

Wrench time is defined as the actual amount of time a crafts person spends doing value-added work. A wrench time study, or work sampling study, is aimed at identifying and then eliminating or mitigating the time spent on non-value-added tasks.

- World-class wrench time is 55–65%
- Most companies have a wrench time of 18–30%
- Wrench time can be increased by effective planning.

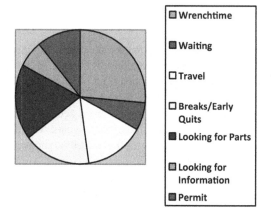

Your system is perfectly designed to deliver precisely the results you're getting.
W. Edwards Deming

Planning

Planning is the identification of all of the resources required to schedule and execute maintenance work effectively and efficiently. Planning is completed before work is scheduled.

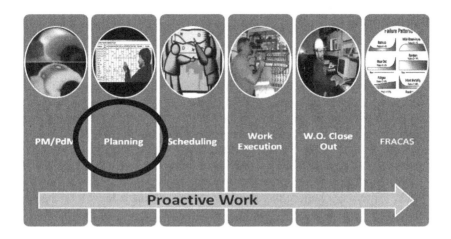

The start of the planning process begins when a planner receives a work request. The planner or a maintenance technician makes a job site inspection in order to identify everything that will be necessary to prevent delays to the job once it is started. The planner should preferably be a fully skilled crafts person who also has the additional skills necessary to execute the responsibilities of planning. Those skills are:

- Broad technical expertise for the craft
- Detail oriented
- Organized
- Communications skills
- Computer skills

If you think you have to get by with something less than a fully qualified maintenance person, then you are selling planning and scheduling (P&S) short. While all of the skills, in addition to technical expertise, are important, most can be easily learned. Technical expertise will allow a planner to quickly asses a work request and job site. In short order, they will have identified everything that is needed for an effective job plan. I say "effective" because I am not one that believes every job plan has to contain a complete list of information, common to all jobs, in order to qualify as fully planned and ready to schedule. All of the activities that are completed during P&S are also completed when no planning and scheduling takes place. The equipment has to be shut down; parts have to be obtained, etc. A planner does these activities much more efficiently. Knowing this, why would any organization choose not to use a fully qualified planner as defined here?

The Job Plan

A job plan is required for any job that is critical to the operation. A job plan should only contain the information that the average maintenance technician would need to execute the job at the desired quality level; any more than that, and you start running the risk of the job plan information being ignored. If you want to, you can include reams of data with the job plan, but on the official form, only required information should be present. If your maintenance personnel are ignoring what is documented in the job plans, it could be a result of too much information that they do not need. If this is happening, they are likely missing key information such as changes that have been made in parts, procedures, specifications, etc. that could be crucial to improving reliability. Therefore, I believe that planners should only include what is absolutely necessary.

Too many job plans focus on information that the average maintenance technician already knows. If we approached job plans a little differently by focusing just on what the technician needs to know but might not, we may find that the job plans would become more useful. The job summary should use bullets rather than text to list the major steps of the job. Details that a less experienced technician might need can be put in the job package as an attachment, referred to in the bulleted list. When details are put in attachments, this enables most technicians to quickly access the information they need by reducing the amount of information they must go through. Any changes from what the technicians would think of as status quo should be highlighted to draw their attention. With the exception of attachments, you should only use details in a job plan where you are going to require a signature from the technician to show that the work was done exactly as required, and these details can usually be simplified with bullets. You can make things even easier for your technicians by standardizing groupings of information and by having no more than seven bullets in each grouping. Additionally, each bullet should have no more than seven words where possible.

The following are some basics that must be present for the job to qualify as fully planned, including:

Job scope: What needs to be accomplished by this job, and what are the basic activities that will need to be executed for completion? If prework is needed, like insulation removal, an electrical disconnect, or any other activity that must be completed before the main task can be initiated, the planner will note this and start making arrangements to have the need resolved before the primary work is scheduled to start. Any follow-up work should be noted, such as jobs to repair equipment that was replaced with spares.

Labor needs by craft, skill level, and duration: The planner will make estimates for the labor hours, and number of individuals needed, for each craft, as well as an overall estimate for the chronological time to complete the job from start to finish. The minimum skill level required for each craft should also be documented in the job plan. For example, if a second year apprentice is qualified, that information should be documented in order for the maintenance manager to make the most effective labor utilization.

Parts: All of the necessary parts will be identified, and those not stocked in the storeroom, or otherwise unavailable, will be ordered. The job will not be considered as "ready to schedule" until all necessary parts are available. Usually, all parts required to do the job will be acquired and assembled in a "kit" a day or two before the job is scheduled to commence.

Permits: Necessary permits will be identified and initiated prior to the job start time.

Procedures: If a job requires a procedure that is anything other than common practice due to regulatory compliance, reliability, or safe practice rules, it should be included in the job plan package and listed as an attachment on the job plan form. Usually this information will be stored in, or linked to, the functional location or equipment master in the CMMS, so that the next time the equipment has to be worked on, that information will be immediately available and it will require less time for the planner to prepare the job.

Specifications: These should be listed when applicable and in bullet form when possible. For example, if bolts and nuts are stored in bulk in the shop and a particular job requires a bolt of a different grade than normal, that should be listed.

Special tools and equipment: These will be noted and assembled for the job.

Some job plans will be created with much more detail because of the complexity of the job. Job plans should be saved by electronically tying them to the functional location for the equipment in the CMMS system (these are known as "preplanned job packages"). In short order, the planner will have documented plans for the equipment types that fail most often. This will be a big time-saver for the planner in the future.

Many times a planner or maintenance technician may have to visit a site to scope the work. Following is an example of a very simple job site inspection form used by a planner for a mechanical maintenance crew.

JOB SCOPE INSPECTION SHEET

Work Order # _____   Date _____

Equip. # _____   F/L _____

Equipment Type:

Pump____Motor ____Coupling____ Valve____ Tank/Vessel ____Filter ____Dryer____

Piping ____ Other _____

Installation Type: Flanged____Welded ____ Screw____ Bolted____Other_____

Fastener Size: Bolt _____ Nut_____Stud _____

Gasket Type: Spiral Wound _____ Gylon _____ Teflon _____

Fiber _____ Other _____Size _____

Wrench Sizes: _____

Special Tools: _____

Job Description: _____

_____

_____

Job Scope: _____

_____

_____

Pre-Work Preparation Steps: _____

_____

_____

Safety Hazards: _____

_____

Permits: SW _____OF _____ Entry _____ Electrical Hot Work _____

Excavation _____ Other _____

LO/TO: Yes   No   **Lock Box:** Yes   No   Power Supply _____

Equipment Repair Location Tagged? Yes   No

Digital Pics: Yes   No   **Field Sketch:** Yes   No

Tools, in addition to hand and power tools, include machines and equipment: ____

_____

_____

Materials, Description and Quantity:_____

_____

_____

Minimal Skill Requirements: _____

_____

_____

Staffing Requirements: MMs # _____ Hrs _____

CSMs # _____ Hrs _____

Equip. Oper. # _____ Hrs _____

Follow up Work (Rebuilds, Fabrications, etc.): _____

_____

Post Completion Steps:_____

_____

Name Plate Data:

Manufacturer _____ Type _____

Serial # _____ Size _____ Pressure _____

Model # _____ Figure _____Frame _____

Horse Power _____RPM _____ Amps _____ Volts _____

Temperature_____Drawing # _____ NB _____

Procedures, Specifications, Additional Info needed: _____

_____

Prepared by: _____

Key traits of an organization that has an effective planner:

- Necessary planner responsibilities do not take a back seat to other needs.
- The majority of a planner's time is spent working on future work.
- Maintenance personnel seldom have to acquire additional parts on "planned" jobs.
- They have documented the type of jobs that should not be planned in order to increase the effectiveness of their planner(s).
- They define emergency work and track its level.
- Emergency work is 15% or less of total labor hours.
- Planning documentation is valued and reviewed by the field personnel. Field personnel realize the importance of sticking to the plan for consistency and maintaining the schedule.
- The planner effectively executes the following responsibilities:

  Inspecting the job

  Writing a job scope

  Identifying parts

  Ordering parts

  Identifying and assembling necessary procedures

  Identifying required permits

  Electronically maintaining current status of the work orders and backlogs

Because the planner is such a skilled and resourceful person, one of those who can always make things happen, it is all too easy for him/her to be saddled with responsibilities that rob time from necessary planning responsibilities. Management must always be wary of this and protect against sacrificing planning quality for convenience. I am not saying that your planner cannot have additional responsibilities but only that management must ensure that the planning responsibilities have first priority and that any additional responsibilities do not impinge on them. We discussed earlier how a planner's time is leveraged by three or more times; it is unlikely that any other responsibilities assigned to a planner will have anywhere near that level of value. However, in my experience, it is a common problem that planners have either been assigned, or they have assumed, other responsibilities that limit the amount of time they have to do their foremost job. Thus planning quality and/or quantity suffers. Some common responsibilities that I find planners fulfilling that do not require a planner's expertise and could compete with higher value adding activities necessary for well planned jobs are:

- Kitting the parts
- Stocking new spare parts
- Maintaining the CMMS information
- Ordering parts for emergency jobs or jobs already in progress. Avoid this at all cost!

Note: It is the responsibility of someone else to take of emergency work. Pulling a planner into emergency work reduces wrench time and decreases work efficiency.

Scheduling

Maintenance scheduling is the coordination of the schedule for the maintenance resources (labor, materials, tools, and equipment) and that of the assets (production equipment) in order to:

- Minimize interruptions to the production schedule.
- Maximize maintenance work within the opportunities present in the production schedule.
- Maximize utilization of the maintenance labor resources.
- Proactively initiate and execute preventive, predictive, and corrective maintenance work.
- Maximize wrench time for the maintenance organization.

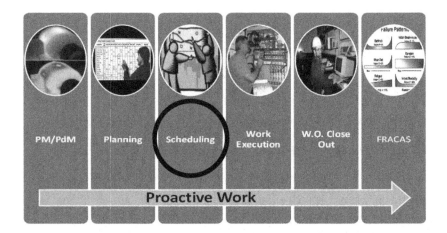

Each of these bullets has already been discussed in detail. So, let us look at how an effective schedule is built. An effective schedule should have an appropriate amount of fill-in work as discussed earlier. Depending on your organization's level of emergency work, you may need more or less fill-in work to protect the remaining planned and scheduled work from interruption. The majority of the labor hours available will be scheduled as routine work that has been planned. The remaining minority of the hours will be scheduled as fill-in work. By design, fill-in work will be sidelined as emergency work necessitates. Forty-six percent of the survey respondents report that their P&S jobs are frequently interrupted by emergency jobs. Interrupting fill-in work does not count as a schedule breaker nor should it require supervision approval. I have seen highly proactive maintenance organizations effectively operate with as little as 5–10% fill-in work, and I have seen those that required 40–50%. As a rule of thumb, if your P&S system is properly designed and executed, in 6–12 months you should need 20% or less fill-in work to protect your higher value work from interruption by emergency jobs.

One of the most effective tools in scheduling maintenance work, particularly in a situation where a single maintenance organization serves more than one production or manufacturing area, is the use of a maintenance coordinator. A maintenance coordinator, who reports through the production department, serves as a point of contact for all information and a funnel for setting priorities for maintenance. I have often seen where each shift supervisor, and each production manager, believes his or her maintenance work is top priority. A maintenance coordinator can minimize this. They should attend all production meetings and be thoroughly capable of leveling work order priorities across department lines. The maintenance coordinator has knowledge of all production schedule issues such as order ship dates, production start-up and shut-down dates, production delays, and product priorities. A representative from the production department, armed with this information, can much more effectively prioritize, help schedule work orders, minimize priority changes, minimize schedule changes, and find opportunities in the production schedule for maintenance work, than all of the maintenance managers talking with the production supervisors in the plant. I strongly believe that the maintenance coordinator role is just as critical as a planner and scheduler in all but the smallest P&S efforts. In the case of a maintenance coordinator, justification for the position should not come from the maintenance department but from the production department. If your P&S system is able to deliver positive gains to your maintenance department, a maintenance coordinator position will be paid for by the gains delivered to the production department. The cost associated with an hour for the production department is much more valuable than an hour of maintenance department time. Reducing the time required to perform a maintenance job will in turn save the same amount of time for production. If you have questions about justifying a maintenance coordinator, keep this section in mind.

A maintenance coordinator working with a maintenance scheduler can create a synergy that will cause a paradigm change in the reactiveness of a maintenance organization. This is the result of merging the production and maintenance schedules, along with adjusting priorities, which will enable more proactive work to be scheduled and completed than without a maintenance coordinator.

The last benefit of having a maintenance coordinator that I will mention is that of an advocate in the production management meetings. The maintenance coordinator's allegiance should be to the operating department, but his or her participation in the P&S process with maintenance will instill a high degree of ownership in the maintenance schedule. Having his or her input in the production scheduling meetings will be invaluable to the maintenance organization in getting support for proactive work and eliminating delays to maintenance caused by production. Forty-seven percent of our survey respondents reported that maintenance frequently waits for equipment to be shutdown and/or prepared for work that was scheduled.

*Key principle: If your maintenance organization has to struggle with competing priorities from the production department and/or a lack of support for the P&S effort, make the case for a maintenance coordinator.

Following are some of the key maintenance coordinator responsibilities:

- Level work priorities across the department.
- Communicate all production schedules to maintenance.

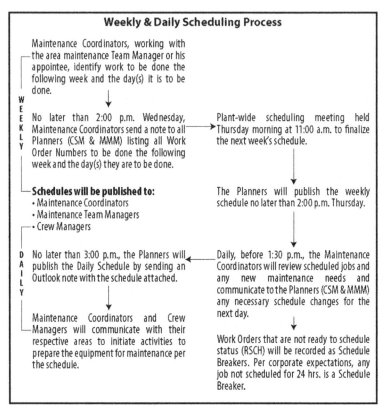

Figure 20.1 The scheduling process.

- Ensure timely equipment preparation for maintenance.
- Delete duplicate and unnecessary work orders.
- Help maintenance find opportunities to complete all proactive work on time.

A process that I have seen work very well is for all maintenance work requests from the operating department to go through the maintenance coordinator before going to the planner. This gives the maintenance coordinator an opportunity to assign a more equitable priority to each work order given his broader perspective across the entire production department. Going through the maintenance coordinator before the planner also helps the planner respond to true priorities.

The scheduling meeting should be attended by the maintenance coordinator, the scheduler(s), the maintenance supervisor, and the planner(s). If the scheduler has properly communicated with the maintenance coordinator and the maintenance supervisor prior to the meeting, the weekly scheduling meeting should be simply a final approval of the schedule. I have sat in scheduling meetings for a large maintenance organization that were very well prepared for and took only 15 min. It was very impressive. The people in those meetings were very good at what they did and they came to the meeting prepared. They had a well-defined process and everyone followed it. I have also sat in scheduling meetings that were either unorganized, ill-prepared for, involved turf wars between maintenance and production, or were simply games being played out because management required the meeting. I have seen these meetings take from one to four hours.

Daily versus weekly scheduling, which is better? I believe that for the most effective scheduling, both should be used. That does not mean that you have daily scheduling meetings. I think you can do both with only a single weekly scheduling meeting.

This allows a weekly schedule, agreed on in advance, that provides the overall target for the following week, then daily updates as needed. Daily changes, if any, are usually small. What this will do for your organization is provide increased flexibility to meet needs unforeseen when the weekly schedule was built. Inflexibility is one of the key downfalls of a rigid scheduling process. Keep in mind, the scheduling process should be very rigidly adhered to. However, that process can have some degree of flexibility designed into it. For example, let us assume it is Tuesday morning and an operator just reported a blower making an odd noise. A work order is written, and the planner inspects the job. The planner determines that a bearing needs to be replaced. Also, since there are several of these blowers in the plant, there is a documented job plan that was previously created. For this job to be fully planned, all the planner has to do is order the bearings and set the status to "ready to schedule," and all of the remaining information will already exist in the preplanned documentation. Using the process described in Figure 20.1, the job can be placed on Wednesday's schedule as long as everything is ready by 1:30 p.m. This is making the scheduling process work for you rather than punishing the organization that does not have the flexibility to count this as a planned and scheduled job. The planner was not rushed abnormally, and the job plan has the necessary elements of a fully planned and scheduled job.

Figure 20.1 is an actual scheduling process from a plant that has a mechanical maintenance crew and a control system/electrical crew, each with their own maintenance team manager. These two crews maintain three separate production areas that operate around the clock seven days per week. Each production area has its own maintenance coordinator. There is a crew manager for each shift to manage production issues. The entire group (about eight people) gets together for their weekly scheduling meeting. With these clarifications, the process is fairly self-explanatory. In this particular organization, the two planners also served as the schedulers.

Some situations that you want to prevent from being counted as scheduled include, for example, when a rush to meet a deadline results in a poor quality job plan, or when your planner is still creating or amending the job plan while work on the job is underway. In my opinion, this is a cardinal sin. When your planner is planning the job while work is in progress on the job, that is not future work—his time is not being leveraged. You should only use the planner's time where it will be leveraged (more about this in the next chapter). If work is in progress, a regular maintenance person should be executing all necessary activities from turning wrenches, ordering parts, and finding procedures. Once work starts, you have missed the opportunity to leverage a planner's time. This is an essential principle that must be understood and practiced.

Scheduling cutoff times is a critical piece of an effective scheduling process.

In Figure 20.1, you will notice that there are two separate cutoff times. The first is the cutoff to get work on the next week's schedule, which is 2:00 p.m. Wednesday. The second cutoff time is 1:30 p.m. daily for updates to the next day's schedule. Cutoff times are necessary in order to "publish" a schedule. With a firm day and time, the scheduler can electronically issue an official schedule for either the next week or the next day, whichever the case may be. There are no separate schedules; the weekly schedule is updated daily. Once published, the production department can start planning to make the necessary preparations to the equipment before maintenance arrives at the job site. Depending on your type of industry, these activities may include:

- Shutting the equipment down
- Cleaning/decontaminating the equipment
- Preparing permits to work on the equipment (lockout/tag out), etc.

Maintenance can also begin a myriad of activities. For example:

- The parts clerk can start kitting the parts for jobs a day or two before execution time.
- Maintenance can look at the schedule and start planning their personal activities.
- The planner can change the status of the work to "scheduled" and discontinue tracking those jobs; they will now effectively be in progress.

Without a cutoff time, it would be impossible to publish a schedule, causing an almost endless amount of back-and-forth communication between all involved in the process to execute a job. Forty-eight percent of survey respondents report using a formal cutoff time for creating a schedule. Only 41% of respondents report publishing a schedule electronically, available to all. Notice in Figure 20.1 that this organization had a firm rule that in order for work to be counted as scheduled, it had to be on the schedule for at least 24 h. This particular organization had historically "punched the card" by regularly listing work as planned and scheduled when it clearly had not been, thus management created the 24 h requirement in an attempt to improve the quality of planning and scheduling. Normally, I would not suggest a 24 h requirement such as this.

*Key principle: Publish a weekly schedule electronically, which is available to any who may need the information, and then update it daily as results and demands change.**

Can planning and scheduling be treated separately, or must they be implemented together? Yes to both questions!

- Planning prevents delays by identifying the needs of the job.
- Planning enables scheduling, by quantifying the resource needs in order for a block of time to be set aside to do the job.
- Scheduling enables effectiveness by maximizing resource utilization in both maintenance and production.
- Scheduling alone can improve effectiveness with only an estimate of labor needs by reducing the down time between maintenance jobs, eliminating false starts when production cannot free up the equipment, and by reducing the down time when maintenance has to wait while production shuts down and prepares equipment.
- Together, P&S are much more effective on certain jobs.
- Not all jobs are candidates for P&S (this will be detailed in the next section).

Ultimately, you will want the majority of your work to be both planned and scheduled because this is how you will get the most effectiveness increase.

Some work, however, is not well suited for planning and thus will only be scheduled. One example is fill-in work and inspections; planning would not be done beyond a basic labor estimate, which should always be documented to enable effective scheduling and backlog management. The major consideration would be if the work was very straightforward and only required "free issue" parts (parts stocked in the shop) or no parts at all. Then the job would only need to be scheduled. Running jobs such as this through the planning process would not add value. In this case, more time would be required to plan than would be saved by delays.

The characteristics of a job that present the opportunity for savings are complexity and predictability. Complexity can come in the form of technical difficulties, multiple crew involvement, special permits, special equipment or tools, special procedures, and multiple part/material needs. It is complexity that creates the opportunity, if not the likelihood, of delays. And, it is on the prevention of delays that we want our planner focused. The more labor that is required by a particular job, the more value a half hour savings can be worth. If you save one person a half hour by having a part ready before the job is started, versus saving 10 people a half hour, it can make the difference in whether a job should be planned or not. If the planner would spend more than 30 min inspecting the job, updating status, ordering parts, etc. then he has used more than the 30 min saved, in the example of a one-person job. You must protect your planner from these types of errors. Very simple and/or short jobs do not have much value that can be added via planning, so they would normally not go through the planning process. This practice frees the planner to focus on jobs where planning can leverage his time by three or more times.

Predictability is a term I use to describe if the parts, person-hours, technical skills, and other resources the job will need can be accurately identified. Jobs that have low predictability are not good candidates since a planner would not be able to accurately predict the resource and part needs for the job or may not be able to estimate the amount of time a given job would require to be completed. A good example of this sort is some jobs for an instrumentation crew. Take, for instance, a level transmitter work order. In inspecting the job, a planner will not be able to determine if the fuse is blown, the probe is bad, or if an amplifier or communication card needs to be replaced. In this case, the planner would not be able to specify the parts for the job or how long the repair might take. The best that can be done is to make an "average" time-required estimate and stop the planning there.

Reasons why Maintenance and Planning Fail

1. **Not knowing what effective planning and scheduling looks like.**

    Visit an organization that has proven effective planning and scheduling with key maintenance and production leadership. Talk to and interview key maintenance technicians and planners.

2. **Planner/Schedulers are not trained by true maintenance planning and scheduling professionals.**

Identify organizations whose primary focus is on training planners and schedulers. Bring the instructor back to your site to train site personnel in effective planning and scheduling. The instructor should help the planner with on-site coaching. If wrench time for maintenance staff of 20 maintenance techs is 20–40%, you just doubled the amount of proactive work conducted at your site.

3. **The culture must be changed from reactive to proactive.**

Patience and discipline is required to move from reactive to proactive.
If their thinking does not change maintenance planning will always be reactive.

4. **Maintenance planning and scheduling metrics are not used to track and lead performance.**

Use the Society for Maintenance and Reliability Professional (SMRP) Metrics and Definitions to move your maintenance planning and scheduling to a proactive state. (www.smrp.org) These metrics were developed by maintenance and reliability professionals worldwide and validated by a large number of professionals. A few SMRP metrics to use for planning and scheduling are:

- Planned Work
- Unplanned Work
- Actual Cost to Planning Estimate
- Actual Hours to Planning Estimate
- Reactive Work
- Proactive Work
- Schedule Compliance Hours
- Schedule Compliance Work Orders
- Standing Work Orders
- Ready Backlog
- Total Backlog
- PM & PdM Compliance

Proactive maintenance planning and scheduling takes time to create a proactive culture, however, patience and perseverance will pay off.

Appendix A

The Scoreboard for Maintenance Excellence—Version 2015

1. Top leaders' support to maintenance and physical asset management

Item#	Rating: Excellent—10, Very Good—9, Good—8, Average—7, Below Average—6, Poor—5	Rating
1.	The organization has a strategic plan that is the starting point for development of the asset management and maintenance strategy, policy, objectives, and plans.	
2.	Top leaders realize that the overall maintenance strategy must be holistic and that just one best practice alone, such as Computerized Maintenance Management Systems (CMMS), alone is not the answer.	
3.	Top leaders understand maintenance in regard to its broader sense as systematic and coordinated activities and practices. This is the way an organization optimally and sustainably manages its assets and asset systems, their associated performance, risks and expenditures over their life cycles to achieve its organizational strategic plans.	
4.	Top leaders consider and optimize the conflicting priorities of asset utilization and maintenance, of short-term performance opportunities and long-term sustainability, and between capital investments and subsequent operating costs, risks, and performance.	
5.	Top leaders have provided an organizational structure that facilitates the implementation of the key principles of asset management with clear direction and leadership.	
6	Top leaders should arrange for the creation of the asset management strategy. Key staff in the organization shall establish, document, implement, and maintain a long-term asset management strategy that shall be approved and authorized by top leaders.	
7.	Top leaders should appoint a member(s) of management at C-level whose responsibility is to ensure that the assets and assets systems deliver the requirements of the asset management policy, strategy, objectives, and plans and who have the authority to achieve this.	
8.	Top leaders should ensure that adequate resources are available for establishing and maintaining the asset management system, including equipment, human resources, expertise and training. Resources can be considered adequate if they are sufficient to deliver asset management plan(s) and activities, including performance measurement and required throughput to meet business plans.	

Continued

1. Top leaders' support to maintenance and physical asset management—cont'd

Item#	Rating: Excellent—10, Very Good—9, Good—8, Average—7, Below Average—6, Poor—5	Rating
9.	Top leaders communicate to the organization and to relevant third parties the importance of meeting its asset management requirements in order to achieve its organizational strategic plan.	
10.	There must be a clear understanding that the ultimate responsibility for the identification, assessment and management of risks associated with the physical asset management and maintenance process rests with top leaders.	
1. Top leaders support maintenance and physical asset management score Subtotal Score Possible: 100		

2. Maintenance strategy, policy and total cost of ownership

Item#	Rating: Excellent—10, Very Good—9, Good—8, Average—7, Below Average—6, Poor—5	Rating
	Your current maintenance strategy provides answers to the following questions:	
1.	Do we know what (existing) assets we have, where they are, what condition they are in, what function they perform, their inherent risk to operate, and their contribution to value?	
2.	Do we know the quality of the existing asset information to include top three databases? (a) equipment asset register, (b) parts and material inventory, and (c) preventive maintenance (PM)/predictive maintenance (PdM)/condition-based maintenance (CBM) database for reliability analysis.	
3.	Do we know what we want from our assets in the short, medium, and long term in their current operating context?	
4.	Can our assets deliver our asset management objectives cost effectively within their current operating context?	
5.	Are we getting the most value from our assets and trying to gain more value from them?	
6.	Do we have enough capability (or overcapacity) in our current asset portfolio?	
7.	Have some assets or asset systems become redundant, underused, unprofitable, or too expensive?	
8.	Are we confident that the risks of our assets causing harm to people and the environment are tolerable and at organizational/legally accepted levels of risk?	
9.	Is our asset-related expenditure (capital investment and costs) insufficient, excessive or optimal and correctly assigned across the asset portfolio? 10 = optimal, 5 = insufficient, excessive = 6, 7, 8, 9.	
10.	Can we readily evaluate the benefits (performance, risk reduction, compliance, sustainability) of proposed work or investment and, conversely, quantify the total impact to the organization of not performing such work, i.e., not investing or delaying such capital, shutdown, or repair actions.	

2. Maintenance strategy, policy and total cost of ownership—cont'd

Item#	Rating: Excellent—10, Very Good—9, Good—8, Average—7, Below Average—6, Poor—5	Rating
11.	Are we allowing future problems to develop and increase maintenance requirements/backlogs (such as performance deterioration, avoiding risks, and decreasing expenditure requirements) in our efforts to obtain short-term gains?	
12.	Have we given due consideration to the other aspects of the organization that affect our asset management plan(s), such as people, knowledge, finance and intangible assets such as synergistic team-based activities via leadership-driven, self-managed teams?	
13.	Do we review the appropriateness of our asset management strategy in light of changes in the operating, regulatory, the financial environment, and existing operating context?	
14.	Are we continually improving our asset management system performance and realizing the benefits of the improvements and know what and where improvements will be most effective?	
15.	Do we have the necessary asset management policy, strategy, and plan to ensure that we manage our assets in a sustainable way?	
16.	Does our approach to sustainable management of the assets take appropriate account of the needs of our stakeholders and are we open in our communication with internal and external stakeholders?	
17.	Are the working conditions, skills and well-being of our employees and contracted service providers given appropriate consideration?	
18.	Are we optimizing our asset management process(es) and/or procedures in light of the latest developments in technology and innovation?	
19.	Can we answer all of the above questions confidently, with a clear audit trail, and demonstrate the answers to our internal and external stakeholders?	
20.	The steps to forming, implementing, and maintaining the asset management policy are typically as follows: • Identify the requirements of the organizational strategic plan in terms of how it will be achieved • Identify current successes and gaps in application of currently accepted best practices as included in this Scoreboard document • The asset management policy should clearly define how it facilitates, supports, and enables achievement of the organization's vision, mission, and business objectives and align with the organization's physical asset management policies • It should consider risks, objectives, strategies, constraints, boundaries, timescales, and responsibilities • Identify all legal, regulatory, statutory, and other top leader–designated mandatory asset management requirements	
21.	Whole life asset management is being used and includes risk exposures and performance attributes, and considers the asset's economic life as the result of an optimization process (depending upon the design, utilization, maintenance, obsolescence, and other factors).	

Continued

2. Maintenance strategy, policy and total cost of ownership—cont'd

Item#	Rating: Excellent—10, Very Good—9, Good—8, Average—7, Below Average—6, Poor—5	Rating
22.	The asset life cycle costs, risks and performance are considered and optimized, which requires definition of clear asset boundaries for measuring performance, life cycle expenditures and attributing associated risks and consequences.	
23.	There is a clear understanding about the relationship between asset management activities and their actual or potential effect upon short-term and long-term costs, risks, performance, and asset life cycle cost or asset system sustainability.	
24.	Information for decisions on total cost of ownership is available for decisions to be made in regard to the optimal mix of life cycle activities such as design/selection, acquisition/construction, utilization, maintenance, renewal, modification, enhancement, decommissioning and disposal.	
25.	The asset management policy should be regularly reviewed, at a frequency determined by the organization, and following significant changes to the operational context of the organization. Issues identified should be addressed, and changes, where appropriate, should be implemented.	
26.	Reviews should ensure that the asset management policy is current and effective and ensures that the policy is continuously improved in light of developments in appropriate fields such as technology, operations, asset care techniques, etc.	
27.	Communication of the policy has been made to all relevant stakeholders, including contracted service providers, where there is a requirement that these persons are made aware of their asset management strategy-related obligations.	
28.	The asset management plan(s) should include a long-term asset replacement program to provide an overview of future asset replacement requirements and associated capital-funding needs.	
29.	Top leaders have a clear understanding as to how the operating context factors influence life cycle costs.	
30.	Major changes in top leaders and maintenance leaders should trigger an internal or external audit of current state of overall physical asset management as a due diligence step for the newcomer.	
2. Maintenance strategy, policy and total cost of ownership score Subtotal Score Possible: 300		

3. The organizational climate and culture

Item#	Rating: Excellent—10, Very Good—9, Good—8, Average—7, Below Average—6, Poor—5	Rating
1.	The organization's vision, mission, and requirements for success include physical asset management and maintenance as a top priority.	
2.	Senior management is visible and actively involved in continuous maintenance improvement and is obviously committed to achieving maintenance excellence with key resources needed.	

3. The organizational climate and culture—cont'd

Item#	Rating: Excellent—10, Very Good—9, Good—8, Average—7, Below Average—6, Poor—5	Rating
3.	The organization's business strategy and the plan for total operations success is known to all in maintenance and includes a strategy for maintenance process improvement.	
4.	Maintenance is kept well informed of changing business conditions, strategies, and long-range plans.	
5.	The organization's culture and the maintenance environment results in innovation, *pride* in maintenance, trust, and an obvious spirit of continuous improvement.	
6.	Open communication exists within maintenance and the overall organization to ensure interdepartmental cooperation, idea sharing, and basic teamwork.	
7.	The organization process of gaining employee involvement in continuous improvement includes the use of leadership driven, self-managed teams formally chartered for addressing a specific site challenge needing ideas or solutions from a cross-functional group of employees.	
8.	The team-based approach process being used such as Six Sigma et al. is well accepted, and employees are open to participation.	
9.	Crafts understand that the company's strength is fast, small-quantity shipments for critical customer reasons. This in turn requires frequent changes of tooling and setups rather than large production runs. Example: Contract packaging of cancer trial medicine packaged and shipped directly to a doctor or hospital for a cancer patient.	
	3. The organizational climate and culture score	
Subtotal Score Possible: 90		

4. Maintenance organization, administration and human resources

Item#	Rating: Excellent—10, Very Good—9, Good—8, Average—7, Below Average—6, Poor—5	Rating
1.	The maintenance organization chart is current and complete with fully defined areas of responsibility.	
2.	The maintenance organization is sufficiently staffed to respond to cover three key areas: (a) emergencies, (b) reliable routine work, and (c) staffing to accomplish timely backlog relief.	
3.	Maintenance leaders do not allow extensive craft resources to be consumed with emergency repairs, and have performance metrics that strive to limit emergency work to only 10–20% of available craft hours.	
4.	There are a number of ways to organize maintenance. Does your organization have sufficient resources to achieve the three main types of work that are listed in Item 2 above?	

Continued

4. Maintenance organization, administration and human resources—cont'd

Item#	Rating: Excellent—10, Very Good—9, Good—8, Average—7, Below Average—6, Poor—5	Rating
5.	The maintenance organization of supervisors and planners is organized such there is direct liaison at the level of supervisor and most importantly the planner. Indirect liaison is viewed as maintenance leader to operational leader and can be the wrong approach. Operations whose customers must be in direct communication with the planning and supervisory function within maintenance operations.	
6	Having total responsibility for the business of maintenance includes not only the craft resources supervisors and planners it should but also maintenance engineering and storeroom operations. Does your organization have complete responsibility for parts inventory control? Yes = 10, No = 5, 6, 7, 8, or 9 depending on service level being received.	
7.	Clear-cut craft job descriptions have been developed that completely define job responsibilities and skill levels required for each craft.	
8.	Craft personnel are provided copies of their job descriptions and counseled periodically on job performance, job responsibilities, and craft skills development needs.	
9.	One single head of maintenance operations is a capable maintenance leader that is supported by adequate clerical and technical staff of planners, first-line supervisors, stores personnel, maintenance engineering, plant engineering, and technical training support.	
10.	The maintenance department head (top leader) has high visibility within the organization and reports to a level such as the plant manager or site top leader.	
11.	The first-line supervisors are responsible for the performance of 12–15 crafts people. Responsible for 12–15 = 10, 8–11 = 8, 16–20 = 8; less than 8 = 5, and over 20 = 5.	
12.	A time-keeping system is in place to charge craft time to each job or task on each work order. 10 = work orders; assets and parts are bar coded to facilitate tracking total time on a work order, wrench time to work order, and all other non-tool time.	
13.	Monthly or weekly reports are available to show distribution of maintenance labor in critical categories: HSSE-related work, true emergency repairs, corrective work, PM/PdM work, project-type work, work generated from PM/PdM/CBM-identified repairs before major failure etc. and other categories per organizational needs.	
14.	Monthly or weekly reports are available to monitor backlog status to include total backlog and ready backlog ready for scheduling and priority of planned or project work, etc.	
15	Backlog trend data is available to highlight the need for craft increases, scheduled overtime, or subcontracting. If multiple shops are on a site, the workload at each site is monitored and cross-leveling can occur upon planner determination to move crafts from shop to shop on a temporary basis.	

4. Maintenance organization, administration and human resources—cont'd

Item#	Rating: Excellent—10, Very Good—9, Good—8, Average—7, Below Average—6, Poor—5	Rating
16.	Guidelines on the level of accepted backlog are established to determine the need for overtime or subcontracting as well as to identify potential problem areas.	
17.	Total maintenance requirement of a maintenance operation are facts based and can be shown to top leaders with priorities, to-be-competed dates, reliable estimates of job duration, and total Person-hours required.	
18.	Sufficient and reliable man-hour data is available that allows valid decisions as to which jobs must be delayed if new jobs or projects need to be added to the schedule. Example: Parts come in for a critical repair on a critical machine, and if this job can be integrated into a weekly schedule is very proactive scheduling.	
4. Maintenance organization, administration and human resources score Subtotal Score Possible: 180		

5. Craft skills development and technical skills

Item#	Rating: Excellent—10, Very Good—9, Good—8, Average—7, Below Average—6, Poor—5	Rating
1.	The types and levels of craft skills required for an effective maintenance operation have been identified, and sufficiently trained staff are in place.	
2.	Job descriptions include well-defined standards for job knowledge and skill levels required for each craft area.	
3.	An assessment of the current job knowledge and skill level of each crafts person has been made to determine individual training needs.	
4.	The overall training needs for the maintenance staff have been developed with a plan of action and cost.	
5.	The organization has committed to providing the necessary resources for maintenance training and skills development.	
6.	A program for craft skills development has been designed to address priority training needs and is being implemented.	
7.	Results of training are determined by a competency-based approach that ensures demonstrated capability to perform on newly trained craft tasks.	
8.	A policy to pay for skills gained is available or is being developed as part of the craft skills development program. This would be a key element for Item #9 below.	
9.	The benefits of developing multicraft capabilities within maintenance have been evaluated and incorporated into the craft skills training program as applicable. Craft performance when effectively planned/scheduled and being measured by a reliable method has shown steady improvement.	
10.	Individual training plans for each crafts person are being used to document compliance-type training as well as training that the crafts person and supervisor see as being needed.	

Continued

5. Craft skills development and technical skills—cont'd

Item#	Rating: Excellent—10, Very Good—9, Good—8, Average—7, Below Average—6, Poor—5	Rating
11.	Actual hands-on job competency is being documented (for example by the work order system) to help validate overall craft skill levels.	
12.	The overall craft workforce has shown the initiative for continuous craft skills development.	
	5. Craft skills development and technical skills score	
Subtotal Score Possible: 120		

6. Operator-based maintenance and pride in ownership

Item#	Rating: Excellent—10, Very Good—9, Good—8, Average—7, Below Average—6, Poor—5	Rating
1.	Operators are responsible for cleaning their equipment and trained to perform selected levels of operator-based maintenance.	
2.	Operator training and certification is a top priority before an operator is placed on specific equipment or processes to monitor.	
3.	An initial cleaning to bring equipment to an optimal or "as new" status has been planned or has been completed for critical equipment where operator-based maintenance is to be performed.	
4.	Operators have been trained to perform periodic inspections on their equipment and report problems.	
5.	Operators have been trained and have proper tools and equipment to do selected lubrication, tighten bolts and fasteners, and to detect symptoms of deterioration.	
6.	The process of transferring maintenance tasks and skills to operators has been well coordinated between maintenance, operations, engineering, and human resources staff.	
7.	Operators have developed greater "pride in ownership" and understand their expanded role in detecting and preventing maintenance problems.	
8.	Results from operator-based maintenance have been measured with such things as improved production levels, quality of output and cost avoidance from reporting small problems and repairing them before major failures occur.	
9.	The organization has a formal operator training program that includes safe start-up, operation, and proper shutdown before operators are assigned to a machine.	
10.	The organization can contribute (or document) increased uptime and asset reliability in part to successful operator training	
	6. Operator-based maintenance and pride in ownership score	
Subtotal Score: 100		

7. Maintenance leadership, management and supervision

Item#	Rating: Excellent—10, Very Good—9, Good—8, Average—7, Below Average—6, Poor—5	Rating
1.	This organization has shown maintenance leadership rather than just maintenance 'management. The maintenance leader is a well-respected member of a top leader staff and has the leadership and technical ability to lead maintenance forward and to support total operations excellence.	
2.	The maintenance leader has taken the leadership role and helped top leaders understand the critical importance of overall physical asset management and has contributed greatly to the organization's strategic plan for asset management.	
3.	The maintenance leader has responsibility for the total business of maintenance and support to maintenance. In addition, he has managed areas such as the maintenance storeroom and inventory as a cornerstone for maintenance excellence described in Best Categories 16 and 17.	
4.	Nonsupervisory work is minimized as a result of supervisors having adequate clerical support, responsive storeroom support, effective planner support, craft training support etc.	
5.	Supervisors primarily perform direct supervision of maintenance to include scheduling work assignments, verifying quality of completed work, evaluating performance, etc.	
6.	Supervisors actively support good housekeeping and the safety program by conducting/attending meetings, providing ideas and having an attitude that creates greater safety awareness and overall HSSE compliance.	
7.	An effective supervisory and leadership development program is available to increase supervisory capabilities as well special technical skills.	
8.	Maintenance supervisors are team players and are able to gain cooperation and support from operations management staff for the improvement of overall customer service.	
9.	Supervisors actively support continuous maintenance improvement with personal ideas and suggestions.	
10.	Supervisors promote *pride* in maintenance and encourage, listen to and get ideas from craft employees.	
11.	Supervisors have the technical background to identify training needs of their craft workforce and create positive support to craft skills development.	
12.	Supervisors manage and lead their work groups as if "they owned their section of the company."	
7. Maintenance leadership, management and supervision score Subtotal Score Possible: 120		

8. Maintenance business operations, budget and cost control

Item#	Rating: Excellent—10, Very Good—9, Good—8, Average—7, Below Average—6, Poor—5	RATING
1.	The maintenance budget is based on a realistic projection of actual needs rather than past budget levels and considers critical repairs that have been deferred.	
2.	Maintenance expenditures are charged to back to assets within work centers within operating departments in a way similar to zero-based budgeting where maintenance allocates most of its labor back to the customer.	
3.	Customers of maintenance are measured based on budget compliance and unit productivity but do not hold back on their maintenance budget when nontypical events occur.	
	Budget variances are monitored at customer level as well overall maintenance to highlight problem areas and evaluate variances early on that can lead to increased total cost of ownership (TOC).	
4.	Deferred maintenance repairs to operating equipment and facilities-related assets are identified and presented to management during the budgeting process with an evaluation as to the increased future cost impact of deferring maintenance.	
5.	Maintenance is analyzing the TOC and advising top leaders when replacements are the most economical option rather than continuously repairing aged equipment.	
6.	Maintenance has been assigned authority to make purchases at a level that ensures minimal delay in purchase order creation, minimal approval levels, and confidence that maintenance will stay within established purchasing guidelines.	
7.	Maintenance provides key input and support to long-range budget planning for new equipment, equipment overhaul and retrofit, facility expansions, rearrangements, and repairs.	
8.	Labor and material costs are established for all work orders accumulated to the equipment history file along with well-defined descriptions of actions taken. The planner has responsibility for the labor and material costs while the supervisor and tradesperson have responsibility for well-defined work accomplished.	
9.	An equipment history file is maintained for major pieces of equipment to track life-cycle cost, types of repairs, and causes of failure and repair trends.	
10.	The equipment history file is reviewed periodically to analyze repair trends and define root causes on critical equipment as means to evaluate recurring problem and to improve reliability. The planner who is most knowledgeable about work planned, scheduled, and completed may be a key resource in this important activity to support the reliability improvement effort.	
11.	Labor and material costs are estimated prior to the start of major planned repair work and projects.	

8. Maintenance business operations, budget and cost control—cont'd

Item#	Rating: Excellent—10, Very Good—9, Good—8, Average—7, Below Average—6, Poor—5	RATING
12.	Major work order and project are established and planned with well-defined cost variances limits within 5–10% and are investigated when extreme variances occur and explained to person authorizing the work. Planner scope of work and estimates are also reviewed for accuracy, as well as change orders needed and authorized.	
13.	Cost approval guidelines are established for large or special repair jobs as compared to normal repair.	
14.	The cost of downtime is known and published for major pieces of equipment or work centers and is used in determining priorities for repair as well positive cost impact when uptime and value of all throughput increases and can be sold.	
15.	Maintenance operations can operate as an internal business with current financial, budgeting, and cost accounting systems.	
8. Maintenance business operations, budget and cost control score Subtotal Score Possible: 150		

9. Work management and control: maintenance and repair (M/R)

Item#	Rating: Excellent—10, Very Good—9, Good—8, Average—7, Below Average—6, Poor—5	Rating
1.	A work management function is established within the maintenance operation generally crafted along functionality of the CMMS.	
2.	Written work management procedures that governs work management and control per the current CMMS are available.	
3.	A printed or electronic work order form is used to capture key planning, cost, performance, and job priority information. 10 = Bar-coded assets, parts, and work order.	
4.	A written procedure that governs the origination, authorization, and processing of all work orders is available and understood by all in maintenance and operations.	
5.	The responsibility for screening and processing of work orders is assigned and clearly defined.	
6.	Work orders are classified by type, e.g., emergency, planned equipment repairs, building systems, PM/PdM, project-type work, planned work created from PM/PdMs.	
7.	Reasonable "date required" is included on each work order with restrictions against "ASAP," etc.	
8.	The originating departments are required to indicate equipment location and number, work center number, and other applicable information on the work orders.	
9.	A well-defined procedure for determining the priority of repair work is established based on the criticality of the work and the criticality of equipment, safety factors, cost of downtime, etc.	
10.	Work orders are given a priority classification based on an established priority system.	

Continued

9. Work management and control: maintenance and repair (M/R)—cont'd

Item#	Rating: Excellent—10, Very Good—9, Good—8, Average—7, Below Average—6, Poor—5	Rating
11.	Work orders provide complete description of repairs performed, type labor and parts used and coding to track causes of failure.	
12.	Work management system provides info back to customer: backlogs, work orders in progress, work completed, work schedules, and actual cost chargebacks to customer.10 = Real time system.	
9. Work management and control: maintenance and repair score Subtotal Score Possible: 120		

10. Work management and control: shutdown, turnarounds, and outages (STO)

Item#	Rating: Excellent—10, Very Good—9, Good—8, Average—7, Below Average—6, Poor—5	Rating
1.	Work management and control is established for major overhaul repairs, shutdowns, turnarounds and outages (STO) and includes effective work management and control by in-house staff and contracted resources.	
2.	Work management and control of major projects provide means for monitoring project costs, schedule compliance and performance of both in-house and contracted resources with a robust project management system.	
3.	Work orders are used to provide key planning info, labor/material costs and performance info for all major STO and overhaul work.	
4.	Equipment history is updated with info from work orders generated from major overhaul repairs and SATO work.	
5.	The responsibility for screening and processing of work orders for major repairs is assigned to one person or unit.	
6.	Change order procedures and control are clear to all and approved at the appropriate level based on company requirements.	
7.	Change orders are reviewed by planners just as they review all jobs: scope of work, key job steps, equipment required, and total additional cost and impact on total STO duration and appropriate approvals received before work execution.	
8.	Work orders for major repairs, shutdown, and overhauls are monitored for schedule compliance, overall costs, and performance info including both in-house staff and contracted services.	
9.	Cost variances are measured at key milestones with cost variance info so extreme variance can be investigated sooner than later when it is too late. A 5–10% variance is set as maximum with clear reasons for increased scope of work.	
10.	Has the current level of plant maintenance/asset management achieved the desired reliability to make an STO (a) needed at longer time frame than normal, (b)needed at a shorter time frame, or (c) needed at the appropriate period based on age and state of asset capabilities in their operation context.? a = 10, 9, 8; b = 5; c = 7, 6.	

10. Work management and control: shutdown, turnarounds, and outages (STO)—cont'd

Item#	Rating: Excellent—10, Very Good—9, Good—8, Average—7, Below Average—6, Poor—5	Rating
11.	The organization has the capability to manage the turnaround program and be cost-effective as compared to the best in the sector, has a strategy for reducing costs in the face of an ageing plant and rising manpower and material costs, and where can we get high level technical advice?	
12.	The organization knows what manpower is available in house, the competence levels where to get additional resources, who will lead the site team, who will do the work plus the cross-functional team to design, monitor, and control the event organization?	
13.	Have STOs received significant level of attention of companies, do companies have a history of tolerating higher than necessary down-time, have older age of plants, see STO as a "necessary evil," are striving to lengthen the STO intervals from 12 to 24 months and even 4, 5, and sometimes 8 years?	
14.	The organization's history of planning and preparation for STOs has been: carried out more carefully, alignment of capital programs, has been scrutinizing and challenging scope of work	
15.	A process of assessing plant equipment deterioration is in place, the likely impact on reliability is known, the planning stocks are safe-guarded and have partnered with major plant overhaul engineering contractors and have a learning organization from past history to manage STOs effectively.	
16.	The plant beginning an STO has personnel available when required and capable of performing design specifications economically and (a) safely for life of plant, (b) knows sum of activities performed to protect reliability of the plant, (c) helps provide consistent means of produc-tion, and (d) helps generate profit, all while (e) reducing the TOC.	
17.	Top leaders clearly understand that STO is a significant maintenance and engineering event during which new plant is installed, existing plant overhauled, and redundant plant removed, which has a direct connection between successful accomplishment and the company's profitability.	
18.	The company includes profit lost during period of STO, which is con-sidered part of turnaround cost because they know the total true cost of event and the real impact can be assessed.	
19.	All involved with STOs realize the potential hazard to plant reliability or can diminish or destroy reliability if not: properly planned, prepared, executed, and there are poor decisions made by managers and engi-neers, bad workmanship, use of incorrect materials and damage done while plant is being shut down, overhauled, and restarted.	
20.	Technical uncertainty due to occurrence of unforeseen problems can be accurately reported, knowing when cost estimates are being exceeded, event's duration must be extended, how both cost and duration increases can be justified. Are reasonable cost and time contingencies built into an STO plan with accurate loss of revenue/profit considered?	

Continued

10. Work management and control: shutdown, turnarounds, and outages (STO)—cont'd

Item#	Rating: Excellent—10, Very Good—9, Good—8, Average—7, Below Average—6, Poor—5	Rating
21.	Have top leaders created their business strategy to manage the STO basic objectives to eliminate STOs altogether unless proven it is absolutely necessary?	
22.	If an STO is proven to be necessary, the top leader ensures that it will align with maintenance objectives, production requirements, and business goals.	
23.	When beginning an STO, the top leader has formed a chartered leadership driven, self-managed (not a committee), forming a cross-functional staff to help a committed company get the best STO value.	
24.	The STO team has senior managers, who are responsible for long-term strategy and meet at regular intervals throughout year to review current performance and formulate high-level strategies for management of events such as a long-term STO program.	
25.	Is an STO truly aligned to overall business strategy which include an evolution of asset management's driven search for change to preventive/predictive maintenance, being driven by technical considerations and a philosophy of maintenance prevention and continuous reliability improvement?	
26.	STOs are driven by business needs and question every maintenance practice to determine if it can be eliminated by addressing the cause that generated the need and examining the largest maintenance initiatives first during an STO.	
10. Work management and control: shutdowns, turnarounds, and outages (STO) score **Subtotal Score Possible: 260**		

11. Shop-level reliable planning, estimating and scheduling (M/R)

Item#	Rating: Excellent—10, Very Good—9, Good—8, Average—7, Below Average—6, Poor—5	Rating
1.	A formal maintenance planning function has been established and staffed with qualified planners in an approximate ratio of one planner to 20–25 crafts people.	
2.	The screening of work orders, reliable estimation of repair times, coordination of repair parts, and planning of repair work is performed as a support service to the supervisor.	
	Planner/schedulers realize their primary scope and role of planning and scheduling is to improve craft labor productivity and quality through the elimination of unforeseen obstacles such as potential delays coronation parts machine time and available resources.	
	Planner/schedulers clearly understand the scope of their defined roles and responsibilities within your organization and are in an organizational structure that promotes close coordination, cooperation and communication with their customer in operations.	

11. Shop-level reliable planning, estimating and scheduling (M/R)—cont'd

Item#	Rating: Excellent—10, Very Good—9, Good—8, Average—7, Below Average—6, Poor—5	Rating
3.	The planner uses the priority system in combination with parts and craft labor availability to develop a start date for each planned job to be scheduled.	
4.	A daily or weekly maintenance work schedule is available to the supervisor who schedules and assigns work to crafts personnel with multiple week "look aheads" if required.	
5.	The maintenance planner develops reliable and well-accepted estimated times for planned repair work and includes on work order for each craft to allow performance reporting, backload levels, and even documentation of work competency for selected jobs.	
6.	A day's planned work is available for each crafts person, keeping at least a half a day ahead (KAHADA) during the working day known in advance.	
7.	A master plan for all repairs is available indicating planned start date, duration, completion date, and type of crafts required to define "total maintenance requirements."	
8.	The master plan is reviewed and updated by maintenance, operations, and engineering as required with project-type work expected from maintenance. Care is taken not to overload maintenance with project work that causes PM/PdM and other work to be neglected.	
9.	Total maintenance requirements are a total of Total Backlog + Ready Backlog that has all resources (except labor or equipment availability) available to be scheduled.	
10.	A firm rule of thumb is never to put anything on the schedule without parts in house. There must be a contingency plan if needed parts do not arrive for critical equipment.	
11.	When parts arrive for critical equipment and can be inserted to the current schedule, this is considered very proactive maintenance cooperation with operations.	
12.	Scheduling/progress meetings are held periodically with operations to ensure understanding, agreement, and coordination of planned work, backlogs, and problem areas.	
13.	Operations cooperate with and support maintenance to accomplish repair and PM schedules.	
14.	Operations staff sign off on the agreed-upon schedule and are responsible to approve change in schedules and are accountable to top leaders for adverse results.	
15.	Setups and changeovers are coordinated with maintenance to allow scheduling of selected maintenance repairs, PM inspections, and lubrication services during scheduled downtime or unexpected "windows of opportunity" to insert ready backlog jobs into the schedule.	
16.	Planned repairs are scheduled by a valid priority system, completed on time and in line with completion dates promised to operations, and measured accordingly.	

Continued

11. Shop-level reliable planning, estimating and scheduling (M/R)—cont'd

Item#	Rating: Excellent—10, Very Good—9, Good—8, Average—7, Below Average—6, Poor—5	Rating
17.	Deferred maintenance is clearly defined on the master plan, and increased costs are identified to management as to the impact of deferring critical repairs, overhauls, etc.	
18.	Maintenance planners and production planners work closely to support planned repairs, adjust schedules, and ensure schedule compliance in a mutual goal.	
19.	The planning process directly supports the supervisor and provides means for effective scheduling of work, direct assignment of crafts, and monitoring of work in progress by the supervisor.	
20.	Planners' training has included formal training in planning/scheduling techniques, super user training on the CMMS, report-generating software or via Excel, and on-the-job training to include developing realistic planning times for craft work being planned. Understand use of MS project or the company's larger project management system such as Primavera 6.	
21.	Benefits of planning/scheduling investments are being validated by various metrics that document areas such as reduced emergency work, improved craft productivity, improved schedule compliance, reduced cost, and improved customer service.	
22.	Planning and scheduling procedures have been established defining work management and control procedures, the planning/scheduling process, the priority system, etc.	
23.	A reasonable number of backup planner/schedulers are selected and properly trained and used to cover for the full-time staff. The number is based on the size and type of work being planned. Ideally, just like the full-time planner, they should have good shop experience and sound technical experience.	
24	If a maintenance coordinator is assigned within a unit of a large refinery or any production, the planner will coordinate with that person about the job request, location within the unit, the problems to be repaired, and related risks. In many cases, this is an engineer or experienced operator who should be able to define complete requirements or a work request and in some cases prepare a risk assessment for the planner to use for the job.	
25.	If issues of any nature arise that are not readily resolved by the planner and operations, the maintenance leader should be the next person that the planner/scheduler goes to for resolution.	
26.	If planned work orders involve participation by several shops or functional crews, they are crossed over to a planner in that area. However, a single planner/scheduler must plan and then coordinate various functional crews with the respective supervisors during the scheduling process.	
27	Planners are in an excellent position to ensure critical spares by asset are accounted for as well as to recommend items to consider for including within the storeroom as critical spares.	

11. Shop-level reliable planning, estimating and scheduling (M/R)—cont'd

Item#	Rating: Excellent—10, Very Good—9, Good—8, Average—7, Below Average—6, Poor—5	Rating
29.	Planners see what is repetitively coming up for nonstock item purchasing as well as what is being repaired over and over again, can your planners who are active in this area support improving reliability and uptime?	
30.	Planners help ensure that warranted parts or equipment are denoted in the equipment file and that work orders for warranted parts or equipment are flagged during the planning process to document supplier reimbursements.	
11. Shop-level reliable planning, estimating and scheduling M/R score Subtotal Score Possible: 300		

12. STO and major planning/scheduling with project management

Item#	Rating: Excellent—10, Very Good—9, Good—8, Average—7, Below Average—6, Poor—5	Rating
1.	The planning and scheduling function includes major repairs, overhauls, and STO-type work not considered as part of normal maintenance work and any work requiring an STO event.	
2.	The planner team is a resource (or member) for the STO team of senior managers and the maintenance leader. Planners should meet at regular intervals to review current jobs awaiting a planned STO event.	
3.	Are your total backlog jobs coded and planned effectively to await an STO event? In large plants and refineries, planners support the plant schedulers with normally detailed job packages for estimates of all required resources for an STO job.	
4.	Schedulers from Item #3 coordinate parts/materials, develop daily or weekly schedules, monitor status of work along with onsite observations, from the supervisor input and from a planner's job package, which could include several crews, defined job steps, and estimated time for each step. With real-time reporting to a project management system or CMMS status, including costs can be readily determined from progress reporting.	
5.	Current planning/scheduling manpower is available with the competency levels needed to support the site team during an STO.	
6.	All planners and schedulers involved with STOs must realize the potential hazard to plant reliability or can diminish or destroy reliability if not properly planned, prepared, and executed.	
7.	There may even be poor decisions by managers and engineers, bad workmanship, use of incorrect materials and damage done while the plant is being shut down, overhauled, or restarted. Planners realize that properly planned, well-prepared work, and work executed to all health, safety, security, and environmental (HSSE) requirements is essential.	
8.	The use of work orders, estimation of repair times, coordination and staging of repair parts/materials and planning/scheduling of internal resources and contractor support is also included for major work and STO work not considered day-to-day maintenance and repair.	

Continued

12. STO and major planning/scheduling with project management—cont'd

Item#	Rating: Excellent—10, Very Good—9, Good—8, Average—7, Below Average—6, Poor—5	Rating
9.	A project work schedule or formal project management system is used to manage status and cost variance for STO work.	
10.	The current CMMS is integrated and linked to the project management system in real time when STO work orders or a change order work is approved.	
11.	Estimated labor and materials are established prior to project start using work orders with effective labor and material reporting to track overall cost, work progress, schedule compliance, etc.	
12.	The master plan for all major STO repairs, overhauls and new installation is available indicating planned start date, duration, completion date, and type crafts required.	
13.	Resources required for day-to-day maintenance work are not compromised by having to perform major repair type work, installation, modifications, etc., consuming in-house resources required for PMs and other day-to-day-type work.	
14.	Scheduling/progress meetings are held periodically with operations to ensure understanding, agreement, and coordination of major work and problem areas such as asset being ready for scheduled work.	
15.	Major work performed by contractors is preplanned, scheduled, and includes measuring performance of contracted services.	
16.	Planning and scheduling procedures have been established for project-type work.	
12. STO and major maintenance planning/scheduling with project management score Subtotal Score Possible: 160		

13. Contractor management

Item#	Rating: Excellent—10, Very Good—9, Good—8, Average—7, Below Average—6, Poor—5	Rating
1.	Contracted work is clearly defined because the better the definition at the early stages, the better the job will go.	
2.	Loose specifications for both materials and work to be done are avoided.	
3.	Communication of your ideas to a contractor is included in the scope of work, and make sure they understand with meeting of the minds at kickoff and status update meetings.	
4.	Ensure that the contractor understands the quality of materials needed from clear specifications.	
5.	Negotiation and award of the contract has had key due diligence by the key owner's representative.	
6.	For larger jobs, owners may check out finances, credit, insurance, and staff.	
7.	Owners may visit other jobs to see contractor quality of work and call references.	

13. Contractor management—cont'd

Item#	Rating: Excellent—10, Very Good—9, Good—8, Average—7, Below Average—6, Poor—5	Rating
8.	Maintain at shop level a copy of the contract documents and keep a fair and complete set of contract info including requests for changes of scope.	
9.	Be aware of and avoid, if possible, lowball bids and negotiate a schedule of extras if applicable (i.e., "the high cost of low bid buying").	
10.	Avoid a common ploy where lowballing the bid to get the job floods the company with small extras.	
11.	Always add in for clauses like "all extras not included in the original price have to be agreed to in writing prior to the commencement of the work."	
12.	Are there deduction clauses in the contract that spell out what you will charge back and when you will charge it? Examples would be debris removal, cleanup, missing firm completion dates.	
13.	Negotiate cancellation clauses and spell out how and why you can cancel the contract. Otherwise, you could find yourself with a mechanic's lien against you over an inadequate job after you did not pay the final payment.	
14.	For ongoing service bids, avoid both too short of a contract term and too long of a contract term for two reasons: 1. If the term is too short, then the contractor will charge excessively for mobilization costs 2. If the term is too long, you might be stuck with a barely adequate vendor with no easy way to improve the situation	
15.	Is the contract as clear as possible about responsibilities on who supplies what, where to unload, site rules (safety, owner contact, cleanup, security, keys, etc.)?	
16.	Are there statements about how the site is to be left at the end of each workday?	
17.	Ensure who is responsible for locking up, barricades, traffic management, cleaning, and debris removal.	
18.	Does the agreement also include who is responsible for municipal permits, job plans, and all HSSE issues?	
19.	Are contractors' insurance policies reviewed with an agreement about what happens when (if) the contractor damages your property?	
20.	Damage to a neighbor's property that then might sue you or might spoil a good relationship is included as required.	
21.	Are all insurance certificates up to date covering: general liability, casualty (property damage), workmen's compensation, and auto liability?	
22.	If the contractor did a design build, then malpractice and errors and omissions are included.	
23.	Define performance as to what a good job would look like.	
24.	Add clause like "all work is expected to be done in a professional and workmanship-like manner and all work will be in compliance with applicable codes."	

Continued

13. Contractor management—cont'd

Item#	Rating: Excellent—10, Very Good—9, Good—8, Average—7, Below Average—6, Poor—5	Rating
25.	Owners should prepare the area to be worked on and remove as much as possible to avoid breakage/theft and isolate area so contractors have no reason to wander around the plant or a large multioperational site such as refinery.	
26.	Does the owner manage the contractor and keep a record of the job/project as it unfolds and provides feedback?	
27.	Does owner perform frequent inspections and document results with a functional planned schedule and compare progress to projections with problems being identified as early as possible for resolution?	
28.	Clear agreements have been made about when and amounts of progress or final payments are to be made, etc.	
29.	To avoid sloppy record keeping, all contractor work is documented on the owner on their CMMS/EAM.	
30.	Owner requires copy of paid receipts to prove subcontractors and material vendors have been paid.	
31.	Owner should get a "release of all liens form" signed before last payment because: 1. You could have paid off the general contractor and still be hit with liens from unpaid jobs 2. Consult with your legal department about lien laws in your state or country and be sure you are covered	
	13. Contractor management score	

Subtotal Score Possible: 310

14. Manufacturing facilities planning and property management

Item#	Rating: Excellent—10, Very Good—9, Good—8, Average—7, Below Average—6, Poor—5	Rating
1.	The equipment asset inventory system provides an accurate and complete record of asset information for both plant and facility assets.	
2.	New facilities planning, equipment additions and renovations are well coordinated with both plant and facility maintenance staff.	
3.	Maintenance staff provides input into the engineering planning process for new facilities, major facility additions and new production equipment additions.	
4.	An effective procedure for adding new facilities and new equipment info to the asset inventory is used as well as the deletion of equipment being removed from the facility.	
5.	Maintenance requirements are clearly designated as to responsibilities for both plant maintenance and facility maintenance, and the site is effectively organized and staffed to accomplish both.	
6.	The overall property management function within the organization provides close coordination with the site's maintenance operation when planning new facilities, renovations, or major production equipment additions.	

14. Manufacturing facilities planning and property management—cont'd

Item#	Rating: Excellent—10, Very Good—9, Good—8, Average—7, Below Average—6, Poor—5	Rating
7.	Facilities planning includes adequate planning for future maintenance requirements.	
8.	Consideration is given to life cycle costing of systems and/or subsystems when designing new facilities, renovations, or major production equipment additions.	
9.	Maximum standardization of facility systems and subsystems is planned for within new facilities, major renovations, and production equipment additions.	
14. Manufacturing facilities, planning and property management score Subtotal Score Possible: 90		

15. Production asset and facility condition evaluation program

Item#	Rating: Excellent—10, Very Good—9, Good—8, Average—7, Below Average—6, Poor—5	Rating
1.	A process is in place to periodically evaluate the current operational status of production equipment and, in turn, defines repair, overhaul, or replacement recommendations for life cycle cost reduction opportunities.	
2.	A facility condition assessment process is in place to evaluate current operational status of facility-type equipment and, in turn, defines repair, overhaul, or replacement recommendations for life cycle cost reduction opportunities.	
3.	The PM and PdM programs provide key info as to overall state of maintenance for production asset.	
4.	The PM and PdM programs provide key info as to overall state of maintenance for facility-type assets.	
5.	The current evaluation processes for production and facility assets support defining the true maintenance requirements of the operation and support STO planning.	
6.	The current evaluation processes for production and facility assets provide an effective baseline for replacement planning, total cost of ownership, and capital justification of new assets, both production and facilities related.	
15. Production asset and facility condition evaluation program score Subtotal Score: 60		

16. Storeroom operations and internal MRO customer service

Item#	Rating: Excellent—10, Very Good—9, Good—8, Average—7, Below Average—6, Poor—5	Rating
1.	The parts inventory system provides an accurate and complete record of information for each stock item. Parts "where used" is included in the master database along with usage, vendor information, and warranty information.	

Continued

16. Storeroom operations and internal MRO customer service—cont'd

Item#	Rating: Excellent—10, Very Good—9, Good—8, Average—7, Below Average—6, Poor—5	Rating
2.	The "ABC" classification of stock items is known and proper storage methods and accountability are established for each.	
3.	"A" and "B" items have valid reorder points, EOQ, and safety stock levels established.	
4.	"C" items (50% of stock items with 5% of total inventory value) are identified and use two-bin system or floor issue.	
5.	Inventory accuracy is determined by an effective cycle counting program. Cycle counting used = 10; count once per year = 7, count occasionally = 5, do no inventory counts = 0.	
6.	Inventory accuracy is regularly measured and is 95% or above. 95% or above = 10; 90–95% = 9; 80–89% = 8; 70–79% = 7; <70% = 5	
7.	An up-to-date storeroom catalogue is readily available to crafts (hard copy or electronic) and includes all stock items, storage locations, stock numbers, etc. Stock items are bar coded to include parts number, storage location, and reference back to parts data defined in Item #1 above.	
8.	Parts usage history is continually reviewed to determine proper stock levels, excess inventory items, and obsolete items.	
9.	A critical spares listing is available for critical assets, and critical spares (insurance items) are denoted in the parts inventory database.	
10.	Spare parts for critical instrumentation system components are listed with inventory tracking numbers by control loop or controlled equipment and stored in a separate area. Remotely stored items (for safety or quick access) must be listed in main inventory as to storage location.	
11.	Storeroom procedures are in place that define issues, receipts, inventory control, access control, parts accountability, reserving of parts, staging, quality control of parts, etc.	
12.	An operations assessment has been conducted for the storeroom operations to provide overall evaluation of facilities, storage and handling of equipment, staffing levels, inventory levels, systems, and procedures.	
16. Storeroom operations and internal MRO customer service score Subtotal Score: 120		

17. MRO materials management and procurement

Item#	Rating: Excellent—10, Very Good—9, Good—8, Average—7, Below Average—6, Poor—5	Rating
1.	Procedures and evaluation criteria for adding new maintenance parts and materials to stores are used.	
2.	Stores requisitions and issues are tied to the maintenance work order and charged directly to the repair job.	
3.	Maintenance planners and the storeroom personnel coordinate to reserve repair parts and material for planned work. "Kitting" or staging for pickup and direct delivery to the job site is done whenever possible.	

17. MRO materials management and procurement—cont'd

Item#	Rating: Excellent—10, Very Good—9, Good—8, Average—7, Below Average—6, Poor—5	Rating
4.	Purchasing has an effective program to evaluate vendor performance and quality.	
5.	Purchasing has developed partnerships with selected vendors and suppliers and is committed to purchasing based on fast delivery, quality parts, and service.	
6.	Maintenance storeroom staff is well trained, customer oriented, and provide a high level of customer service to maintenance. A program to cross-train each person on a specific task such as receiving, issuing, order picking, cycle counting, or running the recommended reorder report.	
7.	Maintenance storeroom and MRO procurement performance indicators such as % inventory accuracy or number of parts requests not available from stocked inventory items not filled are evaluated and reported on a monthly basis.	
8.	Maintenance and purchasing work together to coordinate planned reduction in stock items based on improved vendor delivery capabilities. Indiscriminate cuts to inventory are avoided.	
9.	An effective control method is in place for emergency purchases.	
10.	Purchasing and maintenance work together to ensure procurement of quality parts and materials and the "high cost of low bid buying is avoided."	
11.	Stock outs are being monitored as a performance measure of customer service from the storeroom operation and the MRO procurement process.	
12.	Standardization of parts and components is being pursued by maintenance and procurement staff.	
17. MRO materials management and procurement score		
Subtotal Score: 120		

18. Preventive maintenance and lubrication

Item#	Rating: Excellent—10, Very Good—9, Good—8, Average—7, Below Average—6, Poor—5	Rating
1.	The scope and frequency of PM and lubrication services have been established on applicable equipment.	
2.	Operations staff support and agree with the frequency and scope of the PM and lubrication program.	
3.	Optimum routes for PM inspections and lubrication services are established.	
4.	PM checklists with clear, concise instructions have been developed for each piece of equipment. Lubrication checklists and charts are available for each asset included under the lubrication program.	
5.	Inspection intervals and procedures are periodically reviewed for changes/improvements and updated as required as part of well-defined PM change process.	
6.	Planned times are established for all PM and lubrication tasks.	

Continued

18. Preventive maintenance and lubrication—cont'd

Item#	Rating: Excellent—10, Very Good—9, Good—8, Average—7, Below Average—6, Poor—5	Rating
7.	The total craft labor requirements by craft type to accomplish the overall PM program has been established to validate staffing needs for effective PM.	
8.	The required level of manpower is being committed to achieve the total scope of PM services.	
9.	Actual craft time devoted to PM is known and evaluated as a percentage of total craft time available.	
10.	Goals for PM compliance are established, and overall compliance and results are measured against the company benchmark.	
11.	All noncompliance with scheduled PM and lube services is aggressively evaluated and corrected. Lubrication services are viewed as a key part of preventive maintenance and are not neglected or overlooked.	
12.	Maintenance and operations work with close communication, coordination, and cooperation to schedule PM and lube services.	
13.	The success of PM is measured based on multiple factors: Reduced emergency repairs, increased planned maintenance work, reduced downtime costs, the elimination of the root cause of problems, and improved product throughput and quality, etc.	
14.	Preventive maintenance is a highly visible function within maintenance and is well received as an operational strategy.	
15.	The PM staff are well-qualified crafts people who serve as good maintenance ambassadors and "customer service representatives."	
16.	A PM master schedule is developed to evaluate the weekly or monthly plan and to level load tasks when required.	
17.	Corrective repair work orders are generated and become planned work as a result of PM inspections, and provide one measure of PM success.	
18.	PM manpower needs are adjusted to satisfy changing PM inspection requirements.	
	18. Preventive maintenance and lubrication score	
Subtotal Score Possible: 180		

19. Predictive maintenance and condition monitoring technology applications

Item#	Rating: Excellent—10, Very Good—9, Good—8, Average—7, Below Average—6, Poor—5	Rating
1.	Equipment has been evaluated for the application of PdM technology and the scope and frequency of PdM services has been established on all applicable equipment.	
2.	A plan for using current PdM technology is being developed or is now being put in action. Predictive maintenance techniques used on all applicable equipment = 10; application of PdM in progress based on plan = 9; PdM plan developed, no progress = 6; no plan for PdM = 0.	

19. Predictive maintenance and condition monitoring technology applications—cont'd

Item#	Rating: Excellent—10, Very Good—9, Good—8, Average—7, Below Average—6, Poor—5	Rating
3.	Maintenance, engineering, and others have technical knowledge and necessary skills for using PdM techniques.	
4.	Optimum routes for PdM inspections are established.	
5.	Inspection and testing intervals and procedures are periodically reviewed for changes/improvements and updated as required as part of well-defined PdM change process.	
6.	The required level of manpower is being committed to achieve the total scope of PdM services.	
7.	Goals for increased reliability are established, and overall PM/PdM compliance and results are measured against the company benchmark.	
8.	All noncompliance to scheduled PdM services is aggressively evaluated and corrected.	
9.	Maintenance and operations work with close communication, coordination, and cooperation to schedule PdM services.	
10.	The success of PdM is measured based on multiple factors: reduced emergency repairs, increased planned maintenance work, reduced downtime costs, the elimination of the root cause of problems, and improved product throughput and quality, etc.	
11.	Preventive/predictive maintenance is a highly visible function within maintenance, is well received as an operational strategy.	
12.	Corrective repair work orders are generated and become planned work as a result of PdM inspections and provide one measure of PdM success.	
13.	Critical equipment has been evaluated for the application of continuous monitoring technology.	
14.	A process for evaluating and eliminating the root causes of failure is in place.	
15.	Positive results from PM/PdM and overall reliability improvements are being communicated throughout the entire operation.	
19. Predictive maintenance and conditioning monitoring technology applications score		
Subtotal Score Possible: 150		

20. Reliability-centered maintenance (RCM)

Item#	Rating: Excellent—10, Very Good—9, Good—8, Average—7, Below Average—6, Poor—5	Rating
1.	A process such as RCM is used to determine the maintenance requirements of any critical physical asset in its operating context.	
2.	Criticality analysis to define top candidates for review has been conducted regarding those factors in production consequence (pd), safety/env. cons. (sf), service level (sl), redundancy (rf), frequency of failure (ff), and downtime (dt).	

Continued

20. Reliability-centered maintenance (RCM)—cont'd

Item#	Rating: Excellent—10, Very Good—9, Good—8, Average—7, Below Average—6, Poor—5	Rating
3.	The maintenance team strives to ensure any physical asset continues to do whatever its users want it to do in its present operating context.	
4.	Operations context is clearly defined and understood by maintenance and operations to include the type of process that is the most important feature of the "operating context." Some areas must be very clear before starting the RCM process. • Continuous processing where failure may stop entire plant or failure could significantly reduce output • Batch processing or discrete manufacturing	
5.	Other factors related to operating context of the asset include: redundancy factors, quality standards, environmental standards, actual physical location, safety hazards, shift arrangements, work-in-process levels, repair time and overall mean time to repair (MTTR), spares availability, market demand, and raw material supply.	
6.	Strategies and operating principles are defined and characterized, the key steps of which are: • A focus on the preservation of system function • Identification of specific failure modes to define loss of function or functional failure • Prioritization of the importance of the failure modes, because not all functions or functional failures are equal • Failure consequences • Identification of effective and applicable PM/PdM tasks for the appropriate failure modes • Applicable means that the task will prevent, mitigate, detect the onset of, or discover the failure mode Effective means that among competing candidates the selected PM/PdM task is the most cost-effective option.	
7.	Maintenance can define ways the asset can fail to fulfill its function where a functional failure is defined as the inability of any asset to fulfill a function to a standard of performance that is acceptable to the user.	
8.	Maintenance understands that it is more accurate to define failure in terms of a specific function rather than the asset as a whole. Categories of functional failures may include partial and total failure, upper and lower limits, gauges and indicators, along with the operating context. The RCM process looks to define/record all functional failures with each function.	
9.	The organization's work order system captures functional failures, cause of failures, and frequencies.	
10.	When defining functions, the function statement should consist of a verb, object, and a desired standard of performance, e.g., "to pump water from Tank X to Tank Y at not less than 300 gallons per minute."	
11.	Performance standards are defined in two ways: Desired Performance: What the user wants the asset to do; and Design Capability: What the asset can do.	
12.	Function statements may include the appropriate but different types of performance standards depending on the asset such as multiple, quantitative, qualitative, absolute, variable, or upper and lower limit types of standards.	

20. Reliability-centered maintenance (RCM)—cont'd

Item#	Rating: Excellent—10, Very Good—9, Good—8, Average—7, Below Average—6, Poor—5	Rating
13.	A failure mode that causes functional failures is being documented by the work order system by helping maintenance define the options for a maintenance strategy or development of caring for equipment plans.	
14.	Based upon an analysis of work completed, a reactive maintenance strategy of dealing with failure events after they occur is prevalent.	
15.	Proactive maintenance that deals with events before they occur or deciding, planning, and scheduling how they should be dealt with is a solid strategy.	
16.	Failure mode categories are designated when (a) capability falls below desired performance, and (b) desired performance rises above capability and when asset is not capable from the start per design or operating procedures.	
17.	Failure mode categories capability falling below desired performance can be identified as deterioration (all forms of wear and tear, lubrication failure, dirt/dust, disassembly where integrity of the assemblies (welds, rivets, bolts) decline due to fatigue or corrosion and human errors.	
18.	When desired performance rises above capability, the following can be identified: 1. Sustained, deliberate overloading 2. Sustained, unintentional overloading 3. Incorrect process materials	
19.	The level of detail in defining failure of a mode provides enough detail to select a suitable failure management strategy.	
20.	Failure modes that might reasonably be expected to occur in the current operating context are defined and will include: failures that have occurred on the same or similar assets; failure modes that are already under PM/PdM; any other failure modes that are considered real possibilities and where consequences are very severe if a failure occurs.	
21.	Failure effects and failure consequences' are not the same. Failure effects answer the question, "What happens?" whereas failure consequences answer the question "How does it matter?" Failure consequences are factored as one element of component/equipment criticality rating method being used.	
22.	Describing the effects of a failure strives to answer this question "What evidence is there that the failure occurred?" part of which may be answered by a maintenance planner: • In what way did the failure pose a threat to safety and the environment? • In what ways did the failure affect production or operations? • What physical damage is caused by the failure? • What must be done to repair the failure?Is this reported correctly on a work order, is part of a standard procedure, or becomes an after action report completed for total cost of downtime?	
23.	Completing the failure mode and effects analysis (FMEA) utilizes steps 1–20 above that point maintenance toward the best maintenance strategy to use.	

20. Reliability-centered maintenance (RCM)—cont'd

Item#	Rating: Excellent—10, Very Good—9, Good—8, Average—7, Below Average—6, Poor—5	Rating
24.	Failure consequences answer the question, "How does it matter?" There are many consequences of failure that: • Impact on output, quality and customer service • Impact personal safety and environmental issues • Increase operating costs, energy consumption • Are related to nature and the severity of the effects are governed by whether the users of the asset really believe that a failure matters	
25.	The focus on consequences starts the RCM process of task selection • Assessment of the effects of each failure mode • Classifying into four basic categories of consequences • Safety and environmental consequences • Operational consequences • Nonoperational consequences • Hidden failure consequences	
26.	What can be done to predict or prevent each failure? • Proactive tasks: Tasks undertaken before a failure occurs to prevent asset from going into a failed state include: preventive, predictive, scheduled restoration/overhaul, scheduled discard/replacement, condition monitoring and proactive task is worth doing if it reduces the consequences of failure enough to justify the direct and indirect cost of doing the task	
27.	The organization uses the RCM process to make the best failure management decision for critical assets by considering: • Age-related failures • Non-age-related failures (operator error) • Cost factors for scheduled restoration/overhaul • Cost factors for scheduled discard/replacement • Identifying potential failures and the P–F interval, which is the time that a potential failure begins to the time that a functional failure actually occurs, and the question, "Is the P-F interval enough time to deal with failure/consequences and what condition monitoring options available?"	
28.	Decisions are made on what should be done if a suitable proactive task cannot be found: • For hidden functions that cause multiple failures look for a possible failure finding task; if one is not available maybe redesign? • If safety or environmental issue cannot be resolved by proactive task; redesign or change the process • For operational consequences, if no proactive tasks are available and costs are less than no scheduled maintenance (run to failure). Look to redesign if costs too high • For nonoperational consequences (same as above)	
29.	There is a very clear and solid understanding about the P–F interval curve and the high cost of gambling with maintenance costs when the onsets of failures are known.	

20. Reliability-centered maintenance (RCM)—cont'd

Item#	Rating: Excellent—10, Very Good—9, Good—8, Average—7, Below Average—6, Poor—5	Rating

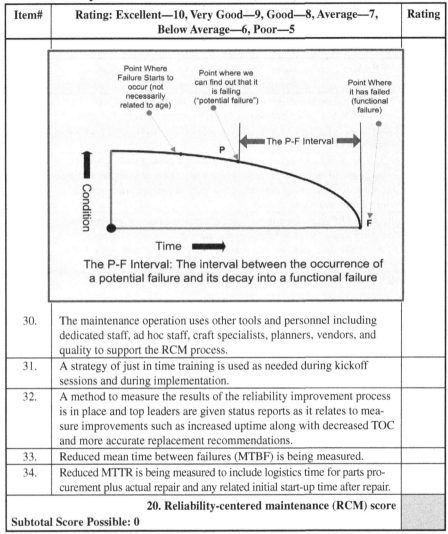

Item#		Rating
30.	The maintenance operation uses other tools and personnel including dedicated staff, ad hoc staff, craft specialists, planners, vendors, and quality to support the RCM process.	
31.	A strategy of just in time training is used as needed during kickoff sessions and during implementation.	
32.	A method to measure the results of the reliability improvement process is in place and top leaders are given status reports as it relates to measure improvements such as increased uptime along with decreased TOC and more accurate replacement recommendations.	
33.	Reduced mean time between failures (MTBF) is being measured.	
34.	Reduced MTTR is being measured to include logistics time for parts procurement plus actual repair and any related initial start-up time after repair.	
	20. Reliability-centered maintenance (RCM) score	
Subtotal Score Possible: 0		

21. Reliability analysis tools: root cause analysis (RCA), root cause corrective action (RCCA), failure modes effects analysis (FMEA), and failure reporting analysis and corrective action system (FRACAS)

Item#	Rating: Excellent—10, Very Good—9, Good—8, Average—7, Below Average—6, Poor—5	Rating
1.	A root cause analysis (RCA) process is in place as an effective process supporting RCM for finding the root causes of an event and facilitating effective corrective actions to prevent recurrence.	

Continued

21. Reliability analysis tools: root cause analysis (RCA), root cause corrective action (RCCA), failure modes effects analysis (FMEA), and failure reporting analysis and corrective action system (FRACAS)—cont'd

Item#	Rating: Excellent—10, Very Good—9, Good—8, Average—7, Below Average—6, Poor—5	Rating
2.	The goal to understand the concepts of RCA and root cause corrective action (RCCA) is to enable and apply the concepts to prevent or eliminate errors, defects and downtime has been communicated to the entire maintenance organization.	
3.	RCCA for nonconformances has been a requirement of the aerospace industry for many years. The organization has a plan or is implementing a process of determining the causes that led to a nonconformance or a failure event of a physical asset.	
4.	RCCA and RCA combine to form and effective method for implementing corrective actions to prevent recurrence and is another tool to support RCM. While these tools are not new, they may not have been aggressively enforced in the past as they are being done currently in your company.	
5.	RCCA when done correctly; this organization strictly enforces the requirements for taking corrective action.	
6.	This organization does not look at root cause analysis only as a way to get through the nonconformance reply response portion of an audit. These tools can be applied to all aspects of business for problem solving and process improvement with the approach varying based upon different functions within the organization and total size of the business.	
7.	An effective tool for problem solving is "a process for identifying the information required" as well as meeting corrective action requirements with steps described that encompass essential elements of any corrective action system that may be accomplished with different tools or called by different names.	
8.	Corrective action establishes and maintains documented procedures with actions taken to a degree appropriate to the magnitude of the problems and commensurate with the risks encountered. Key elements include: • Implementation with record of any changes • Effective handling • Investigating of the cause and recording the results • Determination of action needed to eliminate the cause • Controls to ensure that action taken is effective	
9.	Events are recognized as an all-inclusive term for any of the following: equipment failure, product failure, audit findings, special cause, accidents, customer complaint and failure mode and effects analysis (FMEA).	
10.	Focus is first on containment to "put out the fire," then assess the damage, contain all effects and notify all who are appropriate.	
11.	Assess the damage defines how many parts were impacted? How much were they impacted? and were there any other damages?	

21. Reliability analysis tools: root cause analysis (RCA), root cause corrective action (RCCA), failure modes effects analysis (FMEA), and failure reporting analysis and corrective action system (FRACAS)—cont'd

Item#	Rating: Excellent—10, Very Good—9, Good—8, Average—7, Below Average—6, Poor—5	Rating
12.	The process includes containing all effects, including where all the impacted parts are, segregating all impacted parts, and determining if impacted parts were under warranty or of inferior quality.	
13.	A process for notifying affected customers is in place to include required standard documentation.	
14.	When you have had an event, is it immediately contained and will the problem solution prevent the event from reoccurring?	
15.	The team approach is a leadership driven, self-managed that is formally chartered; these individuals can provide necessary resources to understand the problem, provide additional information, and since they have particular technical expertise, can act as advisors and provide support to top leaders.	
16.	The organization also understands the role of a failure reporting analysis and corrective action system (FRACAS) for helping improve reliability. This is a process in which you identify any reports from your CMMS/EAM or use reliability software that can help you to eliminate causes of equipment breakdown, mitigate or control failure.	
17.	FRACAS is viewed as a way to maximize the information available from your CMMS/EAM for control of losses and to eliminate failures. This is a process in which you identify data and reports from good work order reporting into your CMMS/EAM with special reliability software for data analysis.	
Subtotal Score Possible: 170	**21. Reliability analysis tools score**	

22. Risk-based maintenance (RBM)

Item#	Rating: Excellent—10, Very Good—9, Good—8, Average—7, Below Average—6, Poor—5	Rating
1.	Maintenance in this company has been carried out by integrating analysis, measurement, and periodic test activities to standard preventive maintenance. The gathered information is viewed in the context of the environmental, and has been related to process conditions of the equipment in the system in its operating context.	
	This company has a process and resources to perform the asset condition and risk assessment and define the appropriate maintenance program. All equipment displaying abnormal values is refurbished or replaced. In this way, it is possible to extend the useful life and guarantee over time of high levels of reliability, safety, and efficiency of the plant.	

Continued

22. Risk-based maintenance (RBM)—cont'd

Item#	Rating: Excellent—10, Very Good—9, Good—8, Average—7, Below Average—6, Poor—5	Rating
2.	A risk-based maintenance strategy is in place that prioritizes maintenance resources toward assets that carry the most risk if they were to fail. It is a methodology for determining the most economical use of maintenance resources. This is done so that the maintenance effort across a facility is optimized to minimize the total risk of failure. A risk-based maintenance strategy is based on two main phases: • Risk assessment • Maintenance planning based on the risk	
3.	This company realizes that the objective of maintenance over the last 20 years has steadily shifted from a "prevention" approach to "risk"-based approach. This thinking has been spawned within the oil, gas, and petrochemical sectors by the following factors: • Increasing complexity of systems with relatively high reliability at the component level, but failures have become less predictable at the system level. • Equipment chained together into continuous processes, bigger pieces of equipment with higher capacities. Bigger losses when failures occur. • Lean processes with less in-process storage, lower inventory levels. Short duration downtime affecting output. • Rising expectations of customers wanting higher quality at lower prices in real terms. Organizations unable to adapt to these rising demands go out of business. • Workers, unions, passengers, consumers and the general public demand higher safety and environmental compliance. Industrial action, lawsuits, and consumer resistance more prevalent.	
4.	Physical asset operators and owners within the oil, gas, and petrochemical sectors are increasingly being held to a higher "standard of care" with regard to their physical assets.	
5.	Preventing all failures is generally not feasible from either a technical or an economical point of view. The airline industry attempted failure prevention until the mid-1970s and found that with the increase in technology of modern equipment, their traditional maintenance strategies did not have the desired result. Is your organization still applying these ineffective strategies or has not developed a physical asset management strategy discussed in Bench Mark Area 2? Maintenance strategies must evolve to support the technological requirements of modern equipment and the challenges of a competitive and legislated environment. To the same degree that financial auditors apply "due diligence" to the management of an organization's financial assets, maintainers must apply due diligence in the management of physical assets. To the same degree that the justice system requires "due process" in the implementation of the law, so should physical asset managers apply due process in the management of physical assets? Unfortunately, many physical asset managers are still managing their assets using gut feelings. It is no wonder that:	

22. Risk-based maintenance (RBM)—cont'd

Item#	Rating: Excellent—10, Very Good—9, Good—8, Average—7, Below Average—6, Poor—5	Rating
6.	Your maintenance strategies have evolved to support the technological requirements of modern equipment and the challenges of a competitive and legislated environment, to the same degree that financial auditors apply due diligence to the management of an organization's financial assets. Maintainers are applying due diligence in the management of physical assets to the same degree that the justice system requires due process in the implementation of the law, so should physical asset managers apply due process in the management of physical assets.	
7.	When looking closely at your company, most people will not find a disorientating mismatch between the long-term natures of your liabilities and then see an increasingly short-term nature of your assets management strategy.	
8.	Risk-based maintenance: Your maintenance strategy considers risk-based maintenance (RBM) as an evolution of RCM since RCM is based on the equipment condition and the importance of equipment to the overall process/system; but RCM is limited because it does not solve the quantification of failures.	
9.	Your company has the capability to quantify problems with probability as well level of consequences of failure and has applied RBM successfully and achieved significant direct savings or cost avoidance in oil and gas plants, petrochemical plants, power generation and power distribution networks, etc. And, it achieves important savings.	
10.	API RP 580 standard defines risk as the combination of the probability of an event occurring during a time period and the consequences associated with the event. In mathematical terms: Where Risk Factor = %Probability of failure × consequence of that failure. Does your company define risk factors to obtain an economic value (if the consequence is valued) or a classification by a risk matrix?	
11.	API considers the risk-based inspections (RBI) as the next generation of inspection interval settings, focusing attention specifically on the equipment and associated deterioration mechanisms representing the most risk to the facility. It recognizes the ultimate goal of inspection is the safety and reliability of facilities. Has your company integrated risk factors into PM/PdM inspections or stand-alone RBI tasks for critical processes?	
12.	F.I. Khan and M.M. Haddara propose an RBM methodology broken down into three modules: • Module I: Risk estimation, including a failure scenario development, a consequence assessment and a probability failure analysis; it can be conducted using fault tree analysis. • Module II: Risk evaluation, setting up acceptance criteria and applying these criteria to the estimated risk for each unit in the system. • Module III: Maintenance planning, optimizing the maintenance plan to reduce the probability of failure, reducing the total risk level of the system.	

Continued

22. Risk-based maintenance (RBM)—cont'd

Item#	Rating: Excellent—10, Very Good—9, Good—8, Average—7, Below Average—6, Poor—5	Rating
13.	Some see RBM as a simpler methodology than RCM; it also requires an initial reliability study but includes an economic risk assessment, so it allows doing financial analysis and makes it easier to choose timed-based and on-condition tasks as well as complex actions such as spare parts quantity and location, redesign of equipment, or changes in the process. Does your approach go beyond traditional RCM?	
14.	Your strategy includes maintenance based on equipment performance monitoring and the control of the corrective actions taken as a result. The real actual equipment condition is continuously assessed by the online detection of significant working device parameters and their automatic comparison with average values and performance. Maintenance is carried out when certain indicators give the signaling that the equipment is deteriorating and the failure probability is increasing. This strategy, in the long term, allows reducing drastically the costs associated with maintenance, thereby minimizing the occurrence of serious faults and optimizing the available economic resources management.	
15.	The maintenance frequency and type are prioritized based on the risk of failure. Assets that have a greater risk and consequence of failure are maintained and monitored more frequently. Assets that carry a lower risk are subjected to a less stringent maintenance program. By this process, the total risk of failure is minimized across the facility in the most economical way.	
	The monitoring and maintenance programs for high risk assets are typically CBM programs. Some use signals from wireless input to central data collectors from numerous PdM technologies within control loops or on stand-alone assets.	
16.	Risk-based maintenance is a suitable strategy for all maintenance plants as well as your site. As a methodology, it provides a systematic approach to determine the most appropriate asset maintenance plans. Upon implementation of these maintenance plans, the risk of asset failure will be acceptably low.	
17	A framework for determining risk is in place. Here the risk-based maintenance framework is applied to each system in a facility. A system, for example, may be a high-pressure vessel. That system will have neighboring systems that pass fluid to and from the vessel. The likely failure modes of the system are first determined. Then, the typical risk-based maintenance framework is applied to each risk. Is this framework present at your plant?	
18.	**Collect data**—For each risk that is being identified, accurate data about the risk needs to be collected. This will include information about the risk, its general consequences, and the general methods used to mitigate and predict the risk.	
19.	**Risk evaluation**—At the risk evaluation stage, your operation considers both the probability of the risk and the consequence of the risk, and quantified in the context of the facility/process under consideration.	

22. Risk-based maintenance (RBM)—cont'd

Item#	Rating: Excellent—10, Very Good—9, Good—8, Average—7, Below Average—6, Poor—5	Rating
20.	**Rank risks**—With the risk evaluation complete, the probability and consequence are then combined to determine the total risk. This total risk is ranked against predetermined levels of risk. As a result, your decision on risk is either acceptable or unacceptable.	
21.	**Create an inspection plan**—If the risk is unacceptable, you evaluate a plan to inspect the system using a condition monitoring approach where possible. Or, if it is more cost appropriate, and technically feasible a preventative maintenance program might be selected.	
22.	**Propose mitigation measures**—At this stage, you have a proposal for mitigating the risk, using the condition monitoring and maintenance approach.	
23.	**Reassessment**—Finally, your proposal is evaluated against other factors, such as legal and regulatory requirements. If the proposal needs are deficient, then the process starts again. Otherwise, the maintenance proposal is put into place.	
24.	Risk-based maintenance decision methods are categorized into qualitative, quantitative, and semi-quantitative method. For not long ago, risk-based maintenance is not widely applied, and most of the applications focus on the area of process plant and petroleum transport system.	
	22. Risked-based maintenance score	
Subtotal Score Possible: 240		

23. Process control and instrumentation systems technology

Item#	Rating: excellent—10, very good—9, good—8, average—7, below average—6, poor—5 or less	Rating
1.	List of all instrumentation systems sorted for criticality.	
2.	Instrument systems assigned machine unit (tracking) number along with other vital info such as location, contract service vendor, type (process, regulatory compliance, condition monitoring, etc.).	
3.	Operational instructions and data, installation drawings, catalog info, maintenance documents, and parts info on file and indexed.	
4.	Instrument system components tagged and identified according to standards.	
5.	Calibration procedures and maintenance tasks written and referenced in work order system.	
6.	Calibration records up to date, filed, and indexed. Historical records dated through previous two years or according to insurance carrier and/or legal counsel.	
7.	Critical system operational and emergency procedures written, up to date, and available to operators at control panels or other essential locations.	
8.	All hand-held and portable instruments used for operational checks and maintenance functions listed with tracking numbers and with service/calibration schedules entered into work order system.	

23. Process control and instrumentation systems technology—cont'd

Item#	Rating: Excellent—10, Very Good—9, Good—8, Average—7, Below Average—6, Poor—5	Rating
9.	All laboratory measurement equipment identified with tracking numbers, calibration stickers, and calibration schedules entered in work order system.	
	23. Process control and instrumentation systems technology score **Subtotal Score Possible: 90**	

24. Energy management and control

Item#	Rating: Excellent—10, Very Good—9, Good—8, Average—7, Below Average—6, Poor—5	Rating
1.	Energy costs are reflected in facility budget via monthly tracking of energy usage, demand, and cost by each customer location/building. Designated individual analyzes usage, demand, and costs on a monthly basis to identify and correct problems.	
2.	An energy team (or designated staff position) has been formed and chartered to deal with energy management issues and is actively involved in advising on utility conservation; its recommendations and results are published periodically.	
3.	Steam trap surveys are performed to look at operating status, inlet and outlet pressures, replacement needs, etc.; this is done on a routine basis as part of the RWP/PM program.	
4.	Air compressions have been properly sized and air leaks are routinely corrected.	
5.	Overall systems analysis of the heating and cooling systems of existing facilities is being conducted to optimize the efficiency and operation of the entire heating and cooling system.	
6.	Boilers in the facility are well maintained and periodically inspected and trimmed for maximum efficiency.	
7.	A water management strategy is in place to control process water, monitor usage and address corrosion.	
8.	A comprehensive energy audit has been conducted for the facility based upon federal energy management guidelines.	
9.	Chronic facility breakdowns and problems impacting energy are aggressively investigated to determine root causes.	
10.	New technologies such as infrared analysis are introduced to check the condition of roofs, switch gears etc.	
11.	Energy-efficient motors, lights, ballasts, etc. are used throughout the facility or are being planned as part of new facilities and major renovations.	
12.	Facility automatic energy control systems are in place and being planned for use in new facilities.	
	24. Energy management and control score **Subtotal Score: 120**	

25. Maintenance engineering and reliability engineering support

Item#	Rating: Excellent—10, Very Good—9, Good—8, Average—7, Below Average—6, Poor—5	Rating
1.	Engineering and maintenance work closely during the design and specification stages to improve equipment reliability and maintainability.	
2.	Purchase of new equipment and modifications to existing equipment are subject to maintenance review prior to final approval.	
3.	Engineering provides key support to maintenance and operations for improving equipment effectiveness.	
4.	Engineering provides key support to maintenance during installation and start-up of new equipment to ensure that operating specifications are achieved.	
5.	Engineering supports maintenance as required to troubleshoot chronic equipment problems and to define/eliminate root causes.	
6.	Engineering and maintenance work closely to develop an effective equipment and spare parts standardization program.	
7.	Capital additions, building systems changes, and facility layout changes are subject to maintenance review before final approval.	
8.	Up-to-date prints, parts/service manuals and other documentation for equipment and facility assets are available to maintenance.	
9.	Engineering coordinates material requisitioning with maintenance for project work, major overhauls, and machine building.	
25. Maintenance engineering and reliability engineering support score Subtotal Score: 90		

26. Health, safety, security, and environmental (HSSE) compliance

Item#	Rating: Excellent—10, Very Good—9, Good—8, Average—7, Below Average—6, Poor—5	Rating
1.	Maintenance leaders have created a broad-based awareness and appreciation for achieving a safe maintenance operation.	
2.	Maintenance employees attend at least one safety meeting per month.	
3.	Maintenance has shown a continual improvement in its safety record over the past five years.	
4.	All permits and safety equipment are available and prescribed for each job that requires them.	
5.	All cranes, hoists, lift trucks, and lifting equipment are inspected as part of the preventive maintenance program.	
6.	Good housekeeping within maintenance shops and storerooms is a top priority.	
7.	Maintenance tools, equipment, and leftover materials are always removed from the job site after work completion.	
8.	Maintenance continually evaluates areas throughout the operation where safety conditions can be improved.	
9.	The total scope of regulatory compliance issues within the organization has been defined and a prioritized plan of action established.	

Continued

26. Health, safety, security, and environmental (HSSE) compliance—cont'd

Item#	Rating: Excellent—10, Very Good—9, Good—8, Average—7, Below Average—6, Poor—5	Rating
10.	Maintenance responsibilities related to regulatory compliance have been well defined.	
11.	Maintenance has the technical knowledge and experience to support the organization's regulatory compliance action.	
12.	Maintenance works closely with other staff groups in the organization for a totally integrated approach to regulatory compliance.	
13.	A complete security assessment has been conducted so that accountability of all persons entering and exiting the site is well controlled.	
14.	Recommended physical barriers to unauthorized large or small vehicle entrance is in place.	
15.	Cyber security issues have been reviewed and actions for preventing system hackers are in place.	
26. Health, safety, security, and environmental compliance score Subtotal Score: 120		

27. Maintenance and quality control

Item#	Rating: Excellent—10, Very Good—9, Good—8, Average—7, Below Average—6, Poor—5	Rating
1.	Quality control has included maintenance processes within its span of factors impacting quality and has included maintenance factors within its baseline measurement of quality.	
2.	Major repairs and setups impacting quality have clearly defined procedures and specifications established.	
3.	Documentation of all equipment conditions, factors, and settings that contribute to quality performance is available.	
4.	Quality of maintenance repairs is evaluated and used as a key performance indicator and as a means to validate crafts skills.	
5.	Maintenance and quality work together with close coordination and cooperation to resolve quality issues related to maintenance processes.	
6.	Optimum machine speeds have been established and included in setup procedures and operator training.	
7.	All machine-related quality defects are aggressively evaluated and corrected.	
8.	Losses due to minor stoppages, idling, and minor equipment failures are addressed by operations and maintenance for corrections.	
9.	Chronic equipment breakdowns and problems are aggressively investigated as to cause.	
27. Maintenance and quality control score Subtotal Score: 90		

28. Maintenance performance measurement

Item#	Rating: Excellent—10, Very Good—9, Good—8, Average—7, Below Average—6, Poor—5	Rating
1.	Maintenance performance measurement includes a wide range of performance indicators in order to evaluate the total effectiveness and impact of maintenance service throughout the operation to include craft labor, planning/scheduling, P.M., asset reliability, equipment effectiveness, and cost, etc.	
2.	Maintenance labor and material costs are reported monthly and reviewed against previous costs or budgeted costs to evaluate current trends.	
3.	Equipment downtime attributable to maintenance is monitored. The cost of downtime for major pieces of equipment or processes is known and used to measure value of increased equipment uptime.	
4.	Realistic labor performance standards/estimates have been developed and used for all planned work and recurring tasks.	
5.	Maintenance labor performance is reported monthly or weekly to evaluate actual performance against established performance standards.	
6.	The measurement of craft utilization is available from the labor reporting system to evaluate productive hands-on, wrench time versus nonproductive craft time.	
7.	Periodic reviews are done to evaluate the maintenance operation by determining overall craft utilization and the nature of delays and nonproductive time such as waiting for parts, instructions, unbalanced crew, or waiting for equipment, etc.	
8.	The effectiveness of maintenance planning is evaluated by factors such as percent work orders planned versus total work orders, percent work orders completed as planned versus total planned work orders, and percent work orders with estimates versus total work orders completed.	
9.	Baseline performance factors and information is available to evaluate all ongoing improvements against past performance. Periodic reports to summarize and highlight the tangible benefits from continuous maintenance improvement are provided.	
10.	A method to measure performance of contracted services is in place.	
11.	The craft workforce understands the need for improved craft labor productivity and the challenge to remain competitive with contract service providers.	
12.	Maintenance performance measures are linked to operational performance and support total operations success and profit optimization.	
	28. Maintenance performance measurement score	
Subtotal Score: 120		

29. Computerized maintenance management systems (CMMS) and business systems

Item#	Rating: Excellent—10, Very Good—9, Good—8, Average—7, Below Average—6, Poor—5	Rating
1.	The identification of specific CMMS functional requirements has been clearly defined, and a complete definition of system capabilities has been determined based on the size and type of maintenance operation.	
2.	Equipment (asset) history data complete and accuracy is 95% or better.	
3.	Spare parts inventory master record accuracy is 95% or better.	
4.	Bill of materials for critical equipment includes listing of critical spare parts.	
5.	Preventive maintenance tasks/frequencies data complete for 95% of applicable assets.	
6.	Direct responsibilities for maintaining parts inventory database are assigned.	
7.	Direct responsibilities for maintaining equipment/asset database are assigned.	
8.	Initial CMMS training for all maintenance employees with ongoing CMMS training program for maintenance and storeroom employees.	
9.	Adequate support from supplier and consultants is budgeted to ensure a successful start-up.	
10.	Customization of the CMMS is planned to accommodate specific needs for part numbers, equipment numbers, work orders, and management report formats, etc.	
11.	Training for CMMS is a top priority and will be established as an ongoing process for new and existing users of the system.	
12.	System outputs have been developed into a maintenance information system that provides management reports to monitor a wide range of factors related to labor, material, equipment costs, etc.	
13.	A CMMS systems administrator (and backup) is designated and trained.	
14.	Inventory management module fully utilized and integrated with work order module.	
15.	Reorder notification for stock items is generated and used for reorder decisions.	
16.	CMMS provides MTBF, MTTR, failure trends, and other reliability data.	
17.	Engineering changes related to equipment/asset data, drawings, and specifications are effectively implemented.	
18.	Maintenance standard task database is available and used for recurring planned jobs.	
29. Computerized maintenance management systems (CMMS) and business systems score Subtotal Score Possible: 180		

30. Shop facilities, equipment, and tools

Item#	Rating: Excellent—10, Very Good—9, Good—8, Average—7, Below Average—6, Poor—5	Rating
1.	Maintenance shop facilities are located in an ideal location with adequate space, lighting, and ventilation.	

30. Shop facilities, equipment, and tools—cont'd

Item#	Rating: Excellent—10, Very Good—9, Good—8, Average—7, Below Average—6, Poor—5	Rating
2.	Standard tools are provided to craft employees and accounted for by a method that ensures good accountability and control.	
3.	An adequate number of specialty tools and equipment are available and easily checked out through a tool control procedure.	
4.	All personal safety equipment necessary within the operation is provided and used by maintenance employees.	
5.	Safety equipment for special jobs such as confined space entry, electrical system lockout, etc., is available and used.	
6.	Maintenance achieves a high level of housekeeping in its shop areas.	
7.	Maintenance maintains a broad awareness of new tools and equipment to improve methods, craft safety, and performance.	
8.	Maintenance continually upgrades tools and equipment to increase craft safety and performance.	
9.	An effective process to manage special tools and equipment inventory is in place.	
	30. Shop facilities score	

Subtotal Score: 90

31. Continuous reliability improvement

Item#	Rating: Excellent—10, Very Good—9, Good—8, Average—7, Below Average—6, Poor—5	Rating
1.	Continuous reliability improvement is recognized as an important strategy as evidenced by the current status of maintenance and the ongoing improvement activities.	
2.	The current level of commitment to continuous reliability improvement is based on results of overall assessment.	
3.	Maintenance improvement opportunities have been identified with potential costs and savings established.	
4.	Improvement priorities have been established based on projected benefits and valid economic justifications.	
5.	Top leaders have reviewed, modified, and/or approved maintenance improvement priorities and have made a commitment to action.	
6.	Sufficient resources (time, dollars, and staff) have been established to address priority areas.	
7.	Implementation plans and leaders for each priority area are established.	
8.	A team-based approach is used to identify and implement practical solutions to maintenance improvement opportunities.	
9.	Continuous reliability improvement for the maintenance resource of physical asset and equipment resource is rated as:	
10.	Continuous reliability improvement for the maintenance resource of craft labor resources is rated as:	

Continued

31. Continuous reliability improvement—cont'd

Item#	Rating: Excellent—10, Very Good—9, Good—8, Average—7, Below Average—6, Poor—5	Rating
11.	Continuous reliability improvement for the maintenance resource of MRO material resources is rated as:	
12.	Continuous reliability improvement for the maintenance resource of information resources is rated as:	
13.	Continuous reliability improvement for the maintenance resource of technical skills and knowledge is rated as:	
14.	Maintenance employees participate on functional teams within maintenance and on cross-functional teams with other department employees to develop maintenance improvements.	
15.	Written charters are established for each team to outline reasons for the team, process to be used, resources available, constraints, expectations, and results expected.	
Subtotal Score: 150	**31. Continuous reliability improvement score**	

32. Critical asset facilitation and overall equipment effectiveness (OEE)

Item#	Rating: Excellent—10, Very Good—9, Good—8, Average—7, Below Average—6, Poor—5	Rating
1.	Overall equipment effectiveness (OEE) ratings have been established for major equipment assets or processes to provide a baseline measurement of availability, performance, and quality.	
2.	Priorities have been established with a plan of action for improving OEE.	
3.	The OEE factor of availability is being measured and methodology rated as:	
4.	The OEE factor of performance is being measured and methodology is rated as:	
5.	The OEE factor of quality is being measured and methodology is rated as:	
6.	Equipment improvement teams have been established to focus on improving equipment effectiveness based on established priorities for critical equipment.	
7.	Improvements in OEE are evaluated against baseline OEE measurements to determine progress.	
8.	Optimum machine speeds have been established and included in setup procedures and operator training.	
9.	All machine-related quality defects are aggressively evaluated and corrected.	
10.	A process for critical asset facilitation has been implemented.	
11.	Critical asset facilitation for condition-based maintenance includes CBM technologies and hardware requirements, durations, and/or hours for each routine, frequencies for each routine, necessary tools and equipment for each routine, and perform/record baseline of test and measures.	

32. Critical asset facilitation and overall equipment effectiveness (OEE)—cont'd

Item#	Rating: Excellent—10, Very Good—9, Good—8, Average—7, Below Average—6, Poor—5	Rating
12.	Critical asset facilitation includes developing preventive maintenance plan for each craft/system, inspections/service requirements in sequence of events, schedules for lube and filter routes, duration for each routine by subsystem level and by shift, necessary parts, tools, and equipment for each task, and identification of facilitated asset in CMMS.	
13.	Critical asset facilitation includes a complete review of required documents, manuals, and drawing required to maintaining and operating the asset.	
14.	Critical asset facilitation includes visual management such as labeling of signal or alarm functions/operations, direction of rotation (drives, chains, motors) and replacement parts #, proper control adjustments (pressure, temp., speed, level, voltage, etc., energy lockouts (electrical, hydraulics, compressed air, etc.), electrical conductors, functions/designations of source on cabinets, panels, boxes, switches, valves, buttons, light, etc, all fixable, adjustable, or critical fasteners, filters functions (hydraulic, lube, air, etc.) and replacement part #, normal operating ranges/levels/readings and label function monitored, lube point with product number, major component functions (coolant pump, exhaust fan, etc.), motor function being powered (pump, axis drive, screw, conveyor, etc.), fluid type/direction of flow/pressure on pipes/hoses/lines.	
15.	Critical asset facilitation includes facilitated asset documentation material with the following: list of systems, subsystems and components, CBM routines, CBM baselines, PM routines, lube and filter routes, list of prints, manuals, etc. for PM and CBM routines, visual management digital images, job safety analysis, MSDS, lockout/tagout documentation, craft asset specific training/certification requirements, part list and/or bill of material for PM routines, start-up and shut-down procedures, final operator checklist, documented alignment and test procedures.	
32. Critical asset facilitation and overall equipment effectiveness (OEE) score Subtotal Score Possible: 150		

33. Overall craft effectiveness (OCE)

Item#	Rating: Excellent—10, Very Good—9, Good—8, Average—7, Below Average—6, Poor—5	Rating
1.	Improvement of craft productivity and overall craft effectiveness (OCE) is being established as an important element of the overall maintenance improvement process.	
2.	Priorities have been established with a plan of action for improving the OCE factor for craft utilization (wrench time).	
3.	Priorities have been established with a plan of action for improving the OCE factor for craft performance.	

Continued

33. Overall craft effectiveness (OCE)—cont'd

Item#	Rating: Excellent—10, Very Good—9, Good—8, Average—7, Below Average—6, Poor—5	Rating
4.	Priorities have been established with a plan of action for improving the OCE factor for craft service and quality.	
5.	Improvements in OCE and craft productivity are evaluated against baseline measurements to determine progress.	
6.	A team effort is being used that involves the craft workforce to provide a cooperative effort for improving OCE and eliminating a reactive, fire-fighting strategy that wastes valuable craft resources.	
	33. Overall craft effectiveness (OCE) score	
Subtotal Score: 50		

34. Sustainability

Item#	Rating: Excellent—10, Very Good—9, Good—8, Average—7, Below Average—6, Poor—5	Rating
1.	The organization understands that the organizing principle for sustainability is sustainable development, which includes the four interconnected domains: ecology, economics, politics, and culture.	
2.	Ways of reducing negative human impact are environmentally friendly chemical engineering, environmental resources management, and environmental protection that have been demonstrated or being planned.	
3.	Sustainability is derived from the Latin *sustinere* (*tenere*, to hold; *sub*, up). *Sustain* can mean "maintain," "support," or "endure." Sustainable development is present in regard to maintenance and physical asset management to meet the needs of the current operation without compromising the ability of future site operations to meet their own need to maintain physical assets.	
4.	Top leaders fully understand that sustainability within physical asset management concerns the specification or strategy with a set of best practices to be undertaken by a present operations staff. It is a strategy that will not diminish the prospects of future operations to enjoy levels of throughput, profits, uptime, utility, or welfare comparable to those enjoyed by the present operation.	
5.	The organization's philosophical and analytic framework of sustainability draws on and connects with many different disciplines and fields and has emerged as a sustainability science with internal or external staff to support or be directly responsible for it.	
6.	There is evidence that the organization is developing fewer carbon-hungry technology and transport systems and attempts by individuals to lead carbon-neutral lifestyles. There is monitoring of the fossil fuel use embodied in all the goods and services they use. Engineering of emerging technologies such as carbon-neutral fuel and energy storage systems such as power to gas, compressed air energy storage, and pumped-storage hydroelectricity are necessary to store power from transient renewable energy sources including emerging renewables such as airborne wind turbines.	

34. Sustainability—cont'd

Item#	Rating: Excellent—10, Very Good—9, Good—8, Average—7, Below Average—6, Poor—5	Rating
7.	Water efficiency is being improved on an organizational scale by increased demand management, improved infrastructure, improved plant control, and reuse of water such as condensate water. People are becoming more self-aware, self-sufficient by harvesting rainwater and reducing use of their water system.	
8.	Sustainable use of some materials has targeted the idea of dematerialization, converting the linear path of materials (extraction, use, disposal in landfill) to a circular material flow that reuses materials as much as possible via recycling.	
9.	Top leaders understand that the effects of some chemical agents need long-term measurements to avoid legal battles in regard to their danger to human health per classifications of toxic carcinogenic agents defined by the International Agency for Research on Cancer.	
10.	Top leaders realize and strive for a green economy defined as one that improves economic impact, human well-being and social equity, while significantly reducing environmental risks and ecological scarcities while not favoring one political perspective over another but works to minimize excessive depletion of natural capital.	
11.	Top leaders have benchmarked with research from areas such the World Council for Sustainable Development focused on progressive corporate leaders who have embedded sustainability into commercial strategy, thus yielding a leadership competency model for sustainability. The expansion of sustainable business opportunities has contributed to job creation through the introduction of green-collar workers.	
	34. Sustainability score	
Subtotal Score Possible: 110		

35. Traceability

Item#	Rating: Excellent—10, Very Good—9, Good—8, Average—7, Below Average—6, Poor—5	Rating
1.	*Traceability.* Your organization understands traceability to mean the property of the result of a measurement or the value of a standard whereby it can be related to stated references, usually national or international standards, through an unbroken chain of comparisons all having stated uncertainties.	
2.	*Measuring and testing equipment.* These devices are calibrated against certified measurement standards, which have known valid relationship/*traceability* to national standards at established periods to assure continued accuracy.	

Continued

35. Traceability—cont'd

Item#	Rating: Excellent—10, Very Good—9, Good—8, Average—7, Below Average—6, Poor—5	Rating
3.	Your organization uses International Laboratory Accreditation Cooperation (ILAC) to allow labs to use suppliers accredited by other organizations that have reciprocity agreements. Here different assessment organizations assess each other on a regular basis and the web of reciprocal organizations is large. In the United States, there are a number of organizations that have wide reciprocity with both the European (EA) and Pacific (APLAC) mutual recognition bodies.	
4.	In order to have reciprocity, you have an essential agreement of what the standard (ISO 17025) means, and this includes traceability. Traceability has gone from being a U.S. government contractual term to an internationally recognized regulatory term, defined in detail by ILAC.	
5.	Elements of Traceability—Traceability is characterized by a number of essential elements: • An unbroken chain of comparisons going back to a standard acceptable to the parties, usually a national or international standard • Measurement uncertainty; the measurement uncertainty for each step in the traceability chain must be calculated according to defined methods and must be stated so that an overall uncertainty for the whole chain may be calculated*Documentation.* Each step in the chain must be performed according to documented and generally acknowledged procedures; the results must equally be documented.	
6.	Does your organization realize that the inevitability of traceability in the oil and gas sector is now following a reputational crisis following the BP Deepwater Horizon oil spill, the WikiLeaks disclosures and recent events around the Keystone XL oil pipeline, and controversy in the United Kingdom over the European fuel quality regulation.	
7.	Has your site viewed traceability as a key feature of the rising tide of transparency and accountability, as businesses, customers, and consumers become more discerning in their choice of fuel.	
8.	The growth of traceability within the industry looks set to focus on so-called "unconventional oil production," which has greater environmental and social impacts than conventional fossil fuels.	
9.	The evidence is already there that the trend of traceability is playing out in the purchasing decisions of some leading businesses. Retailers such as Timberland, Walgreens, and Bed Bath & Beyond have pledged to avoid using unconventional oil derived from oil sands.	
10.	To date, the oil and gas industry has taken a rather predictable line of defense: crude oil is fungible and traded as a commodity, and it is not practical to trace a final product back to its source.	
11.	The position has been enabled by the lack of disclosure from the oil and gas companies themselves regarding the derivation of their products.	

35. Traceability—cont'd

Item#	Rating: Excellent—10, Very Good—9, Good—8, Average—7, Below Average—6, Poor—5	Rating
12.	But this looks set to change as nongovernmental organization campaigns gather speed, developments in science and technology unfold, and regulations kick in.	
13.	Forest Ethics is helping companies trace the fuel their shipping suppliers use back to specific refineries. Players like Greenpeace, World Wildlife Fund, Friends of the Earth, and the Pembina Institute are highlighting the damaging environmental and social consequences of unconventional oil, and are pressuring both businesses and consumers on the oil sands issue.	
14.	In parallel, emerging technology is enabling the traceability of crude oil back to its origin based on the product's very specific chemical composition. The science has already been proven in the Gulf of Mexico where the chemical "fingerprinting" of oil following Deepwater Horizon enabled investigators to determine that it indeed came from the Macondo well.	
15.	Regulatory action is another potential challenge to the sector. The European fuel quality regulation is set to designate transport fuel from tar sands as resulting in 22% more greenhouse gas emissions than from conventional fuels. According to the *Guardian* newspaper, this would "make suppliers, who have to reduce the emissions from their fuels by 10% by 2020 very reluctant to include in their fuel mix."	
16.	Low-carbon fuel standards are emerging in markets around the world.	
17.	The overriding risk for oil companies is that, as traceability develops through market or regulatory action, they will be caught on the back foot, defensively attempting to minimize the reputational and financial loss that can come from investment in unconventional oil.	
18.	In the worst case, unconventional assets will be downgraded by investors or even entirely stranded if markets discriminate against them.	
19.	The key message is to jump before you are pushed, and competitive advantage is likely to emerge for those companies that sell "oil sands free" fuels with appropriate branding and verified sourcing. In order to have reciprocity, there must be essential agreement of what the standard (ISO 17025) means, and this includes traceability. Traceability has gone from being a U.S. government contractual term to an internationally recognized regulatory term, defined in detail by ILAC.	
	35. Traceability score	
Subtotal Score Possible: 190		

36. Process safety management (PSM) and management of change (MOC)

Item#	Rating: Excellent—10, Very Good—9, Good—8, Average—7, Below Average—6, Poor—5	Rating
1.	Several PSM regulations (for example, OSHA process safety management (PSM) in the United States, SEVESO SMS in Europe) and industry guidelines drive PSM practices. Your company has in-house staff or contracted services to help your company comply with these PSM regulatory requirements as well as established industry sector standards.	
2.	Your PSM fits seamlessly into your business models and operations, and PSM is embraced and has become a real part of your company's culture with it being successfully implemented throughout the lifecycle of operations to lower risk.	
3.	Your in-house staff or PSM consultants are process, chemical, and safety engineers with a strong industry background complemented by in-depth expertise in process safety.	
4.	How do you rate your PSM development, implementation, and improvement?	
5.	How do you rate your PSM auditing and gap identification?	
6.	How do you rate your PSM training and process safety culture?	
	How do you rate your process hazard analysis? (HAZOP, LOPA, HAZID, What-if, FMEA, Fault Tree)	
7.	How do you rate your quantitative risk assessment, consequence modeling, and blast effects?	
	How do you rate your process safety information and laboratory testing?	
8.	How do you rate your pre start-up safety reviews?	
9.	How do you rate your permitting and emergency plans?	
10.	How do you rate handling your major hazards safety cases and SEVESO compliance?	
11.	How do you rate your technical and organizational prevention measures?	
12.	How do you rate your incident investigation, expert witness, and litigation support?	
13.	How do you rate your company having in-house PSM staff or consulting staff for PSM services that are available in a wide variety of languages as required, including English, Spanish, French, German, Portuguese, Italian, Arabic, and Hindi.	
Note: All evaluation items starting at #14 may be included in your management of change (MOC) process procedure, and some may not be included depending on the scope of a change.		
14.	How well does your company define a proper description and the purpose of the change?	
15.	How well does your company define the technical basis for the change?	
16.	How well does your company define health, safety security, and environmental considerations?	
17.	Documentation of changes for the operating procedures.	
18.	How well does your company define proactive maintenance procedures?	

36. Process safety management (PSM) and management of change (MOC)—cont'd

Item#	Rating: Excellent—10, Very Good—9, Good—8, Average—7, Below Average—6, Poor—5	Rating
19	How well does your company define inspection and testing requirements?	
20.	How well does your company define piping and instrument diagrams (P&ID) and update after changes?	
21.	How well does your company define electrical classification of changes?	
22.	How well does your company define the true training needs for operations and maintenance and ensure good communications on changes across the entire site or overall organization?	
23.	How well does your company define prestart-up inspection and ensure they are enforced by operations?	
24.	How well does your company define duration if it is a temporary change?	
25.	How well does your company redefine and update risk management plans?	
26.	How well does your company define how changes must be approved and authorization levels required and any other requirements due to the operating context of the change?	
36. Process safety management (PSM) and management of change (MOC) score Subtotal Score Possible: 260		

37. Risk-based inspections (RBI) and risk mitigation

Item#	Rating: Excellent—10, Very Good—9, Good—8, Average—7, Below Average—6, Poor—5	Rating
1.	The organization's chief reason for implementing an RBI program is to help manage the risk of a complex and potentially dangerous system.	
2.	Has your organization had a major event that was dangerous to HSSE requirements in the last five years. If No = 10; If Yes rate based on scope i.e., 5, 6, 7, 8, or 9. Remember that equipment failures cost just the U.S. refining industry over $4 billion per year. Two-thirds of those costs are associated with failures of static equipment; the average refinery has a risk of $25 million per year due to failures of static equipment and two-thirds of those costs are associated with piping failures.	
3.	A good RBI program can also help an owner/operator understand if the plant is not being run as designed or if a new feedstock is causing more harm than expected to remaining life of equipment.	
4.	By using RBI as a tool, your organization is helping owner/operators increase turn-around schedules and avoid shutdown inspection activities while maintaining production.	
5.	Your organization understands why predicting any failure prior to occurrence, whether catastrophic or minor, can have a large impact on reputation in the market space and the community, and using tools like RBI can help owner.	

Continued

37. Risk-based inspections (RBI) and risk mitigation—cont'd

Item#	Rating: Excellent—10, Very Good—9, Good—8, Average—7, Below Average—6, Poor—5	Rating
6.	Primary cost savings are moving from a time-based to a risk-based approach often with less inspection but also applying more effective inspection techniques to look for the appropriate damage mechanism. Your organization is focusing on resources on the 5–10% of the assets that are driving the bulk of the risk.	
7.	You have seen the cost benefit of RBI require 50–90% fewer inspection points than a traditional API inspection program while reducing the risk of failures by 80–95%.	
8.	You have seen other benefits of RBI that extend plant shutdown intervals and reduce the number of equipment items being opened by 30–60%. And over a five-year period, the benefit of RBI is generally 5–20 times the cost of implementation and management.	
9.	From a HSSE standpoint, a good RBI approach will force an owner/operator to have a better understanding of their key hazardous processes and provide a solid mitigation program. Has this been shown in your company?	
10.	You have seen and have achieved greater HSSE awareness and compliance by the multidisciplinary approach to the RBI process when operations and process engineers, along with maintenance teams, provide their input on how the plant works. This allows for the development of inspection plans that truly address where corrosion and cracking is likely to occur.	
11.	Your approach to good RBI has helped top leaders down to the operators understand better the corrosion mechanism and deterioration rates and in doing so predict leaks prior to happening.	
12.	An extension of a good RBI program supports your physical asset integrity operating windows and also helps the mechanical integrity departments understand process changes that accelerate through-wall corrosion events.	
13.	One of the primary objectives of PSM is to minimize the risk of release of highly hazardous chemicals, which has not occurred at your site in the last five years. 5 years+ = 10.	
14.	All categories of equipment that contain or control a hazardous process, including piping systems, must be covered in a mechanical integrity program.	
15.	Your organization is going through a process implementing a good RBI program and will allow your organization to: (a) develop a systematic process for understanding hazardous processes, (b) determine mitigation plans, and (c) create an inventory and analysis for all PSM-related equipment and have a good documentation trail of inspections, decisions, and nonconformances.	
16.	Your organization is going through implementing a good RBI program that will allow your organization to develop a systematic process for understanding hazardous processes and determining mitigation plans.	

37. Risk-based inspections (RBI) and risk mitigation—cont'd

Item#	Rating: Excellent—10, Very Good—9, Good—8, Average—7, Below Average—6, Poor—5	Rating
17.	You have a good documentation trail of inspections, decisions, and nonconformances.	
18.	You have created an inventory and informed analysis for all PSM-related equipment and have a well-documented audit trail of all inspections, decisions, and nonconformances.	
19.	Your organization is in compliance with the new OSHA and EPA documents that require inspection and test procedures following recognized and generally accepted good engineering practices defined in Appendix C of the OSHA regulations, good engineering practice areas that are described.	
20.	The codes and standards are relied upon to establish good engineering practice within your organization. These codes and standards are published by organizations as the American Society of Mechanical Engineers, American Petroleum Institute, American National Standards Institute, National Fire Protection Association, American Society for Testing and Materials, National Boiler and Pressure Vessel Inspectors, National Association of Corrosion Engineers, American Society of Exchanger Manufacturers Association, and model building code.	
21.	Various engineering societies issue technical reports that impact process design. For example, the American Institute of Chemical Engineers has published technical reports on topics such as two-phase flow for venting devices. This type of technically recognized report is being used by your organization for guidelines toward good engineering practice.	
22.	The recently enacted Environmental Protection Agency regulation EPA 40 CFR 68 for Accident Release Prevention is being used and closely parallels the OSHA regulation, but adds a new section (Subpart G) titled Risk Management Plan. Are you in compliance with this section by having an offsite consequence analysis to calculate worst-case release scenarios for the site? The EPA has defined a consequence analysis of the same type used in risk-based inspection.	
23.	Worst-case release is defined and used as being "the release of the largest quantity of a regulated substance from a vessel or process line failure that results in the greatest distance to an endpoint…"	
24.	The shortcoming of a worst-case scenario approach is understood because of its overemphasis on the failure consequence without regard to the probability of the event occurring. Your organization realizes that true risk management evaluates both the consequence and probability of the event, and the combination determines the risk.	
25.	The API codes covering the inspection of fixed equipment and piping are clearly recognized and generally accepted as good engineering practices in their respective areas. They have recently been revised to acknowledge the use of risk-based methods in inspection programs. The API-570 Piping Inspection Code was the first API recommended practice to take a rudimentary risk management approach, by categorizing piping based on the flammability or toxicity of its contents.	

Continued

37. Risk-based inspections (RBI) and risk mitigation—cont'd

Item#	Rating: Excellent—10, Very Good—9, Good—8, Average—7, Below Average—6, Poor—5	Rating
26.	Class 1 is the highest consequence category, and Class 3 is the lowest. The required coverage and frequency of inspections is based on the fluid categories. In a refinery, most of the process lines are Class 2, based on flammability. The weakness of the API-570 classification scheme is that it uses contents as the sole basis for determining categories. Your organization understands the weaknesses of certain areas of API-570.	
27.	As an example, the API-570 classification does not discriminate between a 10-inch line full of liquid propane (Class 1) and a 2-inch propane vapor line (Class 1). The RBI approach based on the API Risk Based Inspection methodology (API 590 and 581) does provide significant discrimination between the two propane lines, however.	
28.	API 580 requirements for an RBI program are, from a management system point of view: vision/mission, policies, procedures, risk acceptance criteria, personnel competencies, document and data management of change (MOC), etc., documented method for probability of failure determination, and documented method for consequence of failure determination.	
29.	Documented methods are in place for managing risks through inspections and other mitigation activities (e.g., operating conditions, inventory control or release containment, etc.)	
	37. Risk-based inspections and risk mitigation score Subtotal Score Possible: 290	

38. Pride in maintenance

Item#	Rating: Excellent—10, Very Good—9, Good—8, Average—7, Below Average—6, Poor—5	Rating
1.	Overall, has your crafts workforce displayed evidence that they were indeed proud of their profession?	
	If your craft employees were independently surveyed, would they reveal that their positions were essential to company success and are fully a part of the mission-vision-values of the organization?	
2.	Your organization has held events to recognize employees on their job achievement as well as outstanding service to their community.	
3.	Your organization has held or sponsored activities such as a softball team, a bowling league, golf league, or soccer team or any events for all of maintenance to enjoy off-site as teams outside of work hours.	
4.	Turnover related to technicians moving to another company has been normal compared to your expectations.	
5.	Subject matter experts, well before official retirement, are able to influence and train new employees with a positive attitude and with special recognition from the company.	

38. Pride in maintenance—cont'd

Item#	Rating: Excellent—10, Very Good—9, Good—8, Average—7, Below Average—6, Poor—5	Rating
6.	Technicians at other plants desire to work at your site due to its positive culture and environment as well as its pay scales being competitive in your area.	
7.	Craft employees are all given opportunities to contribute their skills and ideas for improving site reliability.	
8.	*Pride in maintenance* is a basic philosophy embraced and practiced by a majority of the staff in the maintenance and storeroom teams.	
	38. Pride in maintenance score	
Subtotal Score Possible: 80		

Appendix B

Acronyms and Glossary of Maintenance, Maintenance Repair Operations Stores/ Inventory, and Oil and Gas Terms

List of Acronyms

API   American Petroleum Institute
APL   Applications parts list
BBL   Barrel
BCF   Billion cubic feet of natural gas
BITE   Built-in test equipment
BOE   Barrel of oil equivalent
BOED   Barrels of oil equivalent per day
BOM   Bill of materials
BTU   British thermal unit
CAGR   Compound annual growth rate
CBM   Coal bed methane
CBM   Conditioned-based maintenance
CCF   Centum cubic feet
CM   Corrective maintenance
CMMS   Computerized maintenance management system
CMO   Chief maintenance officer
CNG   Compressed natural gas
CO$_2$e   Carbon dioxide equivalents
CPM   Critical path method
CRI   Continuous reliability improvement
CSS   Carbon capture and storage
DT   Downtime
E&P   Exploration and production
ECN   Engineering change notice
EOQ   Economic order quantity
EOR   Enhanced oil recovery
EUR   Estimated ultimate recovery
EWO   Engineering work order
FEED   Front-end engineering and design
FIFO   First in–first out
FISH   First in still here
FMEA   Failure modes and effects analysis
FMECA   Failure modes, effects, and criticality analysis

FPSO   Floating production, storage, and offloading
FTA   Fault tree analysis
GCR   Gas cost recovery
GSE   General support equipment
GWP   Global warming potential
HSSE   Health, safety, security, and environment
IOR   Improved oil recovery
JIT   Just-in-time
JOA   Joint operating agreement
KPI   Key performance indicator
LCA   Life cycle analysis
LCC   Life cycle cost
LDC   Local distribution company
LIFO   Last in–first out
LNG   Liquefied natural gas
LOR   Level of repair
LSA   Logical support analysis
MBBL   One thousand barrels of crude oil, bitumen, condensate, or natural gas liquids
MBD   One thousand barrels per day
MBOE   One thousand barrels of oil equivalent
MCF   One thousand cubic feet
MIL-STD   United States military standard
MMBBL   One million barrels
MMBOE   One million barrels of oil equivalent
MMBTU   One million British thermal units
MMCF   One million standard cubic feet of natural gas
MOC   Management of change
MRO   Maintenance repair operations
MTBF   Mean time between failures
MTPA   Millions of tons per annum
MTTR   Mean time to repair
NDT   Nondestructive testing
NGLs   Natural gas liquids
NGV   Natural gas vehicle
NYMEX   New York Mercantile Exchange
OBM   Operator-based maintenance
OCE   Overall craft effectiveness
OEE   Overall equipment effectiveness
PCCM   Profit and customer-centered maintenance
PCM   Profit-centered maintenance
PdM   Predictive maintenance
PERT   Projection evaluation and review technique
PM   Preventive maintenance
PMRPW   Preventive maintenance–related planned work
PRA   Probabilistic risk assessment
PSA   Probabilistic safety assessment
Psi   Pounds per square inch
PSM   Process safety management
PSC   Production sharing contract
RBI   Risk-based inspection

RBM   Risk-based maintenance
RCM   Reliability-centered maintenance
RIME   Ranking index for maintenance expenditures
ROCE   Return on capital employee
ROI   Return on investment
ROMI   Return on maintenance investment
ROP   Reorder point
RPI   Reliability performance indicator
S-LCA   Social life cycle analysis
SAGD   Steam-assisted gravity drainage
SKU   Stock keeping unit
TCF   One trillion cubic feet of natural gas
TCO   Total cost of ownership
ToSS   Total system support
TPM   Total productive maintenance
TRR   Total recordable rate
UM   Unscheduled maintenance
WO   Work order
YTD   Year-to-date

Glossary

ABC inventory policy   A collection of prioritizing practices to give varied levels of attention to different classes of inventories. For example, Class A items typically make up 15 to 25% of stock items but 75 to 85% of inventory value. Class C items, in turn, might be 60% of the stock items but only 10% of the inventory value. Class B items would be somewhere in between these two.

ACE system   Today's best methodology for establishing quality repair methods and team-based planning times (see **ACE Team Benchmarking Process**).

ACE Team Benchmarking Process (also known as the **ACE Team Process**)   A propriety system for maintenance work measurement developed by The Maintenance Excellence Institute International in the 1980s. It is a team-based process for using experienced craftspeople, supervisors, and planners to develop a consensus on maintenance repair method improvement, job task times, and total work content/wrench time. It involves a consensus of experts (ACE) for first evaluating a job for improved methods, tools, special equipment, and root cause elimination. Secondly, this process develops a consensus on work content and then spreadsheets for work content comparison (slotting). The process provides a methodology for a planner to include various job-specific allowances to wrench time and then develop reliable planning times for scheduling purposes. The ACE Team process allows for a wide range and number of jobs to be estimated using a relatively small number of benchmark jobs arranged on spreadsheets by craft areas.

Acreage   Land leased for oil and gas exploration and/or land for which a company owns the mineral rights.

Actuarial analysis   Statistical analysis of failure data to determine the age-reliability characteristics of an item.

Adjustments   Minor tune-up actions requiring only hand tools, no parts, and usually lasting less than a half hour.

American Petroleum Institute (API)   The American Petroleum Institute is the oil and gas industry's trade organization. API's research and engineering work provides a basis for establishing operating and safety standard issues and specifications for the manufacturing of oil field equipment and furnishes statistical and other information to related agencies. Visit API at www.api.org.

Anticline   A convex-upward formation of rock layers, which may form a trap for hydrocarbons.

Applications parts list   A list of all parts required to perform a specific maintenance activity, typically set up as a standard list attached to a standard job for routine tasks. Not to be confused with a bill of materials.

Apprentice   A craftsperson in training, typically following a specifically defined technical training program while gaining hands on experience in a craft area to gain a specified number of hours experience.

Aquifer   An underground layer of water-bearing permeable rock or unconsolidated materials (gravel, sand, silt, or clay) from which groundwater can be extracted using a water well.

Area maintenance   A method for organizing maintenance operations in which the first-line maintenance leader is responsible for all maintenance crafts within a certain department, area, or location within the facility.

Asset care   An alternative term for the maintenance process. A kinder, gentler term but still pure maintenance and physical asset management.

Asset list   A register of physical assets (equipment, facilities, building systems, etc.) usually with information on manufacturer, vendor, specifications, classification, costs, warranty, and tax status.

Asset management   The systematic planning and control of a physical asset resource throughout its economic life; the systematic planning and control of a physical resource throughout its life. This may include the specification, design, and construction of the asset; its operation, maintenance and modification while in use; and its disposal when no longer required.

Asset number   A unique alphanumerical identification of an asset list, which is used for its management.

Asset register   A list of all the assets in a particular workplace, together with information about those assets, such as manufacturer, vendor, make, model, specifications, etc.

Asset utilization   The percentage of total time the equipment/asset is running.

Assets   The physical resources of a business, such as plant equipment, facilities, building systems, fleets, or their parts and components; unlike in the accounting definition, in maintenance this is commonly taken to be any item of physical plant or equipment.

Availability   The probability that an asset will, when used under specified conditions, operate satisfactorily and effectively. Also, the percentage of time or number of occurrences for which an asset will operate properly when called upon; the proportion of total time that an item of equipment is capable of performing its specified functions, normally expressed as a percentage. It can be calculated by dividing the equipment available hours by the total number of hours in any given period.

Available hours   The total number of hours that an item of equipment is capable of performing its specified functions. It is equal to the total hours in any given period, less the downtime hours.

Average life   How long, on average, a component will last before it suffers a failure. Commonly measured by mean time between failures (MTBF).

Backlog   Work orders planned and prioritized, awaiting scheduling and execution; work that has not been completed by the "required by" date. The period for which each work order is overdue is defined as the difference between the current date and the "required by" date. All work for which no "required by" date has been specified is generally included on the backlog. Backlog is generally measured in crew-weeks: the total number of labor hours represented by the work on the backlog, divided by the number of labor hours available to be worked in an average week by the work crew responsible for completing this work. As such, it is one of the common key performance indicators (KPI) used in maintenance.

Bar code   An identification method using symbols for encoding data using lines of varying thickness, with designated alphanumeric characters. It can be used on work orders, physical assets, and parts to form a state-of-the-art information gathering methodology.

Barrel (BBL)   One stock tank barrel of 42 US gallons liquid volume used in reference to crude oil, bitumen, condensate, or natural gas liquids.

Barrel of oil equivalent (BOE)   A measure used to aggregate oil and gas resources or production, with one BOE being approximately equal to 6000 cubic feet of natural gas.

Basin   A large, natural depression on the earth's surface in which sediments, generally brought by water, accumulate.

Benchmarking   The process of comparing performance with other organizations, identifying comparatively high-performance organizations, and learning what it is they do that allows them to achieve that high level of performance.

Bill of materials (BOM)   List of components and parts for an asset, usually structured in hierarchical layers from gross assemblies or major end items to minor items down to component parts; a list of all the parts.

Bitumen   A highly viscous form of crude oil (greater than 10,000 cP) resembling cold molasses (at room temperature). Bitumen must be heated or combined with lighter hydrocarbons for it to be produced. It contains sulfur, metals, and other nonhydrocarbons in its natural form.

Borehole   A hole in the earth made by a drilling rig.

Breakdown   Failure to perform to a functional standard; a specific type of failure where an item of plant or equipment is completely unable to function.

Breakdown maintenance   A maintenance strategy or policy where no maintenance is done until an item fails and no longer meets its functional standard; see No scheduled maintenance.

British thermal unit (BTU)   A unit of measure for heat energy. The quantity of heat necessary to raise the temperature of one pound of water 1 °F under a stated pressure.

Built-in test equipment (BITE)   Diagnostic and checkout devices integrated into equipment to assist operation, trouble shooting and service.

Business vacation   A well-planned event that combines real business and/or training with a fun time either before or after the event. "Killing two birds with one stone" for fun and professional development at a place such as The Maintenance Excellence Institute International's oceanfront site, The Breakwaters on Oak Island in North Carolina.

Calibrate   To verify the accuracy of equipment and assure performance within tolerance, usually by comparison to a reference standard that can be traced to a primary standard.

Call-out   To summon a tradesperson to the workplace during his normal nonworking time so that he can perform a maintenance activity (normally an emergency maintenance task).

CAPEX   Capital expenditures.

Capital   Durable items with long lives or high values that necessitate asset control and depreciation under tax guidelines, rather than being expensed.

Carbon capture and storage (CCS)   Process by which carbon dioxide emissions are captured and removed from the atmosphere and then stored, normally via injection into a secure underground geological formation.

Carbon dioxide equivalents (CO_2e)   The quantity that describes, for a given mixture and amount of greenhouse gas, the amount of CO_2 that would have the same global warming potential (GWP) when measured over a specified timescale (generally 100 years).

Carbon intensity   The quantity of greenhouse gas emissions associated with producing an intermediate or final product. For the oil and gas industry, carbon intensity is commonly expressed in units of tons CO_2e per product volume (e.g., tons CO_2e/BBL or tons CO_2e/ MCF).

Carbon sequestration   The fixation of atmospheric carbon dioxide in a carbon sink through biological or physical processes.

Carbon sink   A reservoir that absorbs or takes up released carbon from another part of the carbon cycle. The four sinks, which are regions of the earth within which carbon behaves in a systematic manner, are the atmosphere, terrestrial biosphere (usually including freshwater systems), oceans, and sediments (including fossil fuels).

Carrying costs   Expense of handling, space, information, insurance, special conditions, obsolescence, personnel, and the cost of capital or alternative use of funds to keep parts in inventory. Also, called *holding costs* and generally in the 30–40% range when all factors are considered.

Casing   Thick-walled steel pipe placed in wells to isolate formation fluids (such as freshwater) and to prevent borehole collapse.

Central maintenance   A method for organizing maintenance operations in which the maintenance leader is responsible for all maintenance and all craft areas operating on call from a central location to support the entire operation.

Centum cubic feet (CFF)   One hundred cubic feet; residential billing units are common in CCF. One cubic foot of natural gas contains about 1000 BTUs; 1 CCF contains about 100,000 BTUs. CCF is a volumetric measurement. Because the exact composition of the natural gas can be different, so can the BTUs it contains. One Therm is exactly 100,000 BTUs of any fuel. Some gas companies therefore use the Therm as their unit of sale. Often, the Therm and CCF are used interchangeably, although technically they are not exactly the same thing unless the natural gas is pure methane.

Change out   The removal of a component or part and the replacement of it with a new or rebuilt one.

Chief maintenance officer (CMO)   The technical leader (or actual leader) of a profit-centered maintenance operation within a large or small corporation or nonmanufacturing organization. First coined, defined and promoted by staff from The Maintenance Excellence Institute International in the late 1990s. The CMO represents the prototype of the New Millennium leader for maintenance and physical asset management and managing the maintenance process as a true profit center.

Checkout   The determination of the working condition of a system.

City gate stations   City gate stations mark the point where natural gas leaves the transmission system and enters the distribution System. City gate stations have equipment that reduces the pressure of the gas, meter/measure it, add odorant, and may include heaters to heat the gas if the pressure drop is large enough to cause it to drop below 32 °F.

Clean   To remove all sources of dirt, debris, and contamination for the purpose of inspection and to avoid chronic losses.

CMMS benchmarking system   A methodology developed by staff from The Maintenance Excellence Institute International to evaluate the effectiveness of a computerized maintenance management system (CMMS)/ enterprise asset management (EAM) installation. A process for evaluating CMMS/EAM implementation progress and full utilization of system functionality to enhance best practices.

Coal bed methane (CBM)   Natural gas extracted from coal beds.

Code   Symbolic designation used for identification, such as failure code, repair code, and commodity code.

Commodity code   Classifications of parts by group and class according to their material content or type of consolidation of procurement, storage, and use.

Completion   The process of making a well ready to produce natural gas or oil. Completion involves installing permanent equipment, such as a wellhead, and often includes hydraulic fracturing.

Component A constituent part of an asset, usually modular and replaceable, that may be serialized and interchangeable; a subassembly of an asset, usually removable in one piece and interchangeable with other, standard components (e.g., truck engine).

Component number Designation, usually structured by system, group, or serial number.

Compound annual growth rate (CAGR) The average year-over-year growth rate of a metric over a specific period of time.

Compressed natural gas (CNG) Natural gas that is compressed into storage cylinders for use in transportation fuel in buses, trucks, cars, and forklifts. CNG is *not* the same as liquid propane gas that is typically contained in cylinders.

Computerized maintenance management system (CMMS) Integrated computer system modules such as work orders, equipment, inventory, purchasing, planning, and preventive maintenance that support asset management and overall maintenance management.

Condensate Mixture of hydrocarbons that are in a gaseous state under reservoir conditions and, when produced, become a liquid as the temperature and pressure is reduced.

Condition-based maintenance (CBM) Maintenance based on the measured condition of an asset. Testing and/or inspection of characteristics that will warn of pending failure and performance of maintenance after the warning threshold but before total failure. Predictive maintenance technologies such as vibration analysis, thermography/infrared, oil analysis, and ultrasonic provide tools and technology for condition-based maintenance. An equipment maintenance strategy based on measuring the condition of equipment in order to assess whether it will fail during some future period, and then taking appropriate action to avoid the consequences of that failure. The terms condition based maintenance, on-condition maintenance and predictive maintenance can be used interchangeably.

Condition monitoring The use of specialist equipment to measure the condition of equipment. Vibration analysis, oil analysis and thermography are all examples of condition monitoring techniques.

Conditional probability of failure The probability that an item will fail during a particular age interval, given that it survives to enter that age.

Confidence Degree of certainty that something will happen. For example, a low confidence of replenishment means repair parts probably will not be readily available and is one reason that maintenance personnel retain excess parts in uncontrolled areas.

Configuration The arrangement and contour of the physical and functional characteristics of systems, equipment, and related items of hardware or software; the shape of a thing at a given time. The specific parts used to construct a machine.

Consumables Supplies such as fuel, lubricants, paper, printer ribbons, cleaning materials, and forms that are exhausted during use in operation and maintenance.

Contingency Alternate actions that can be taken if the main actions do not work.

Continuous reliability improvement (CRI) A process developed and used by staff from The Maintenance Excellence Institute International that goes beyond current reliability-centered maintenance (RCM) approaches to outline a continuous, integrated process for improving total reliability of the following resources:

- Equipment/facility resources (asset care/management and maximum uptime via RCM techniques)
- Craft and operator resources (recognizing the most important resource: craftspeople and equipment/process operators)
- Maintenance repair operations (MRO) resources (establishing effective materials management processes)

- Maintenance information resources (effective information technology applications for maintenance)
- Maintenance technical knowledge/craft skills base (closing the technical knowledge resource gap)
- Synergistic team processes (tapping the value-added resource of effective leadership-driven team, self-managed teams to support total operations success)

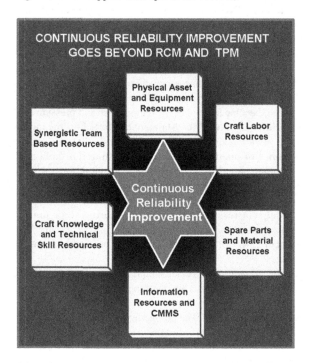

Contract acceptance sheet   A document that is completed by the appropriate contract supervisor and contractor to indicate job completion and acceptance. It also forms part of the appraisal of the contractor's performance.

Conventional resources   Discrete accumulations of hydrocarbons contained in rocks with relatively high matrix permeability, which normally have relatively high recovery factors.

Coordination   Daily adjustment of maintenance activities to achieve the best short-term use of resources or to accommodate changes in needs for service.

Corrective maintenance (CM)   Unscheduled maintenance or repair actions, performed as a result of failures or deficiencies, to restore items to a specific condition. Maintenance done to bring an asset back to its standard functional performance. Any maintenance activity that is required to correct a failure that has occurred or is in the process of occurring. This activity may consist of repair, restoration, or replacement of components.

Craft availability   Percentage of time that craft labor is free to perform productive work during a scheduled working period.

Craft leaders   Subject matter experts (SMEs), crew leaders, and technician specialists.

Craft utilization   Percentage of time that craft labor is engaged in productive work, hands-on during a scheduled working period. The actual wrench time as compared to total time paid.

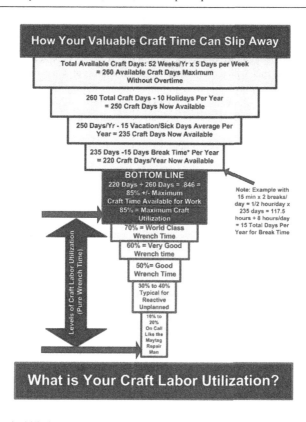

Craftsperson   A skilled maintenance worker who has typically been formally trained through an apprenticeship program. See also **MVP** and **Tradesperson**.

Critical   Describes items that are especially important to product performance and more vital to operation than noncritical items.

Critical equipment   Items especially important to performance, capacity, and throughput and more vital to the operation than noncritical items.

Criticality   The priority rank of a failure mode based on some assessment criteria.

Critical path method (CPM)   A logical method of planning and control that analyzes events, the time required, and the interactions of the considered activities.

Critical spare   Parts and materials that are not used often enough to meet detailed stock accounting criteria but are stocked as "insurance items" because of their essentially or the lead time involved in procuring replacements; similar to safety stocks, except on low-use parts. Normally critical spares require special approval.

Crossdocking   Term for the function capability of an inventory management module to track high-priority inbound orders into receiving, to initiate immediate delivery and to receive/process issue transactions with minimal manual effort. Crossdocking provides quick turnaround at receiving without putting items into storage locations and then having to pick for issue later.

Cycle count   An inventory accountability strategy where counting and verification of stock item quantities is done continuously based on a predetermined schedule and frequency based on the ABC classification of the item. As opposed to an annual physical inventory, cycle counting allows for continuous counting, immediate reconciliation of inventory discrepancies and inventory accuracy of 98% plus.

Dead stocks   Items for which no demand has occurred over a specific period of time.

Defect   A condition that causes deviation from design or expected performance. A term typically used in the maintenance of mobile equipment. A defect is typically a potential failure or other condition that will require maintenance attention at some time in the future, but which is not currently preventing the equipment from fulfilling its functions.

Deferred maintenance   Maintenance that can be or has been postponed from a schedule.

Deterioration rate   The rate at which an item approaches a departure from its functional standard.

Demand   Requests and orders for an item. Demands become issues only when a requested part is given from stock.

Developed acreage   The number of acres that are allocated or assignable to productive wells or wells capable of production.

Developed reserves   Reserves that can be expected to be recovered through existing wells with existing equipment and operating methods or in which the cost of the required equipment is relatively minor compared to the cost of a new well and, if extraction is by means other than a well, through installed equipment and infrastructure operational at the time of the reserves estimate.

Development well   A well drilled within the proved area of an oil or gas reservoir to the depth of a stratigraphic horizon known to be productive.

Direct costs   Any expenses that can be associated with a specific product, operation, or service that is value added.

Directional Drilling   The application of special tools and techniques to drill a wellbore at a predetermined angle. Horizontal drilling is a form of directional drilling where the wellbore is ultimately drilled at ±90° to the vertical direction.

Discard task   The removal and disposal of items or parts. One type of maintenance strategy for selected components.

Disposal   The act of getting rid of excess or surplus property under proper authorization. Such processes as transfer, donation, sale, abandonment, destruction, or recycling may accomplish disposal.

Distribution line   System of low-pressure pipes used to move gas to customers within a service area.

Distribution system   The distribution system moves the natural gas from city gate stations to the users. Distribution systems typically operate in the 2-60 psi range for residential areas, often higher in industrial areas. The operating pressure depends on how old the system is, what it is made of (i.e., cast iron, steel, plastic), what load it needs to serve and how far it is from the city gate station.

Down   Out of service, usually due to breakdown, unsatisfactory condition, or production scheduling.

Downtime (DT)   The time that an item of equipment is out of service, as a result of equipment failure. The time that an item of equipment is available, but not utilized is generally not included in the calculation of downtime.

Drilling rig   The machine used to drill a wellbore.

Dry gas   Dry gas is almost pure methane and occurs in the absence of liquid hydrocarbons or by processing natural gas to remove liquid hydrocarbons and impurities.

Dry hole   A well incapable of economically producing salable hydrocarbons in sufficient quantities to justify commercial exploitation.

Economic life   The total length of time that an asset is expected to remain actively in service before it is expected that it would be cheaper to replace the equipment rather than continuing to maintain it. In practice, equipment is more often replaced for other reasons, including-because it no longer meets operational requirements for efficiency, product quality, comfort etc., or because newer equipment can provide the same quality and quantity of output more efficiently.

Economic order quantity (EOQ)   Amount of an item that should be ordered at one time to get the lowest possible combination of inventory carrying costs and ordering costs.

Economic repair   A repair that will restore the item to a sound condition at a cost less than the value of its estimated remaining useful life.

Economically producible   A resource that generates revenue that exceeds, or is reasonably expected to exceed, the costs of the operation.

Emergency maintenance   A condition requiring immediate corrective action for safety, environmental, or economic risk, caused by equipment breakdown.

Emergency maintenance task   A maintenance task carried out in order to avert an immediate safety or environmental hazard, or to correct a failure with significant economic impact.

Engineering change   Any design change that will require revision to specifications, drawings, documents, or configurations.

Engineering change notice (ECN)   A control document from engineering authorizing changes or modifications to a previous design or configuration.

Engineering work order   The prime document used to initiate an engineering investigation, engineering design activity, or engineering modifications to an item of equipment.

Enhanced oil recovery (EOR)   One or more of a variety of processes that seek to improve recovery of hydrocarbon from a reservoir after the primary production phase.

Environmental assessment   A study that can be required to assess the potential direct, indirect, and cumulative environmental impacts of a project.

Environmental consequences   A failure has environmental consequences if it could cause a breach of any known environmental standard or regulation.

Equipment configuration   List of assets usually arranged to simulate the process, or functional or sequential flow.

Equipment maintenance strategies   The choice of routine maintenance tasks and the timing of those tasks, designed to ensure that an item of equipment continues to fulfill its intended functions.

Equipment repair history   A chronological list of defaults, repairs, and costs on key assets so that chronic problems can be identified and corrected and economic decisions made.

Equipment use   Accumulated hours, cycles, distance, throughput, etc., or performance.

Estimated plant replacement value   The estimated cost of capital required to replace all the existing assets with new assets capable of producing the same quantity and quality of output.

Estimated ultimate recovery (EUR)   The sum of reserves remaining as of a given date and cumulative production as of that date.

Estimating index   The ratio of estimated labor hours required completing the work specified on work orders to the actual labor hours required to complete the work specified on those work orders, commonly expressed as a percentage. This is also a measure of craft performance, one element of craft labor productivity, particularly when there are well-defined estimating standards.

Examination   A comprehensive inspection with measurement and physical testing to determine the condition of an item.

Expediting   Special efforts to accelerate a process. An expediter coordinates and assures adequate supplies of parts, materials and equipment. Typically found in a reactive maintenance operation.

Expense   Those items that are directly charged as a cost of doing business. They generally have a short, nondurable life. Most nonrepairable repair parts are expensed when installed on equipment.

Expensed inventory   Parts written off as a "cost of sales." Material transferred from ledger inventory to expensed inventory is to be used within 12 months.

Expert system   Decision support software with some ability to make or evaluate decisions based on rules or experience parameters incorporated in the database; a software-based system that makes or evaluates decisions based on rules established within the software. Typically used for fault diagnosis.

Exploratory well   A well drilled to find a new field or to find a new reservoir in a field previously found to be productive of oil or gas in another reservoir.

Failsafe   An item is failsafe if, when the item itself incurs a failure, that failure becomes apparent to the operating workforce in the normal course of events.

Failure   Termination of the ability of an item to perform its required function to a standard; an item of equipment has suffered a failure when it is no longer capable of fulfilling one or more of its intended functions. Note that an item does not need to be completely unable to function to have suffered a failure. For example, a pump that is still operating, but is not capable of pumping the required flow rate, has failed. In reliability-centered maintenance terminology, a failure is often called a functional failure.

Failure analysis   The logical, systematic examination of an item or its design, to identify and analyze the probability, causes, and consequences of real or potential malfunction. A study of failures to analyze the root causes, to develop improvements, to eliminate or reduce the occurrence of failures. (See also **Failure modes, effects, and critical analysis.**)

Failure cause   See **Failure mode.**

Failure coding   Identifying and indexing the causes of equipment failure on which corrective action can be based, such as lack of lubrication, operator abuse, material fatigue, etc.; a code typically entered against a work order in a CMMS that indicates the cause of failure (e.g., lack of lubrication, metal fatigue).

Failure consequences   A term used in reliability-centered maintenance. The consequences of all failures can be classified as being eitherhidden, safety, environmental, operational, or nonoperational. Very important to a risk-based maintenance approach that considers probability of failure and failure consequences.

Failure effect   A description of the events that occur after a failure has occurred as a result of a specific failure mode. Used in reliability-centered maintenance and FMEA/FMECA analyses.

Failure finding interval   The frequency with which a failure finding task is performed. It is determined by the frequency of failure of the protective device, and the desired availability required of that protective device.

Failure finding task   Used in reliability-centered maintenance terminology. A routine maintenance task, normally an inspection or a testing task, designed to determine, for hidden failures, whether an item or component has failed. A failure finding task should not be confused with an on-condition task, which is intended to determine whether an item is about to fail. Failure finding tasks are sometimes referred to as functional tests.

Failure mode   Any event that causes a failure.

Failure modes and effects analysis (FMEA)   A structured method of determining equipment functions, functional failures, assessing the causes of failures and their failure effects. The first part of a reliability-centered maintenance analysis is a FMEA.

Failure modes, effects, and critical analysis (FMECA)   A logical, progressive method used to understand and assess the root causes of failures and their subsequent effect on production, safety, cost, quality, etc.

Failure pattern   The relationship between the conditional probability of failure of an item and its age. Failure patterns are generally applied to failure modes. Research in the airline industry established that there are six distinct failure patterns. The type of failure pattern that applies to any given failure mode is of vital importance in determining the most appropriate equipment maintenance strategy. This fact is one of the key principles underlying reliability-centered maintenance.

Failure rate   The number of failures per unit measure of life (cycles, time, miles, events, and the like) as applicable for the item.

Farm-in   The acquisition of part or all of an oil, natural gas, or mineral interest from a third party.

Farm-out   The assignment of part or all of an oil, natural gas, or mineral interest to a third party.

Fault tree analysis (FTA)   A review of failures, faults, defects, and shortcomings based on a hierarchy or relationship to find the root cause.

Field   An area consisting of a single hydrocarbon reservoir or multiple geologically related reservoirs all grouped on or related to the same individual geological structure or stratigraphic condition.

Fill rate   Service level of a specific stock point. An 85% fill rate means that if 100 parts are requested, then 85 of them are available and issued. This also means that 15% of parts requests are not filled due to stock outs. Goal for service levels should be 98% plus.

First in–first out (FIFO)   Use the oldest item in inventory next. FIFO accounting values each item used at the cost of the oldest item in inventory. Contrasts with LIFO (last in–first out).

First in still here (FISH)   A fun term for obsolete parts not identified and still taking up space.

Flaring   The burning of natural gas for safety reasons or when there is no way to transport the gas to market or use the gas for other beneficial purposes (e.g., enhanced oil recovery, reservoir pressure maintenance). The practice of flaring is being steadily reduced as pipelines are completed and in response to environmental concerns. One major refining company calls flaring "the flames of incompetency."

Floating production, storage, and offloading (FPSO)   Provides alternative to pipeline to store oil production and load vessels for movement to markets.

Peng Bo FPSO in Bohai Bay, China

Forecast   To calculate or predict some future event or condition, usually as a result of rational study and analysis of pertinent data. The projection of the most probable, as in forecasting failures and maintenance activities.

Formation   A rock layer that has distinct characteristics (e.g., rock type, geologic age).

Forward workload   All known backlog work and work which is due or predicted to become backlog work within a pre-specified future time period.

Fossil fuel   A fuel source (e.g., oil, condensate, natural gas, natural gas liquids or coal) formed in the earth from plant or animal remains.

Frequency   Count of occurrences during each time period or event. A typical frequency chart for inventory plots demand versus days.

Front-end engineering and design (FEED)   Part of a project's life cycle.

Fugitive emissions   Emissions of gases or vapors from pressurized equipment, including pipelines, due to leakage, unintended, or irregular releases of gases.

Function   A separate and distinct action required to achieve a given objective, to be accomplished by the use of hardware, computer programs, personnel, facilities, procedural data, or a combination thereof; or an operation a system must perform to fulfill its mission or reach its objective; the definition of what we want an item of equipment to do, and the level of performance that the users of the equipment require when it does it. An item of equipment can have many functions, commonly split into primary and secondary functions.

Functional failure   Used in reliability-centered maintenance terminology. The inability of an item of equipment to fulfill one or more of its functions. Interchangeably used with "failure."

Functional levels   Rankings of the physical hierarchy of a product. Typical levels of significance from the smallest to the largest are part, subassembly, assembly, subsystem, and system.

Functional maintenance structure   A method for organizing the maintenance operation where the first-line maintenance leader is responsible for conduction a specific kind of maintenance, for example, electrical maintenance, pump maintenance, HVAC maintenance, etc.

Gantt chart   A bar chart format of scheduled activities showing the duration and sequencing of activities.

Gas cost recovery (GCR)   Most gas utility companies bill their customers based on the actual cost of gas as a separate charge from the distribution charge and other charges on the bill. The GCR factor is the cost for the gas itself, before delivery and other charges are added on.

Gathering system   Pipelines, compressors, and additional equipment used to move gas from the wellhead to a processing facility.

General support equipment (GSE)   Equipment that has maintenance application to more than a single model or type of equipment.

Global warming potential (GWP)   The relative measure of how much heat a greenhouse gas traps in the atmosphere. It compares the amount of heat trapped by a certain mass of the gas in question to the amount of heat trapped by a similar mass of carbon dioxide. GWP is calculated over a specific time interval, commonly 100 years. GWP is expressed as a multiple of that for carbon dioxide (whose GWP is standardized to 1).

Go-line   Used in relation to mobile equipment. Equipment that is available but not being utilized is typically parked on the go-line. This term is used interchangeably with ready line.

Greenhouse gas   Atmospheric gases that are transparent to solar (short-wave) radiation but opaque to long-wave (infrared) radiation, thus preventing long-wave radiant energy from leaving Earth's atmosphere. The net effect of these gases is a trapping of absorbed radiation and a tendency to warm the planet's surface. The greenhouse gases most relevant to the oil and gas industry are carbon dioxide, methane, and nitrous oxide.

Hardware   A physical object or physical objects, as distinguished from capability or function. A generic term dealing with physical items of equipment-tools instruments, components, parts-as opposed to funds, personnel, services, programs, and plans, which are termed "software."

Heavy oil   Crude oil with an API gravity that is less than 20°. Heavy oil generally does not flow easily due to its elevated viscosity.

Hidden failure   A failure that, on its own, does not become evident to the operating crew under normal circumstances. Typically, protective devices that are not fail-safe (examples could include standby plant and equipment, emergency systems, etc.)

Hold for disposition stock   Defective material held at a stock location pending removal for repair or for scrap.

Horizontal drilling   A drilling technique whereby a well is progressively turned from vertical to horizontal so as to allow for greater exposure to an oil or natural gas reservoir. Horizontal laterals can be more than a mile long. In general, longer exposure lengths allow for more oil and natural gas to be recovered from a well and often can reduce the number of wells required to develop a field, thereby minimizing surface disturbance. Horizontal drilling technology has been extensively used since the 1980s and is appropriate for many, but not all, developments.

Hot work   Typically used in welding, a work area that is too hot to handle safely without special personal protective equipment (PPE).

Hydraulic fracturing   Hydraulic fracturing (also referred to as frac'ing or fracking) is an essential completion technique in use since the 1940s that facilitates production of oil and natural gas trapped in low-permeability reservoir rocks. The process involves pumping fluid at high pressure into the target formation, thereby creating small fractures in the rock that enable hydrocarbons to flow to the wellbore.

Hydraulic fracturing fluids   Mixture of water and proppant along with minor amounts of chemical additives used to hydraulically fracture low permeability formations. Water and sand typically comprise up to 99.5% of the mixture.

Hydrocarbons   An organic compound containing only carbon and hydrogen and often occurring in nature as petroleum, natural gas, coal, and bitumens or in refined products such as gasoline and jet fuel.

Identification   Means by which items are named or numbered to indicate that they have a given set of characteristics. Identification may be in terms of name, part number, drawing number, code, stock number, or catalog number. Items may also be identified as part of an assembly, a piece of equipment, or a system.

Improved oil recovery (IOR)   Term used to describe methods employed to improve the flow of hydrocarbons from the reservoir to the wellbore or to recover more oil or natural gas. **Enhanced oil recovery** (EOR) is one form of IOR.

Indirect costs   Expenses not directly associated with specific products, operations, or services; usually considered overhead.

Infant mortality   The relatively high conditional probability of failure during the period immediately after an item returns to service.

Infill wells   Wells drilled into the same reservoir as known producing wells so that oil or natural gas does not have to travel as far through the formation, thereby helping to improve or accelerate recovery.

Inherent reliability   A measure of the reliability of an item, in its present operating context, assuming adherence to ideal equipment maintenance strategies.

In-situ recovery   Techniques used to extract hydrocarbons from deposits of extra-heavy crude oil, bitumen, or oil shale without removing the soil and other overburden materials.

Inspection   A review to determine maintenance needs and priority on equipment. Any task undertaken to determine the condition of equipment, and/or to determine the tools, labor, materials, and equipment required to repair the item.

Insurance items   Parts and materials that are considered as critical spares but not used often enough to meet detailed stock accounting criteria. Insurance items are stocked because of their essentially or the lead time involved in procuring replacements. May be of high dollar value to classify them as capital spares.

Interchangeable   Parts with different configurations and numbers that may be substituted for another part, usually without any modification or different performance, because they have the same form, fit, and function.

Interface   A common boundary between two or more items, characteristics, systems, functions, activities, departments, or objectives. That portion impinges upon or directly affects something else.

Interval-based   Periodic preventive maintenance based on calendar time or hours operated.

Inventory turnover   Ratio of the value of materials and parts issues annually to the value of materials and parts on-hand, expressed as percentage. For maintenance, two to three turns per year is expected, with much more being a sign of excess inventory.

Inventory   Physical count of all items on hand by number, weight, length, or other measurement, as well as any items held in anticipation of future use. Annual physical inventory remains accurate for a short time unless cycle counting is used.

Inventory control   A phase or function of logistics that includes management, cataloging, requirements, determination, procurement, inspection, storage, distribution, overhaul, and disposal of material. Managing the acquisition, receipt, storing and issuance of materials and spare parts; managing the investment efficiently of the store's inventory.

Issues   Stock consumed through stores.

Item   Generic term used to identify a specific entity. Items may be parts, components, assemblies, subassemblies, accessories, groups, parents, components, equipment, or attachments.

Item of supply   An article or material that is recurrently purchased, stocked, distributed, used and is identified by one distinctive set of numbers or letters throughout the organization concerned. It consists of any number of pieces or objects that can be treated as a unit.

Jack-up rig   An offshore rig with retractable steel legs that are placed on the ocean floor to raise the rig above the waterline.

Joint operating agreement (JOA)   An agreement governing the rights and obligations of co-owners in a field or undeveloped acreage, which defines, amongst other things, how costs and revenues are to be shared among the parties and who is the operator.

Just-in-time (JIT)   A buzzword term for proactive planning of many processes, such as JIT inventory service, JIT maintenance services, JIT training conducted as needed for project activities rather train-train-train and then doing projects and activities, etc. Often seen with total productive maintenance (TPM) training that drags out long before an actual start date.

Keep full   Term used for maintaining set levels of shop stock inventory of Class C items (see also **Shop stock**). May involve vendor-supplied C items and periodic refill of "bins" and invoice of amount used to refill.

Key performance indicators (KPIs)   A select number of key measures that enable performance against targets to be monitored.

Knuckle buster   A poor-quality, imitation, adjustable wrench.

Last in–first out (LIFO)   Use newest inventory next. LIFO accounting values each item used at the cost of the last item added to inventory. Contrasts with **First-in-first out** (FIFO).

Lead time   Allowance made for that amount of time estimated or actually required to accomplish a specific task such as acquiring a part. Remote locations may have much longer lead times than operations near a large city or industrialized area.

Lease   A legal document executed between a mineral owner and a company or individual that conveys the right to explore for and develop hydrocarbons and/or other products for a specified period of time over a given area.

Ledger inventory   Items carried on the corporate financial balance sheet as material valued at cost.

Level of repair (LOR)   Locations and facilities at which items are to be repaired. Typical levels are operator, field technician, bench, and factory.

Level of services (stores)   Usually measured as the ratio of stock outs to all stores issues.

Life   That strange experience you have in a maintenance context when an emergency breakdown occurs; you may want to look at equipment life.

Life cycle analysis (LCA)   LCA is an analytical methodology used to comprehensively quantify and interpret the environmental flows to and from the environment (including air emissions, water effluents, solid waste, and the consumption/depletion of energy and other resources) over the life cycle of a product or process. LCAs should be performed in adherence to the International Organization for Standardization (ISO) 14040 series of standards.

Life cycle   The series of phases or events that constitute the total existence of anything. The entire "womb to tomb" scenario of a product from the time concept planning is started until the product is finally discarded.

Life cycle cost (LCC)   All costs associated with the items of life cycle including design, manufacture, operation maintenance, and disposal; a process of estimating and assessing the total costs of ownership, operation, and maintenance of an item of equipment during its projected equipment life. Typically used in comparing alternative equipment design or purchase options in order to select the most appropriate option.

Liquefied natural gas (LNG)   Natural gas that has been converted to a liquid by refrigerating it to −260 °F. Liquefying natural gas reduces the fuel's volume by 600 times, enabling it to be shipped economically from distant producing areas to markets.

Local distribution company (LDC)   A company that delivers natural gas to end users in a geographic area. Could also apply to a local distributor for maintenance parts and materials.

Logistics engineering   The professional art of applying science to the optimum planning, handling and implementation of personnel, materials and facilities. This includes life-cycle designs, procurements, production, maintenance, and supply.

Logistics support analysis (LSA)   A methodology for determining the type and quantity of logistics support required for a system over its entire life cycle. Used to determine the cost-effectiveness of asset-based solutions.

Main   Gas pipe generally laid along street right-of-ways from which extends smaller lateral service lines to individual customer gas meters.

Maintainability   The inherent characteristic of a design or installation that determines the ease, economy, safety, and accuracy with which maintenance actions can be performed. Also, the ability to restore a product to service or to perform preventive maintenance within required limits. The rapidity and ease with which maintenance operations can be performed to help prevent malfunctions or correct them if they occur, usually measures as mean time to repair; the ease and speed with which any maintenance activity can be carried out on an item of equipment. May be measured by mean time to repair; a function of equipment design, and maintenance task design (including use of appropriate tools, jigs, work platforms, etc.)

Maintainability engineering   The application of applied scientific knowledge, methods, and management skills to the development of equipment, systems, projects, or operations that have the inherent ability of being effectively and efficiently maintained; the set of technical processes that apply maintainability theory to establish system maintainability requirements, allocate these requirements down to system elements, and predict and verify system maintainability performance.

Maintenance   The function of keeping items or equipment in, or restoring them to, serviceable condition. It includes servicing, test, inspection, adjustment/alignment, removal, replacement, reinstallation, troubleshooting, calibration, condition determination, repair, modification, overhaul, rebuilding, and reclamation. Maintenance includes both corrective and preventive activities. Any activity carried out to retain an item in, or restore it to, an acceptable condition for use or to meet its functional standard.

Maintenance engineering   Developing concepts, criteria, and technical requirements for maintenance during the conceptual and acquisition phases of a project. Providing policy guidance for maintenance activities, and exercising technical and management direction and review of maintenance programs. A staff function intended to ensure that maintenance techniques are effective, equipment is designed for optimum maintainability persistent and chronic problems are analyzed and corrective actions or modifications are made.

The Reliable Maintenance Excellence Index (RMEI)   An essential component to The Maintenance Excellence Institute International's implementation of profit and customer-centered maintenance. It is a progressive approach to managing the business of maintenance with performance measurement of maintenance operations achieved by integrating multiple metrics into a composite total RMEI value. It includes the comparison of current performance to both the performance goal and baseline value for each metric selected. It is ideally suited to measure progress across multiple sites within a large organization.

Maintenance leaders   Maintenance managers, supervisors, foremen, maintenance engineers, and reliability engineers.

Maintenance policy   A statement of principle used to guide maintenance management decision-making.

Maintenance repair operations (MRO)   Term for maintenance repair operations and generally used as MRO items referring to parts, materials, tools, and equipment used in the maintenance process.

Maintenance requirements   There are two foundational needs for an effective facility management or plant maintenance operation:

Maintenance business process improvement: Business process improvement is what this book strives to help and promote with a profit and customer-centered strategy and related attitudes. If this is truly present, then the plant maintenance leader or facilities management leader in governmental operations at least has a chance to survive. However, regardless of the type of maintenance operation, they must be able to show top leaders they really are maximizing all available maintenance resources and there is a true need for resources to address the next item, the basic maintenance requirements.

Total maintenance requirements

a) *Knowing your total maintenance requirements:* Achieving the total maintenance requirement is the primary mission. It is executing the required maintenance while providing, maintaining, and improving the asset or facilities and related services for production operation and the tenants/customers. This is what maintenance leaders must achieve in addition to many other activities that compete for engineering, craft, and administrative resources. Defining true maintenance requirement to top leaders is extremely important when all resources are maxed out and basic preventive maintenance is being neglected. If you do not know and cannot convince top leaders what your total true maintenance requirements are, then you will experience really mean times between failures with your boss and your boss's boss!

b) **Inspection** is a maintenance requirement when the basic objective is to assure that a requisite condition or quality exists. In order to inspect for the desired condition, it may be necessary to remove the item, to gain access by removing other items, or to disassemble partially the item for inspection purposes. In such cases, these associated actions necessary to accomplish the required inspection would be specific tasks.

c) **Troubleshooting** is a maintenance operation that involves the logical process (series of tasks) that leads to positive identification, location, and isolation of the cause of a malfunction.

d) **Remove** is a maintenance requirement when the basic objective is to separate the item from the next higher assembly. This requirement is usually applied for a configuration change.

e) **Remove and replace** is a maintenance requirement that constitutes the removal of one item and replacement of it with another like item. Such action can result from a failure or from a scheduled action.

f) **Remove and reinstall** is a maintenance requirement when an item is removed for any reason, and the same items reinstalled.

g) **Adjustment/alignment** is a maintenance requirement when the primary cause of the maintenance action is to adjust or align, or to verify adjustment/alignment of specific equipment. Adjustment/alignment accomplished subsequent to repair of a given item is not considered to be a separate requirement and is included as a task in the repair requirement.

h) **Functional test** constitutes a system or subsystem operational checkout either as a condition verification after the accomplishment of corrective maintenance action or as a scheduled requirement on a periodic basis.

i) **Conditioning** is a maintenance requirement whenever an item is completely disassembled, refurbished, tested, and returned to a serviceable condition, meeting all requirements set forth is applicable specifications. It may result from either a scheduled or unscheduled requirement and is generally accomplished at the depot/factory level of maintenance.

Maintenance schedule A comprehensive list of planned maintenance and its sequence of occurrence based on priority in a designated period of time; a list of planned maintenance tasks to be performed during a given time period, together with the expected start times and durations of each of these tasks. Schedules can apply to different time periods (e.g., daily schedule, weekly schedule).

Maintenance shutdown A period of time during which a plant, department, process, or asset is removed from service specifically for maintenance.

Maintenance strategy Principles and strategies for guiding decisions for maintenance management; a long-term plan, covering all aspects of maintenance management that sets the direction for maintenance management, and contains firm action plans for achieving a desired future state for the maintenance function.

Maintenance task routing file A computer file containing skills, hours, and descriptions to perform standard maintenance tasks.

Maintenance, winning   Maintenance that wins most than it loses; maintenance that plays well as part of the total operations team; maintenance that never, never, never gives up on trying to get better; maintenance that performs like Richard Petty's pit crew.

Management of change (MOC)   May refer to the management of organizational changes. From a technical standpoint in a complex continuous operation, this may be defined in a formally written management of change (MOC) process procedures and cover complete documentation of all physical and procedural changes depending on the scope of the change.

Mean time between failures   A measure of equipment reliability. Equal to the number of failures in a given time period, divided by the total equipment uptime in that period.

Mean time to repair   A measure of maintainability. Equal to the total equipment downtime in a given time period, divided by the number of failures in that period.

Mercaptan   A distinctive odorant added to natural gas. The rotten-egg smell ensures that gas escaping into the atmosphere will be detected.

Model work order   A work order stored in the CMMS which contains all the necessary information required to perform a maintenance task (see also **Standard job**).

Modification   Change in configuration; any activity carried out on an asset which increases the capability of that asset to perform its required functions.

Modularization   Separation of components of a product or equipment into physically and functionally distinct entities to facilitate identification, removal and replacement unitization.

Most valuable people (MVP)   Your craftspeople and your storeroom staff.

MRO materials management   The overall management of the process for requisitioning, storage/warehousing, purchasing, inventory management and issue of MRO type items used in the maintenance process.

Natural gas   A fossil fuel composed mostly of methane (CH_4). Naturally occurring hydrocarbon gases found in porous rock formations. Its principal component is usually methane. Nonhydrocarbon gases, such as carbon dioxide and hydrogen sulfide, can sometimes be present in natural gas.

Natural gas liquids (NGLs)   A general term for highly volatile liquid products separated from natural gas in a gas processing plant. NGLs include ethane, propane, butane and condensate.

Natural gas vehicle (NVG)   A vehicle designed or converted to run on compressed natural gas (CNG).

Net acres   The percentage that a company owns in an acreage position with multiple owners. For example, a company that has a 50% interest in a lease covering 10,000 acres owns 5000 net acres.

Nondestructive testing (NDT)   Testing of equipment to detect abnormalities in physical, chemical, or electrical characteristics, using such technologies as ultrasonic (thickness), liquid dye penetrates (cracks), X-ray (weld discontinuities), and voltage generators (resistance).

Nonoperational consequences   A failure has nonoperational consequences if the only impact of the failure is the direct cost of the repair (plus any secondary damage caused to other equipment as a result of the failure).

Nonrepairable   Parts or items that are discarded upon failure for technical or economic reasons.

Nonroutine maintenance   Maintenance performed at irregular intervals, with each job unique, and based on inspection, failure, or condition. Any maintenance task that is not performed at a regular, predetermined frequency.

No scheduled maintenance   An equipment maintenance strategy where no routine maintenance tasks are performed on the equipment. The only maintenance performed on the equipment is corrective maintenance, and then only after the equipment has suffered a failure. Also described as a "run-to-failure" strategy.

Obsolescence   Decrease in value or use of items that have been superseded by superior items.

Obsolete   Designation of an item for which there is no replacement. The part has probably become unnecessary as a result of design change.

Odorant   Any material added to natural gas to impart a distinctive odor to aid in leak detection, commonly mercaptan.

Oil analysis   See **Teratology.**

Oil sands   Geologic formation comprised predominantly of sand grains and bitumen, a highly viscous form of crude oil.

On-condition maintenance   See **Condition-based maintenance.**

One million barrels (MMBBL)   One million barrels of crude oil, bitumen, condensate or natural gas liquids.

One million cubic feet (MMCF)   In the United States, standard conditions are defined as gas at 14.7 psia and 60 °F.

One thousand cubic feet (MCF)   A common volumetric unit of measurement for natural gas. One MCF has about 1 million BTUs in it, when natural gas has about 1000 BTUs per cubic foot.

Operating context   The operational situation within which an asset operates. For example, is it a stand-alone piece of plant, or is it one of a duty–standby pair? Is it part of a batch manufacturing process or a continuous production process? What is the impact of failure of this item of equipment on the remainder of the production process? The operating context has enormous influence over the choice of appropriate equipment maintenance strategies for any asset.

Operating hours   The length of time that an item of equipment is actually operating.

Operational consequences   A failure has operational consequences if it has a direct adverse impact on operational capability (lost production, increased production costs, loss of product quality, or reduced customer service).

Operational efficiency   Used in the calculation of overall equipment effectiveness. The actual output produced from an asset in a given time period divided by the output that would have been produced from that asset in that period, had it produced at its rated capacity. Normally expressed as a percentage.

Operator   The entity responsible for managing operations in a field or undeveloped acreage position.

Operator-based maintenance (OBM)   A maintenance excellence strategy where equipment or process operators are trained and accountable for selected maintenance tasks. Also known as autonomous maintenance.

Order point   Quantity of parts at which an order will be placed when usage reduces stock to that level; also called reorder point (ROP).

Order quantity   Number of items demanded. The economic order quantity (EOQ), also called minimum cost quantity, is a specific number; but the actual order quantity may vary as a result of cost, transportation, discounts, or extraordinary demand.

Outage   A term used in some industries, for example, electrical power distribution, to denote when an item or system is not in use.

Overall craft effectiveness (OCE)   The OCE Factor is a method developed by TMEI founder Ralph "Pete" Peters to measure craft labor productivity that combines three key elements: craft utilization, craft performance and craft methods and quality. Typically, the OCE factor is determined by only the two elements: % Craft Utilization × Craft Performance. Compares to overall equipment effectiveness in basic concepts but applies directly to productivity of craft labor assets.

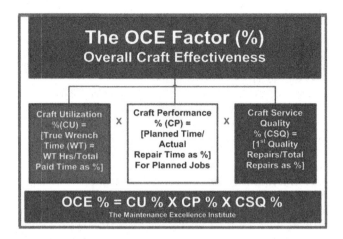

Overall equipment effectiveness (OEE)   The OEE factor is a method to measure overall equipment effectiveness that originated with Japan's total productive maintenance (TPM) strategy for maintenance improvement. The OEE factor combines three key elements; equipment availability, performance, and quality measurement into a common metric that reflects key elements of the manufacturing environment. The OEE factor equals % Availability × % Performance × % Quality. Compares to OCE in basic concept but applies directly to the productivity of physical assets.

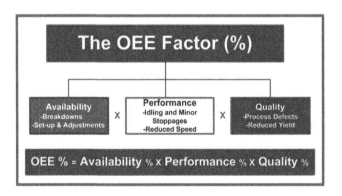

Overhaul   A comprehensive examination and restoration of an item to an acceptable condition.

Pareto's principle   Named for Vilfredo Pareto, a very smart Italian economist whose principle applies to about anything! A critical percentage (often about 20%) of parts or people or users should receive attention before the insignificant many, which are usually about 80%.

Part numbers   Unique identifying numbers and letters that denote each specific part configuration; also called stock numbers or item numbers.

Periodic maintenance   Cyclic maintenance actions carried out at regular intervals, based on repair history data, use, or elapsed time.

Percent planned work   The percentage of total work (in labor hours) performed in a given time period which has been planned in advance.

P–F interval A term used in reliability-centered maintenance. The time from when a potential failure can first be detected on an asset or component using a selected predictive maintenance task, until the asset or component has failed. Reliability-centered maintenance principles state that the frequency with which a predictive maintenance task should be performed is determined solely by the P–F interval.

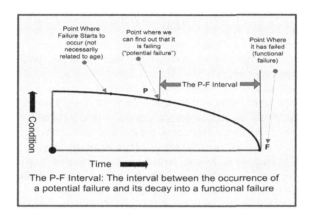

The P-F Interval: The interval between the occurrence of a potential failure and its decay into a functional failure

Permeability The permeability of a rock is the measure of the resistance to the flow of fluid through the rock. High permeability means fluid passes through the rock easily.

Pick list A selection of required stores items for a work order or task.

Planned maintenance Maintenance carried out according to a documented plan of tasks, skills, and resources; any maintenance activity for which a predetermined job procedure has been documented, for which all labor, materials, tools, and equipment required to carry out the task have been estimated, and their availability assured before commencement of the task.

Plant engineering A staff function whose prime responsibility is to ensure that maintenance techniques are effective, that equipment is designed and modified to improve maintainability, that ongoing maintenance technical problems are investigated, and appropriate corrective and improvement actions are taken. Used interchangeably with maintenance engineering and reliability engineering.

Play An area in which hydrocarbon accumulations or prospects with similar characteristics occur, such as the Lower Tertiary play in the deepwater Gulf of Mexico or the Marcellus play in the eastern United States.

Pounds per square inch (Psi) The common method of measuring natural gas pressure. One psi is equal to 28″ of water column. A common delivery pressure for residential applications is 1/4 psi or 7″ water column. A common delivery pressure for industrial customers may be 10 psi. Power plants may get gas pressure delivered in several 100 psi or have a compressor on-site to boost the pressure to run combustion turbines.

Porosity The measure of a rock's ability to hold a fluid. Porosity is normally expressed as a percentage of the total rock that is taken up by pore space.

Possible reserves Additional reserves that are less certain to be recovered than probable reserves.

Potential failure A term used in reliability-centered maintenance. An identifiable condition which indicates that a functional failure is either about to occur, or in the process of occurring.

Predictive maintenance (PdM) Use of measured physical parameters against known engineering limits for detecting, analyzing, and correcting equipment problems before a failure occurs; examples include vibration analysis, sonic testing, dye testing, infrared testing, thermal testing, coolant analysis, teratology, and equipment history analysis. Subset of preventive maintenance that uses nondestructive testing such as spectral oil analysis, vibration evaluation, and ultrasonic with statistics and probabilities to predict when and what maintenance should be done to prevent failures; an equipment maintenance strategy based on measuring the condition of equipment in order to assess whether it will fail during some future period, and then taking appropriate action to avoid the consequences of that failure. The terms condition based maintenance, on condition maintenance, and predictive maintenance can be used interchangeably.

Preventive maintenance (PM) Maintenance carried out at predetermined intervals, or to other prescribed criteria, and intended to reduce the likelihood of a functional failure. Actions performed in an attempt to keep an item in a specific operating condition by means of systematic inspection, detection, and prevention of incipient failure; an equipment maintenance strategy based on replacing, overhauling or remanufacturing an item at a fixed interval, regardless of its condition at the time. Scheduled Restoration tasks and scheduled discard tasks are both examples of preventive maintenance tasks. See also **Scheduled maintenance.**

PRIDE-in-Maintenance Coined originally in 1981 as the theme for a presentation to the craft work force at a manufacturing plant in Greenville, Mississippi. It is about changing the hearts, minds and attitudes about the profession and practice of maintenance. It is about PRIDE and **P**eople **R**eally **I**nterested **in D**eveloping **E**xcellence in maintenance operations of all types. Its foundation starts with the most important maintenance resource, the crafts work force. The goal is to achieve PRIDE-in-Maintenance from within the crafts work force and among their maintenance leaders and to have top leaders realize the true value of their total maintenance operation and take positive action.

Priority The relative importance of a single job in relationship to other jobs, operational needs, safety, etc., and the time within which the job should be done; used for scheduling work orders.

Proactive maintenance A maintenance strategy that is anticipatory and includes a level of planning; any tasks used to predict or prevent equipment failures.

Probable reserves Additional reserves that are less certain to be recovered than proved reserves but which, together with proved reserves, are as likely as not to be recovered.

Probabilistic risk assessment A "top-down" approach used to apportion risk to individual areas of plant and equipment, and possibly to individual assets so as to achieve an overall target level of risk for a plant, site, or organization. These levels of risk are then used in risk-based techniques, such as reliability-centered maintenance, to assist in the development of appropriate equipment maintenance strategies and to identify required equipment modifications.

Probabilistic safety assessment Similar to probabilistic risk assessment, except focused solely on safety-related risks.

Process safety management (PSM) Several PSM regulations and industry guidelines drive PSM practices. Most companies have in-house staff or contracted services to help their companies comply with these PSM regulatory requirements as well as other established industry sector standards, such as those from API.

Processing Processing is the process of removing impurities and byproducts such as water, carbon dioxide, sulfur, helium, and heavy hydrocarbons from the natural gas that is processed.

Processing natural gas   The extraction of impurities such as water vapor, H_2S (hydrogen sulfide), and CO_2 from newly produced gas or gas in storage.

Procurement   Process of obtaining persons, services, supplies, facilities, materials, or equipment. It may include the function of design, standards determination, specification development, and selection of suppliers, financing, contract administration, and other related functions.

Produced water   Water produced in connection with oil and natural gas exploration and development activities.

Production   The process of extracting natural gas from the rocks buried beneath the earth's surface. The wells in the gas field produce by allowing pressured gas in the rock formations to flow or be lifted to the surface in a controlled manner.

Wellhead   is a series of valves, flanges, gages and vertical piping that delivers fluids/gas from the well to the surface of the earth.

Gathering Lines   deliver gas from wells and transport it to the processing plant or directly to the transmission system.

Production sharing contract (PSC)   An agreement between a host government and the owners (or co-owners) of a well or field regarding the percentage of production each party will receive after the parties have recovered a specified amount of capital and operational expenses.

Productive well   A well that is capable of producing hydrocarbons in sufficient quantities to justify commercial exploitation.

Profit-centered maintenance (PCM)   A value-adding business approach to the leadership and management of maintenance and physical asset management. Simply stated, it asks the question, "If I owned this maintenance operation as a business to make a profit, what would I do differently?" On a broader scope, it is the application of world-class maintenance practices, attitudes, and leadership principles. When applied, it makes an in-house maintenance operation equivalent to a profit center with both a financial system and performance measurement process in place to validate results.

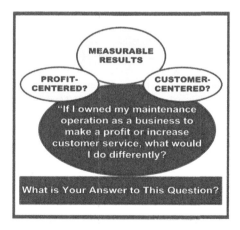

Profit and customer-centered maintenance (PCCM)   On a broader scope, it combines the philosophies of profit-centered with customer-centered into management and leadership of all types of maintenance processes. It is the application of maintenance best practices, attitudes, and leadership principles to both profit and maintenance customer service. When applied, it makes an in-house maintenance operation equivalent to a profit center when both a financial system and performance measurement process in place to validate results.

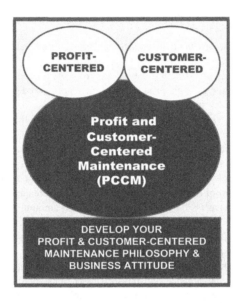

Project evaluation and review technique (PERT) chart   Scheduling tool that shows the interdependencies between project activities in flowchart format.

Proppant   Sand or manmade, sand-sized particles pumped into a formation during a hydraulic fracturing treatment to keep fractures open so that oil and natural gas can flow through the fractures to the wellbore.

Protective device   Devices and assets intended to eliminate or reduce the consequences of equipment failure. Some examples include standby plant and equipment, emergency systems, safety valves, alarms, trip devices, and guards.

Proved developed reserves   Proved reserves that can be expected to be recovered through existing wells with existing equipment and operating methods or in which the cost of the required equipment is relatively minor compared to the cost of a new well.

Proved reserves   Proved oil and gas reserves are those quantities of oil and gas that, by analysis of geosciences and engineering data, can be estimated with reasonable certainty to be economically producible.

Provisioning   Process of determining and selecting the varieties and quantities of repair parts, spares, special tools, and test and support equipment that should be procured and stocked to sustain and maintain equipment for specified periods of time. It includes identification of items of supply; establishing data for catalogs, technical manuals, and allowance lists; and providing instructions and schedules for delivery of provisioned items.

Purchase order   The prime document created by an organization, and issued to an external supplier, ordering specific materials, parts, supplies, equipment or services.

Purchase requisition   The prime document raised by user departments authorizing the purchase of specific materials, parts, supplies, equipment or services from external suppliers.

Quality rate   Used in the calculation of overall equipment effectiveness (OEE) and overall craft effectiveness (OCE). For OEE, the proportion of the output from a machine or process that meets required product quality standards. For OCE, the proportion of repairs completed right the first time. Normally specified as a percentage.

Risk-based maintenance (RBM)   A risk-based maintenance strategy is in place to prioritize maintenance resources toward assets that carry the most risk if they were to fail. It is a methodology for determining the most economical use of maintenance resources. This is done so that the maintenance effort across a facility is optimized to minimize the total risk of failure. A risk-based maintenance strategy is based on the main phases of risk assessment for probability of failure, consequences of failure, and maintenance planning based on the risk.

Reaction time/response time   The time required between the receipt of an order or impulse triggering some action and the initiation of the action.

Ready line   Used in relation to mobile equipment. Equipment that is available but not being utilized is typically parked on the ready line. This term is used interchangeably with go-line.

Reasonable certainty   A high degree of certainty. Much more likely to be achieved than not.

Rebuild   Restore an item to an acceptable condition in accordance with the original design specifications.

Rebuild/recondition   Total teardown and reassembly of a product, usually to the latest configuration.

Recompletion   The process of entering an existing wellbore and performing work designed to establish production from a new zone.

Recordable cases   As related to health, safety and environment (HSE), recordable cases include occupational death, nonfatal occupational illness and those nonfatal occupational injuries which involve one or more of the following: loss of consciousness, restriction of work or motion, transfer to another job or medical treatment (other than first aid).

Redesign   A term that, in reliability-centered maintenance, means any one-off intervention to enhance the capability of a piece of equipment, a job procedure, a management system, or people's skills.

Redundancy   Two or more parts, components, or systems joined functionally so that if one fails, some or all of the remaining components are capable of continuing with function accomplishment; fail-safe; backup.

Refurbish   Clean and replace worn parts on a selective basis to make the product usable to a customer. Less involved than rebuild.

Reliability   The probability that an item will perform its intended function without failure for a specified time period under specified conditions. The ability of an item to perform a required function under stated conditions for a stated period of time; is usually expressed as the mean time between failures. Normally measured by mean time between failures.

Reliability analysis   The process of identifying maintenance of significant items and classifying them with respect to malfunction on safety environmental, operational, and economic consequences. Possible failure mode of an item is identified and an appropriate maintenance policy is assigned to counter it. Subsets are failure mode, effect, and criticality analysis (FMECA), fault tree analysis (FTA), risk analysis, and HAZOP (hazardous operations) analysis.

Reliability-centered maintenance (RCM)   Optimizing maintenance intervention and tactics to meet predetermined reliability goals. A structured process, originally developed in the airline industry, but now commonly used in all industries to determine the equipment maintenance strategies required for any physical asset to ensure that it continues to fulfill its intended functions in its present operating context. A number of books have been written on the subject. The seven key elements of RCM are shown in the following figure.

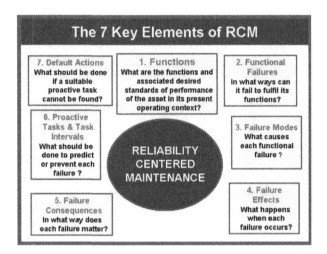

Reliability engineering   A staff function whose prime responsibility is to ensure that maintenance techniques are effective, that equipment is designed and modified to improve maintainability, that ongoing maintenance technical problems are investigated, and appropriate corrective and improvement actions are taken. Used interchangeably with plant engineering and maintenance engineering.

Reorder point (ROP)   The minimum quantity, established by economic calculation and management direction, that triggers the ordering of more items.

Repair   To restore an item to an acceptable condition by the renewal, replacement, or mending of worn or damaged parts. Restoration or replacement of parts or components as necessitated by wear, tear, damage, or failure; to return the facility, equipment, or part to efficient operating condition; any activity which returns the capability of an asset that has failed to a level of performance equal to, or greater than, that specified by its functions, but not greater than its original maximum capability. An activity that increases the maximum capability of an asset is a modification.

Repair parts   Individual parts or assemblies required for the maintenance or repair of equipment, systems, or spares. Such repair parts may also be repairable or nonrepairable assemblies, or one-piece items. Consumable supplies used in maintenance or repair, such as wiping rags, solvents, and lubricants, are not considered repair parts. Repair parts are also service parts.

Repairable   Parts or items that are technically and economically repairable. A repairable part, upon becoming defective, is subject to return to the repair point for repair action.

Replaceable Item   Hardware that is functionally interchangeable with another item but differs physically from the original part to the extent that installation of the replacement requires such operations as drilling, reaming, cutting, filling, or shimming in addition to normal attachment or installation operations.

Reserves   Estimated remaining quantities of oil and gas and related substances anticipated to be economically producible, as of a given date, by application of development projects to known accumulations. In addition, there must exist, or there must be a reasonable expectation that there will exist, the legal right to produce or a revenue interest in production, installed means of delivering oil and gas or related substances to market and all permits and financing required to implement the project.

Reservoir   A porous and permeable underground formation containing a natural accumulation of producible oil and/or gas that is confined by impermeable rock or water barriers and is individual and separate from other reservoirs.

Resources   Quantities of oil and gas estimated to exist in naturally occurring accumulations. A portion of the resources may be estimated to be recoverable, and another portion may be considered to be unrecoverable. Resources include both discovered and undiscovered accumulations.

Restoration   Any activity which returns the capability of an asset that has *not* failed to a level of performance equal to, or greater than, that specified by its functions, but not greater than its original maximum capability. Not to be confused with a modification or a repair.

Return-on-assets   An accounting term that, with regard to maintenance, is the profit attributable to a particular plant or factory, divided by the amount of money invested in plant and equipment at that plant or factory. It is normally expressed as a percentage. As such, it is roughly equivalent to the interest rate that you get on money invested in the bank, except that in this case the money is invested in plant and equipment.

Return on capital employed (ROCE)   ROCE is a measure of the profitability of a company's capital employed in its business compared with that of its peers. ROCE is calculated as a ratio, with the numerator of net income plus after-tax interest expense and the denominator of average total equity plus total debt. The net income is adjusted for nonoperational or special items impacts.

Ranking index for maintenance expenditures (RIME)   A maintenance priority methodology that provides a method to include a ranking of equipment/asset criticality combined with the repair work classification ranking to produce a priority index value.

Reliability performance indicators (RPI)   Also known as key metrics that relate to the measurement of asset reliability. Examples include the following:

• Maximum corrective time (MCT) and maximum preventive time (MPT). The most time expected for maintenance, usually specified at 95% confidence level.
• Mean active maintenance time (MAMT). Weighted average of mean corrective time and mean preventive time but excluding administrative and logistics support time.
• Mean downtime (MDT). Average time a system cannot perform its mission; including response time, active maintenance, supply time and administrative and logistics support time.
• Mean time between failures (MTBF). The average time/distance/events a product or equipment process delivers between breakdowns.
• Mean time between maintenance (MTBM). The average time between corrective and preventive actions.
• Mean time to repair (MTTR). The average time it takes to fix a failed item.

Risk   The potential for the realization of the unwanted, negative consequences of an event. The product of conditional probability of an event, and the event outcomes.

Risk-based inspection   An organization's chief reason for implementing an RBI program is to help manage the risk of a complex and potentially dangerous system. A good RBI program can also help an owner/operator understand of the plant is not being run as designed or if a new feedstock is causing more harm than expected to remaining life of equipment. Also, RBI is tool for helping owner/operators increase turn around schedules and avoiding shutdown inspection activities while maintaining production.

Rotable   A term often used in the maintenance of heavy mobile equipment. A rotable component is one which, when it has failed or is about to fail, is removed from the asset and a replacement component is installed in its place. The component that has been removed is then repaired or restored, and placed back in the maintenance store or warehouse, ready for re-issue.

Routine maintenance task   Any maintenance task that is performed at a regular, predefined interval.

Running maintenance   Maintenance that can be done while the asset is in service.

Run-to-failure   An equipment maintenance strategy where no routine maintenance tasks are performed on the equipment. The only maintenance performed on the equipment is corrective maintenance, and then only after the equipment has suffered a failure. Also known as "no scheduled maintenance."

Safety consequences   A failure has safety consequences if it causes a loss of function or other damage that could hurt or kill someone.

Safety stock   Quantity of an item, in addition to the normal level of supply, required to be on hand to permit continuing operation with a specific level of confidence if the supply is late or demand suddenly increases.

Salvage   The saving of reuse of condemned, discarded, or abandoned property, and of materials contained therein for reuse or scrapping. As a noun, it refers to property that has some value in excess of its basic material content, but is in such condition that it has no reasonable prospect of original use, and its repair or rehabilitation is clearly not practical.

Schedule compliance   The number of scheduled jobs actually accomplished during the period covered by an approved schedule; also the number of scheduled labor hours actually worked against a planned number of scheduled labor hours, expressed as percentage; one of the key performance indicators often used to monitor and control maintenance. It is defined as the number of scheduled work orders completed in a given time period (normally one week), divided by the total number of scheduled work orders that should have been completed during that period, according to the approved maintenance schedule for that period. It is normally expressed as a percentage, and will always be less than or equal to 100%. The closer to 100%, the better the performance for that time period.

Scheduled discard task   Replacement of an item at a fixed, predetermined interval, regardless of its current condition; a maintenance task to replace a component with a new component at a specified, predetermined frequency, regardless of the condition of the component at the time of its replacement. An example would be the routine replacement of the oil filter on a motor vehicle every 6000 miles. The frequency with which a scheduled discard task should be performed is determined by the useful life of the component.

Scheduled maintenance (SM)   Preplanned actions performed to keep an item in specified operating condition by means of systematic inspection, detection, and prevention of incipient failure. Sometimes called preventive maintenance, but actually a subset of PM.

Scheduled operating time   The time during which an asset is scheduled to be operating, according to a long-term production schedule.

Scheduled restoration task   A maintenance task to restore a component at a specified, pre-determined frequency, regardless of the condition of the component at the time of its replacement. An example would be the routine overhaul of a slurry pump every 1000 operating hours. The frequency with which a scheduled restoration task should be performed is determined by the useful life of the component.

Scheduled work order   A work order that has been planned and included on an approved maintenance schedule.

Scoping   A planning activity which outlines the extent/scope and detail of work to be done and defines the resources needed.

Scoreboard for Excellence   Baseline for today's most comprehensive benchmarking guides for maintenance operations. Developed initially in 1981 and enhanced into its present format of five different Scoreboard for Excellence versions. See **Scoreboard for Facilities Management Excellence** and **Scoreboard for Maintenance Excellence.**

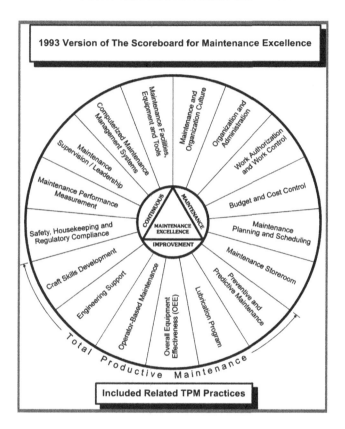

1993 Version of The Scoreboard for Maintenance Excellence

Included Related TPM Practices

Scoreboard for Facilities Management Excellence   Today's most comprehensive benchmarking guide for pure facilities maintenance operations. Developed along the same format as the Scoreboard for Maintenance Excellence, the New Millennium version includes 27 evaluation (best practice) categories and 300 evaluation categories. An excellence benchmarking guide for operations within large physical plant and facilities complexes such as universities, state and municipal building complexes, healthcare facilities, secondary school complexes and retail organizations with nationwide system of sites. Provides an essential benchmarking guide where results become an important external benchmark against recognized best practices and also the user's baseline for Continuous Reliability Improvement.

The Scoreboard for Facilities Management Excellence

CATEGORY	The Scoreboard for Facilities Management Excellence Category Descriptions (Part 1)	Evaluation Items	Total Points in Category
A.	The Organizational Culture and PRIDE in Maintenance	5	50
B.	Facilities Organization, Administration and Human Resources	10	100
C.	Craft Skills Development	10	100
D.	Facilities Management Supervision/Leadership	10	100
E.	Business Operations, Budget and Cost Control	15	150
F.	Work Management and Control: Maintenance and Repair (M/R)	10	100
G.	Work Management and Control: Construction and Renovation (C/R)	5	50
H.	Facilities Maintenance and Repair Planning and Scheduling	15	150
I.	Facilities Construction and Renovation Planning /Scheduling and Project Management	10	100
J.	Facilities Planning and Property Management	10	100
K.	Facilities Condition Evaluation Program	5	50
L.	Facilities Storeroom Operations and Internal MRO Customer Service	15	150
M.	MRO Materials Management and Procurement	10	100

© 2001-2006 The Maintenance Excellence Institute. All Rights Reserved

The Scoreboard for Facilities Management Excellence

CATEGORY	The Scoreboard for Facilities Management Excellence Category Descriptions (Part 2 Continued)	Evaluation Items	Total Points in Category
N.	Preventive Maintenance and Lubrication	20	200
O.	Predictive Maint. and Conditioning Monitoring Technologies	10	100
P.	Building Automation and Control Technology	5	50
Q.	Utilities Systems Management	10	100
R.	Energy Management and Control	10	100
S.	Facilities Engineering Support	10	100
T.	Safety and Regulatory Compliance	15	150
U.	Security Systems and Access Control	10	90
V.	Facilities Management Performance Measurement	15	150
W.	Facilities Maintenance Management System (FMMS) and Business System	15	150
X.	Shop Facilities, Equipment, and Tools	10	100
Y.	Continuous Reliability Improvement	10	100
Z.	Grounds and Landscape Maintenance	15	150
ZZ.	Housekeeping Service Operations	15	150
Total Evaluation Items and Points		**300**	**3000**

© 2001-2006 The Maintenance Excellence Institute. All Rights Reserved

Scoreboard for Maintenance Excellence   TMEII has today's most comprehensive benchmarking guide for plant maintenance operations and now an updated 2015 version for complex continuous processing operations. Developed initially in 1982 and enhanced into its present format of 38 evaluation (best practice) categories and 600 evaluation items. Previous versions have been used for over 200 assessment/audits and has been used by approximately 5000 organizations for benchmarking all types of maintenance operations. Provides an essential benchmarking guide where results become an important external benchmark against recognized best practices and also the user's baseline for Continuous Reliability Improvement.

THE SCOREBOARD FOR MAINTENANCE EXCELLENCE™ 2015 Part 1

MAINTENANCE BENCHMARK EVALUATION SUMMARY

Category Number	Benchmark Category Descriptions &Rating Values for Each Evaluation Criteria 5=Poor, 6=Below Average, 7=Average, 8=Good, 9-Very Good & 10= Excellent New Benchmark or Expanded Category	Number of Criteria	Total Assessment Points per Category
1.	Top Management Support to Maintenance and Physical Asset Management	10	100
2.	Maintenance Strategy, Policy and Total Cost of Ownership	30	300
3.	The Organizational Climate and Culture	9	90
4.	Maintenance Organization, Administration and Human Resources	18	180
5.	Craft Skills Development and Technical Skills	12	120
6.	Operator Based Maintenance and PRIDE in Ownership	10	80
7.	Maintenance Leadership, Management and Supervision	12	120
8.	Maintenance Business Operations, Budget and Cost Control	15	150
9.	Work Management and Control: Shop Level Maintenance Repair (M/R)	12	120
10.	Work Management and Control: Shutdowns, Turnarounds and Outages (STO)	26	260
11.	Shop Level Reliable Planning, Estimating and Scheduling M/R	30	300
12.	STO and Major Planning /Scheduling and Project Management	16	160
13.	Contractor Management	31	310
14.	Manufacturing Facilities Planning and Site Property Management	9	90
15.	Production Asset and Facilities Condition Evaluation Program	6	60
16.	Storeroom Operations and Internal MRO Customer Service	12	120
17.	MRO Materials Management and Procurement	12	120
18.	Preventive Maintenance and Lubrication	18	180
19.	Predictive Maintenance and Condition Monitoring Technology Applications	15	150
20.	Reliability Centered Maintenance (RCM)	34	340

THE SCOREBOARD FOR MAINTENANCE EXCELLENCE™ 2015 Part 2

Category	Benchmark Category Descriptions &Rating Values for Each Evaluation Criteria 5=Poor, 6=Below Average, 7=Average, 8=Good, 9-Very Good & 10= Excellent New Benchmark Category or Expanded Category	Number of Criteria	Total Assessment Points Per Category
21.	Reliability Analysis Tools: Root Cause Analysis (RCA), Root Cause Corrective Action (RCCA) Failure Modes Effects Analysis (FMEA), Root Cause Failure Analysis (RCFA) and Failure Reporting Analysis and Corrective Action System (FRACAS)	17	170
22.	Risk Based Maintenance (RBM)	24	240
23.	Process Control and Instrumentation Systems Technology	9	90
24.	Energy Management and Control	12	120
25.	Maintenance Engineering and Reliability Engineering Support	9	90
26.	Health, Safety, Security and Environmental (HSSE) Compliance	15	150
27.	Maintenance and Quality Control	9	90
28.	Maintenance Performance Measurement	12	120
29.	Computerized Maintenance Management System (CMMS) as a Business System	18	180
30.	Shop Facilities, Equipment, and Tools	9	90
31.	Continuous Reliability Improvement	15	150
32.	Critical Asset Facilitation and Overall Equipment Effectiveness (OEE)	15	150
33	Overall Craft Effectiveness (OCE)	6	60
34.	Sustainability	11	110
35.	Traceability	19	190
36	Process Safety Management (PSM) and Management of Change (MOC)	26	260
37.	Risk Based Inspections (RBI) and Risk Mitigation	29	290
38.	PRIDE in Maintenance	8	80
	Total Evaluation Items:	600	
	Scoreboard Total Possible Points:		6000
	Actual Total Benchmark Value Score of All Ratings:		
	Assessment Performed By:	Date:	

Secondary damage Any additional damage to equipment, above and beyond the initial failure mode, that occurs as a direct consequence of the initial failure mode.

Secondary failures Malfunctions that are caused by the failure of another item.

Secondary function A term used in reliability-centered maintenance. The secondary functionality required of an asset generally not associated with the reason for acquiring the asset, but now that the asset has been acquired, the asset is now required to provide this functionality. For example, a secondary function of a pump may be to ensure that all of the liquid that is pumped is contained within the pump (i.e., the pump does no't leak). An asset may have tens or hundreds of secondary functions associated with it.

Serial number   Number or letters that uniquely identify an item.

Service contract   Contract calling directly for a contractor's time and effort rather than for a specific end product.

Service level   Frequency usually expressed as a percentage, with which a repair part demand can be filled through a particular service stock echelon. A 95% level of service means that 95 out of 100 demands are properly issued. If viewed from the end customer service technician perspective, the service level is the percent of parts received out of those requested, from all levels of the support system.

Service line   The pipe the carries the gas from the distribution mains in the street, across private property to the customer's meter.

Serviceability   Characteristics of an item, equipment, or system that make it easy to maintain after it is put into operation; similar to maintainability.

Servicing   The replenishment of consumables needed to keep an item in operating condition.

Shale   A very fine-grained sedimentary rock that is formed by the consolidation of clay, mud, or silt and that usually has a finely stratified or laminated structure. Certain shale formations, such as the Eagle Ford and the Barnett, contain large amounts of oil and natural gas.

Shelf life   The period of time during which an item can remain unused in proper storage without significant deterioration.

Shop stock   Self-service items, such as nuts, bolts, fitting, etc., that are stored directly in the shop work area. May be on consignment directly from the vendor or vendor may inventory and "keep full" as needed without significant paperwork requirement.

Shutdown   The period of time when equipment is out of service.

Shutdown, turnaround, and outages (STO)   Typically, a well-planned, scheduled, and executed amount of work (all types) with complete or partial shutdown of production activities. Much more complex for oil and gas operations as compared to discrete manufacturing.

Shutdown maintenance   Maintenance done while the asset is out of service, as in the annual plant shutdown.

Social life cycle analysis (S-LCA)   A methodology for assessing internalities and externalities of the production of goods and services based on social and socioeconomic indicators.

Sour gas   Sour gas is natural gas or any other gas containing significant amounts of hydrogen sulfide (H_2S).

Spacing   The distance between wells producing from the same reservoir. Spacing is often expressed in terms of acres (e.g., 80-acer spacing) and is often established by regulatory agencies.

Specifications   Physical, chemical, or performance characteristics of equipment, parts, or work required to meet minimum acceptable standards.

Standard item   Part, component, material, subassembly, assembly, or equipment that is identified or described accurately by a standard document or drawing.

Standardization   Process of establishing the greatest practical uniformity of items and of practices to assure the minimum feasible variety of such items and practices, and affect optimum interchangeability.

Standard job   A work order stored in the CMMS that contains all the necessary information required to perform a maintenance task (see also **Model Work Order**).

Standby   Assets installed or available but not in use.

Standing work order   A work order that is left open either indefinitely or for a predetermined period of time for the purpose of collecting labor hours, costs, and/or history for tasks for which it has been decided that individual work orders should not be raised. Examples would include standing work orders raised to collect time spent at safety meetings, or in general housekeeping activities; a work order that remains open, usually for the annual budget cycle, to accommodate information small jobs or for specific tasks.

Steam-assisted gravity drainage (SAGD)   A process used to recover bitumen that is too deep to mine. A pair of horizontal wells is drilled from a central well pad. In a plant nearby, steam generators heat water and transform it into steam. The steam then travels through above-ground pipelines to the wells. It enters the ground via the steam injection well and heats the bitumen to a temperature at which it can flow by gravity into the producing well. The resulting bitumen and condensed steam emulsion is then piped from the producing well to the plant, where it is separated and treated. The water is recycled for generating new steam.

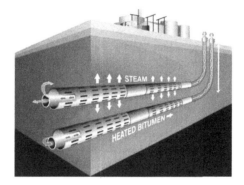

Stock keeping unit (SKU)   A warehouse inventory management term for individual stock items carried in inventory.

Stock number   Number assigned by the stocking organization to each group of articles or material, which are then treated as if identical within the using supply system; also called part number, item number, or part identifier.

Stock out   Indicates that all quantities of a part normally on hand have been used, so that the items are not presently available. Demand for a nonstock part is usually treated as a separate situation and procured via defined procedures and vendors.

Storage   Natural gas can be stored in underground storage fields. Storage fields are typically old gas fields that have been depleted of natural gas, salt-domes that have been hollowed out by solution mining, or other porous rock that is surrounded by impermeable rock.

• **Injection** is the process of pumping natural gas into storage fields, using during off-peak summer months

• **Withdrawal** is the process of pumping natural gas out of storage fields, usually during periods of winter peak demand

Storage underground   The use of subsurface porous rock formations for storing gas. Depleted natural gas production fields are often used for storage. Typically, natural gas is pumped into a field during the summer when demand and commodity prices usually are lower and pumped out of storage during winter when demand is high.

Stores issue   The issue and/or delivery of parts and materials from the store or warehouse.

Stores requisition   The prime document raised by user departments authorizing the issue of specific materials, parts, supplies, or equipment from the store or warehouse.

Sweet Gas   Natural gas that contains little or no hydrogen sulfide.

Supply   Procurement, storage, and distribution of items.

Support equipment Items required to maintain systems in effective operating condition under various environments. Support equipment includes general and special-purpose vehicles, power units, stands, test equipment, tools, or test benches needed to facilitate or sustain maintenance action, to detect or diagnose malfunctions, or to monitor the operational status of equipment and systems.

Sustainability Sustainable development is present in regards to maintenance and physical asset management to meet the needs of the current operation without compromising the ability of future site operations to meet their own need to maintain physical assets. The organization understands that the organizing principle for sustainability is sustainable development, which includes the four interconnected domains of ecology, economics, politics, and culture.

Teamwork works! The proven practice of the synergistic results by uncommon people working toward a common mutual goal. Often, this is ordinary people achieving extraordinary results (see also **Maintenance, winning**)

Technical data and information Includes, but is not limited to, production and engineering data, prints and drawings, documents such as standards, specifications, technical manuals, changes in modifications, inspection and testing procedures, and performance and failure data.

Terotechnology An integration of management, financial, engineering, operating maintenance, and other practices applied to physical assets in pursuit of an economical life cycle; the application of managerial, financial, engineering and other skills to extend the operational life of, and increase the efficiency of, equipment and machinery.

Test and support equipment All special tools and checkout equipment, metrology and calibrations equipment, maintenance stands, and handling equipment required for maintenance. Includes external and built-in test equipment (BITE) considered part of the supported system or equipment.

Thermography The process of monitoring the condition of equipment through the measurement and analysis of heat. Typically conducted through the use of infra-red cameras and associated software. Commonly used for monitoring the condition of high voltage insulators and electrical connections, as well as for monitoring the condition of refractory in furnaces and boilers, amongst other applications.

Throwaway maintenance Maintenance performed by discarding used parts rather than attempting to repair them.

Top leaders C-positions, especially the CFO, VP operations, managing directors, and engineering/maintenance managers.

Total asset management An integrated approach to asset management that incorporates elements such as reliability centered maintenance, total productive maintenance, design for maintainability, design for reliability, value engineering, life cycle costing, probabalistic risk assessment, and others, to arrive at the optimum cost-benefit-risk asset solution to meet any given production requirements.

Total productive maintenance (TPM) A Japanese-based maintenance strategy for companywide equipment management program emphasizing operator involvement in equipment maintenance, continuous improvement in equipment effectiveness, and measurement of overall equipment effectiveness (OEE).

Traceability Traceability is characterized by a number of essential elements: a) an unbroken chain of comparisons going back to a standard acceptable to the parties, usually a national or international standard; b) measurement of uncertainty; the measurement uncertainty for each step in the traceability chain must be calculated according to defined methods; and c) must be stated so that an overall uncertainty for the whole chain may be calculate and have documentation for each step in the chain performed according to documented and generally acknowledged procedures. The results must equally be documented.

Tradesperson   Alternative to "craftsperson." A skilled maintenance worker who has typically been formally trained through an apprenticeship program. TMEI likes to think of craftspeople and tradespeople as our MVPs—today's most valuable people for all types of maintenance operations around the world.

Transmission lines   A gas pipeline that transports large quantities of highly pressurized natural gas over long distances.

Transportation   The movement by a pipeline operator of natural gas owned by another party, typically a distribution company or end-user. Transportation service is a type of tariff many gas companies offer to large users who desire to purchase their gas from someone other than the local distribution company (LDC).

Teratology   The process of monitoring the condition of equipment through the analysis of properties of its lubricating and other oils. Typically conducted through the measurement of particulates in the oil, or the measurement of the chemical composition of the oil (spectrographic oil analysis). Commonly used for monitoring the condition of large gearboxes, engines, and transformers, amongst other applications.

Tight gas   Natural gas produced from relatively impermeable rock. Getting tight gas out usually requires enhanced technology applications like hydraulic fracturing. The term is generally used for reservoirs other than shale.

Total recordable rate (TRR)   The total recordable rate is a measure of the rate of recordable cases, normalized per 100 workers per year. The factor is derived by multiplying the number of recordable injuries in a calendar year by 200,000 (100 employees working 2000 h per year) and dividing this value by the total man hours actually worked in the year.

Total shareholder return (TSR)   Represents share price appreciation and dividends returned to shareholders over a period. It is calculated as follows: [(stock price at the end of the period) − (stock price at the start of the period) + (dividends paid during the calculation period) ÷ (stock price at the start of the period)].

Total system support (ToSS)   The composite of all considerations needed to assure the effective and economical support of a system throughout its programmed life cycle.

Transmission   Transmission is the process of moving gas from producing regions or storage fields to major consuming areas.

- **Compression** is the process of increasing the pressure of the gas moving through a pipeline, so that more gas can be moved a further distance. Large pipelines may operate in the range of 3,000 psi and be up to 60″ in diameter.
- **Compressor** is the device used to add pressure to the gas moving through the pipeline. Compressors are typically powered by natural gas engines and are rated in hundreds or thousands of horsepower per compressor.

Troubleshooting   Locating or isolating and identifying discrepancies or malfunctions of equipment and determine the corrective action required.

Turnaround time   Interval between the time a repairable item is removed from use and the time it is again available if full serviceable condition.

Turnover   Measurement on either numbers of parts or on monetary value that evaluates how often a part is demanded versus the average number kept in inventory. For example, if two widgets are kept in inventory and eight are used each year, then the turnover is 8/2 = 4 times per year. In monetary terms, turnover is cost of inventory sold/average cost of inventory carried.

Unconventional reservoirs   Reservoirs with permeability so low (generally less than 0.1 millidarcy) that horizontal hydraulically fractured stimulated wells or other advanced completion techniques must be utilized to extract hydrocarbons at commercial rates. Shale reservoirs such as the Eagle Ford and Barnett, as well as tight reservoirs like the Bakken and Three Forks, both are examples of unconventional reservoirs.

Undeveloped acreage   Acreage on which wells have not been drilled or completed to a point that would permit the production of commercial quantities of oil and gas regardless of whether or not the acreage contains proved reserves.

Unit   The joining of interests in a reservoir or field to provide for development and operations without regard to separate property interests. Also, the area covered by a unitization agreement.

Unplanned maintenance   Maintenance done without planning or scheduling; could be related to a breakdown, running repair, or corrective work; any maintenance activity for which a predetermined job procedure has not been documented, or for which all labor, materials, tools, and equipment required to carry out the task have been not been estimated, and their availability assured before commencement of the task.

Unscheduled maintenance (UM)   Emergency maintenance (EM) or corrective maintenance (CM) to restore a failed item to usable condition.

Up   In a condition suitable for use.

Uptime   The time that an item of equipment is in service and operating.

Usage   Quantity of items consumed or necessary for product support. Usage is generally greater than the technical failure rate.

Useful life   The maximum length of time that a component can be left in service, before it will start to experience a rapidly increasing probability of failure. The useful life determines the frequency with which a scheduled restoration or a scheduled discard task should be performed.

Utilization   The proportion of available time that an item of equipment is operating. Calculated by dividing equipment operating hours by equipment available hours. Generally expressed as a percentage.

Value engineering   A systematic approach to assessing and analyzing the user's requirements of a new asset, and ensuring that those requirements are met, but not exceeded. Consists primarily of eliminating perceived "non-value-adding" features of new equipment.

Variance analysis   Interpretation of the causes for a difference between actual and some norm, budget, or estimate.

Vibration analysis   The process of monitoring the condition of equipment, and the diagnosis of faults in equipment through the measurement and analysis of vibration within that equipment. Typically conducted through handheld or permanently positioned accelerometers placed on key measurement points on the equipment. Commonly used on most large items of rotating equipment, such as turbines, centrifugal pumps, motors, gearboxes etc.

Warranty   Guarantee that an item will perform as specified for at least a specified time, or will be repaired or replaced at no cost to the user.

Warranty claims   Replacement or reimbursement due to bad parts or equipment from a vendor or original equipment manufacturer (OEM). Sunken gold treasure within most maintenance storerooms where warranty claims never get identified and sent back to the item vendor or OEM. For the most part, it is not the CMMS's fault!

Waterflood   An improved oil recovery technique that involves injecting water into a producing reservoir to enhance movement of oil to producing wells.

Wear out   Deterioration as a result of age, corrosion, temperature, or friction that generally increases the failure rate over time.

Wet gas   Produced gas that contains natural gas liquids.

Working interest   The right granted to the lessee of a property to explore for, produce and own oil, gas or other minerals. The working interest owners bear the exploration, development, and operating costs.

Workload   The number of labor hours needed to carry out a maintenance program, including all scheduled and unscheduled work and maintenance support of project work.

Work order (WO)   A unique control document that comprehensively describes the job to be done; may include formal requisition for maintenance, authorization, and charge codes, as well as what actually was done. The prime document used by the maintenance function to manage maintenance tasks. It may include such information as a description of the work required, the task priority, the job procedure to be followed, the parts, materials, tools and equipment required to complete the job, the labor hours, costs and materials consumed in completing the task, as well as key information on failure causes, what work was performed etc.

Work request   The initial request for maintenance service or work usually as a statement of the problem. The work request provides the preliminary information for creation of the work order. Depending on the cost and scope of a work request an approval process may be required before the work order is created, planned, and scheduled. The prime document raised by user departments requesting the initiation of a maintenance task. This is usually converted to a work order after the work request has been authorized for completion.

X-maintenance   Extreme maintenance challenges: Illustrates four key stages that as assets age and new assets are also added, the total maintenance requirements increase often without a corresponding increase in staffing levels.

Zero-based budgeting   An accounting strategy that, as related to maintenance, is the charge back of all maintenance work to the customer and is based on a computed shop rate per hour plus parts and rental equipment if required.

Appendix C

Maintenance Planner/Scheduler or Maintenance Coordinator: Position Description, Job Evaluation Form

Position Questionnaire

Position Title: **Maintenance Planner/Coordinator**   Date: **mm/dd/yyyyy**

Number of Associates with same title in Department:

Department: Maintenance Operations   Reports to: **Maintenance Manager**

Is this a new position? X Yes __ No

1. **Primary purpose and function of position.** Please explain in two or three sentences. Provides overall maintenance planning and scheduling support to include contract maintenance and the planning of repairs and overhauls during scheduled shutdowns, turnarounds, and outages (STO) in support to engineering. Provides close interaction and support to storeroom operations and is responsible for maintaining the site's preventive maintenance program. Serves as a primary resource for computerized maintenance management system (CMMS) implementation and the maintenance of all related databases.
2. **Duties and responsibilities.** List in order of importance the major duties that are performed and normally assigned to the position. Indicate the percentage of time spent on each duty. Please add additional sheets if needed.

% of time	
5	1. Coordinates maintenance work order (MWO) system, processing of MWOs, and provides overall work control to ensure maximum utilization of internal craft labor and contractor labor.
5	2. Reviews MWOs to confirm priorities with manufacturing operations and maintenance and to establish craft requirements. Reviews drawings/prints, equipment, and parts manuals as required to determine sequence of crafts, scope of work, and job plans for major repairs. Utilizes the site's priority system to support effective planning/scheduling of critical repairs.
10	3. Develops labor estimates for planned work and establishes schedules based on confirmed priorities and craft labor availability. Coordinates with manufacturing operations to establish firm schedules for equipment availability prior to start of repair.

Continued

—cont'd

% of time	
5	4. Develops and maintains a current backlog summary of maintenance repairs and preventive maintenance/predictive maintenance (PM/PdM) inspections due. Evaluates level of backlog and provides recommendations on use of overtime for priority repairs and PM scheduling. Monitors status of ongoing major planned repairs, overhauls, or new system installations. Maintains the site's master plan for major repairs, overhauls, and all maintenance work to be scheduled during scheduled shutdowns or as future capital projects.
10	5. Develops and coordinates modernization of current stores area into an organized storeroom operation with an effective parts inventory management system, access control method, and bin location system along with improved storage methods. Provides support and coordinates completing a physical inventory of existing stock items and developing the initial parts database master file on a new CMMS. Becomes the primary resource for ensuring that the parts database is maintained over the long term with proper internal stock numbers, vendor part information, min/max stock levels, specifications, etc.
10	6. Determines material and spare parts availability prior to scheduling. Works with craft supervisors and crafts workforce to determine parts required on major repair. Coordinates with storeroom to reserve and kit parts prior to major planned repairs. Maintains status of requested materials/parts to meet established repair schedule. Provides recommendations to storeroom on stock levels, additions, and/or deletions from inventory. Identifies critical spares and updates equipment history database with critical spares that are carried in storeroom inventory.
10	7. Develops job plans and standard repair practices during planning process. Coordinates entering of job plans to the CMMS and updating these standard job plans as required.
5	8. Evaluates requirements for contract maintenance against existing craft capability and current backlog levels.
2	9. Ensures that information on contract maintenance repairs, overhauls, and all repairs is added to the equipment history database and that this database is accurately maintained.
10	10. Responsible for coordinating the PM/PdM review team for review and updating of current PM/PdM procedures and the development of new PM/PdM inspection procedures and frequencies. Develops planning times for PM inspection procedures and frequencies for revision and improvements and ensures PM/PdM database is accurately maintained. Coordinates development of predictive maintenance technology applications based upon results of assessment for predictive technologies application at the Las Vegas site. Ensures that current PM/PdM procedures reflect actual operating conditions.

—cont'd

% of time	
2	11. Supports scheduling PdM technology application such as oil analysis/ infrared, etc. and planning of maintenance repairs to correct deficiencies found during both PM and PdM inspections. Coordinates introduction of all new PdM technologies to the operation.
5	12. Develops and provides data (MTTR/MTBF) and recommendations for improving equipment reliability. Reviews equipment repair history and PM/PdM results for failure trends and supports maintenance in identifying and correcting root causes of major failures.
2	13. Coordinates with plant operations staff to establish PM/PdM, lubrication, and parts requirements for any new equipment or systems. Updates equipment history database with information on vendor, equipment specifications, warranty, information, parts lists, etc.
5	14. Responsible for supporting implementation and administration of a future computerized maintenance management system (CMMS) as a prime internal technical resource. Coordinates development and maintenance of databases for equipment history, work orders, and PM. Supervises clerical support for major data entry tasks.
2	15. Provides support to updating engineering drawings and coordinates cross-referencing of engineering drawing numbers to equipment history database. Supports craft areas (and contract maintenance) by providing drawings/prints as required with MWO for major planned repairs and overhauls.
2	16. Works closely with the maintenance manager, all craft supervisors, and manufacturing operations staff to support effective planning for shutdowns and special projects. Develops the overall master plan for maintenance-related activities and coordinates development of materials/parts requirements and requisitioning, estimates of craft labor, and plan for contract maintenance support as required.
5	17. Serves as the primary support to the maintenance manager in developing performance information for the Accuride Maintenance Excellence Index, life cycle costing, and other maintenance cost information.
5	18. Performs other related duties and special projects as assigned. a. Provides training and support for a maintenance planner/ coordinator backup in case the incumbent is absent. b. Provides support to the maintenance manager for other maintenance initiatives such as craft skill development, operator-based maintenance, or vendor partnership for MRO parts and materials.

3. Briefly describe any special functions performed in addition to normal responsibilities.

Serves as a key internal resource for effective CMMS implementation, supports utilization of all CMMS system modules, and supports training of internal users on a new CMMS system.

Are the special functions above performed?

[✓] Weekly   [ ] Monthly   [ ] Quarterly   [ ] Other _____

4. Office machines and clerical skills needed to perform in the position:

Please check one:

	Daily	Weekly	Occasionally	Special Training Required
[ ] Typing	✓			
[ ] Dictation				
[ ] 10-key				
[✓] Personal computer	✓			
[✓] Windows office	✓			MS project & access DB
[✓] Other CRTs	✓			Basic knowledge of energy mgmt system
[ ] Graphics packages and document management system			✓	
[✓] Programming/CMMS language			✓	Future CMMS

In detail, list any special computer packages, programming knowledge, etc., used:
Should understand how company's STO project management operates and how to load deferred maintenance work orders to this system. Capability to use CMMS Report Writer is helpful as well as is being able to use others such as *Crystal Report Writer.*

5. What level of education is required to perform in this position?

[ ] Less than high school   [ ] High school diploma/equivalent

[ ] College degree   [ ] Advanced degree

[✓] Certification in planning, estimating and scheduling   [✓] Technical/trade school plus multi-skilled crafts background

Why do you feel this level of education is required for the position?

a. Needs good craft background (preferably multi-craft experience) for technical knowledge to support planning of maintenance work across multiple trade areas
b. Needs formal maintenance planning & scheduling training and on the job training as well.
c. Must have good computer skills
d. Must be able to develop skills in MRO materials management

6. In detail, give an example of an independent decision made in this position?

a. Work backlog level high –this position recommends scheduling of overtime
b. Contract maintenance work is of poor quality – this position reschedules contractor, recommend new contractor or recommends with holding payment to contractor
c. This position may recommend new vendor/supplier for spare parts

7. Does this position make recommendations to:
[✓]   Policies & procedures   [✓] Vendors   [✓]   Course of action

8. Is this position supervised? [✓] Yes   [ ] No

What amount of supervision and technical?
[✓]   Shop floor supervisors are available on the floor/department at all times for technical support
[✓]   Maintenance manager is available only when needed

9. Does this position have supervisory responsibilities?   [X] Yes   [ ] No

If so, which of the following:
[✓] Instructing   [✓] Assigning work (to craft work groups, First-line supervisors
                                             actually assigns work to individual craft person)
[✓] Review work   [✓] Work schedules

[ ] Performance reviews
__recommends __approves/handles

[ ] Handles disciplinary problems
__recommends __approves/handles

List position titles supervised: i.e., Possible on size of operation

Storeroom Attendant if
there is a direct
responsibility for stores
-Data Entry-Admin Person   _____

_____   _____

10. How does this position use resourcefulness, originality and/or initiative?

This position allows the maintenance department to operate in a planned and controlled environment instead of a high stressed reactive environment. It provides direction to storeroom operations in some case for a small operation. It ensures availability of parts and materials at right time, right quantity, right place and best price for proactive planned. When performed correctly this position can help achieve craft productivity gains equal to a minimum of 5 equivalent craft positions for each planner position. This position is also in a role to help increase asset uptime resulting in greater output and greater profits

11. What kinds of errors could happen in this position?

There are four main areas a) Lost capacity on plant operations equipment, b) low productivity of maintenance man hours and c) excess spares parts inventory cost and d) excessive obsolete parts on hand and high costs of net inventory.

How would such errors ordinarily be checked or discovered?

Maintenance metrics via a system such The Reliable Maintenance Excellence Index will clearly reveal, excessive downtime of critical equipment (low OEE), high % emergency wow, low craft productivity (or OCE),

low PM and Schedule compliance will all show clearly any negative as well as positive results
Direct feedback from associates, maintenance manager, first-line maintenance supervisors and plant operations staff

What would happen if an error is not caught?

Nothing catastrophic, but this position support a business approach to physical asset management and allows a progression toward continuous improvement for the maintenance of valuable plant assets (both human capital & plant equipment)

12. What "people" contact does this position have?

	Continually (at least 2x daily)	Frequently (4-5x week)	Occasionally (1-2x week)	Never	Method (phone, letter, in person)
✓ Other departments	✓				All
✓ Stores (maint)	✓				In person, via a PC
✓ Vendors			✓		All
			✓		Contractor's Staff
☐ General public					
					Auditors on occasion
☐ Govt/state agencies					

What level of contact (clerical, administrative, managers, Vice Presidents, etc.)?

High level of contact with first-line maintenance supervisors and maintenance manager
Regular contact with plant manager, operations manager, VP Operations and other plant operations personnel needing information or maintenance service

Give an example of such contacts:

Coordinate scheduling of repairs or PM tasks where asset must be scheduled out of service for repair or PM
Has major input on deciding the level of repairs made. This person may be asked along with Maintenance Leader to review deferred jobs for shutdowns (STO type work) with STO Project Manager and STO team

ADDITIONAL INFORMATION: Please describe any other aspects of the position that are important and have not been covered above.

This person can provide valuable information such as which vendors/contractors are performing to the company's performance standards based on actual performance and reliability data. This person can have major economic impact affecting such things as craft utilization, the cost of maintenance repairs and parts, timely repair service and status of work being requested by operations.

TASK LIST:

Please check the appropriate section for each task on the following checklist. Indicate whether or not the task is required for the position and the degree to which it is required:
NOT REQUIRED, OCCASIONALLY REQUIRED, FREQUENTLY REQUIRED.
(Occasionally is defined as irregular or infrequent intervals.)

If a task statement does not exactly fit the position, change the statement by crossing out or adding additional words.

	Not Required	Occasionally Required	Frequently Required
➤ Design and determine performance report, layout, format and retrieve data for finished report	☐	☐	✓
➤ Proofread and correct errors	☐	✓	☐
➤ Have experience to operate power equipment, power tools or perform general maintenance	☐	☐	✓
➤ Conduct specialized training sessions related to CMMS to educate department or company associates	☐	✓	☐
➤ Operate computer peripheral equipment	☐	☐	✓
➤ Set up and type financial reports or statistical tabulations	☐	✓	☐
➤ Compose department memos without supervisor approval	☐	✓	☐
➤ Set up appointments independently, schedule meetings, conferences and/or make travel arrangements and reservations	☐	☐	✓

TASK LIST (continued)

Task	Col 1	Col 2	Col 3
Maintain confidential records and files and handle confidential correspondence	✓		
Open, sort and distribute incoming mail. Date stamp papers and documents	✓		
Answer telephone , screen and place calls, and refer callers to appropriate parties		✓	
Gather and analyze data to prepare complex reports		✓	
Audit records for accuracy and make appropriate corrections		✓	
Investigate accidents/injuries to determine cause	✓		
Receive and welcome visitors and refer them to appropriate parties	✓		
Write and test complex computer programs	✓		
Prepare and process expense accounts	✓		
Maintain status of inventoried maintenance parts & supplies and requisition new parts & supplies			✓
Monitor department overtime, sick leave, vacations and holidays	✓		✓
Maintain, sort and update mailing lists and labels	✓		
Handle work requests, complaints, or inquiries from callers/customers			✓
Prepare, process and verify invoices, bills and/or receipts			✓
Coordinate work to complete, process and trace purchase orders		X	
Gather and tabulate data; compile routine reports as directed by supervisor			✓
Record and verify entries in CMMS data base	✓		✓
Balance ledger accounts and reconcile accounting statements			
Arrange for equipment repairs via planning process			✓
Maintain records of cash receipts and disbursements	X		
Prepare complex legal documents	✓		
Photocopy, collate, punch and/or bind documents		X	
Compile routine reports for federal, state and local agencies	✓		
Extract, tabulate, and summarize information for periodic or special reports		✓	
Interpret and enforce Company policies and procedures		✓	
Manually key in and verify data and correct errors		✓	
Negotiate prices with vendors		X	
Troubleshoot equipment for problems or repairs			✓
Support maintenance manager in researching, analyzing and prepare budget for department or Company		✓	
Unload/load materials, unpack cartons	✓		
Compose department memos to be reviewed by supervisor		✓	
Manipulate computer control switches on console panels, storage devices, readers, printers and punches according to detailed instructions		✓	
Type and proofread legal documents	✓		
Research problems by reconstructing past repairs as related to RCM/RCA team work		✓	
Council associates or managers on performance related issues		✓	
Assist others in the company with computer, and CMMS questions		✓	
Write basic computer programs as directed by supervisor	X		
File, retrieve, assemble and distribute records, papers documents and reports		✓	
Make non-routine decisions without a supervisor			✓
Make decisions that impact bottom line of company			✓

Please draw your organizational chart:

Immediate supervisor's position ➤

This position ➤

Any direct reports to this position ➤

Associate's printed name

Associate signature

Phone extension

Date completed

I have reviewed the contents of this questionnaire and feel it is an accurate and complete representation of the position

Appendix D

Charter: Format for a Leadership Driven-Self-Managed Team at GRIDCo Ghana

Ralph W. Peters, Franklin Takyi
The Maintenance Excellence Institute International, Ghana

The GRIDCo CMMS Selection and Best Practices Implementation Team-Draft January 14, 2014

1. Purpose: To provide senior guidance to CMMS (computerized maintenance management system) selection and best practice across the entire GRIDCo operation in Ghana. Support the team with resources of people and capital to complete a very important project for GRIDCo's current operation and future goals for expanded power transmission. This team is to identify areas for improvement and to ensure that total operations success and profit optimization occur as a result of this major project.	Proposed start date February 1, 2014	Projected completion September 30, 2014

2. Expectations (sponsor's):
   a. Team will incorporate the 5 steps of the "Action Cycle" into their work processes (Assess the Situation/Determine the Causes/Target Solutions/Implement Actions/ Make it ongoing)
   b. Team will establish a consistent reporting process to keep their sponsor informed of progress
   c. Team Will escalate issues, that cannot be resolved in the team, to the sponsor
   d. Team Will establish a measurement process on Day One with key baseline data that will validate results over time
   e. Team will use a facilitator throughout the process

3. Importance of this project to GRIDCo:
   a. Customers are serviced faster and more efficiently to restore power transmission
   b. Reduce waiting times for parts to arrive in the field
   c. Control of inventory levels
   d. One central location for parts: Overall and regionally
   e. Meet regulatory requirements and company goals for 2014

4. Team skills:
   a. All GRIDCo areas represented will work have process experience. Combined years of experience exceeds _____ years
   b. Trades knowledge and parts materials management knowledge. Involve a direct trades person on this team
   c. Different leadership levels, styles, and approaches to problem-solving
   d. Understanding of regulatory requirements, system engineering, and grid control system

5. Scope: The scope of the parts team will include making decision on top CMMS to purchase and define a successful implementation plan. Review, modify, and approve resources for a final GRIDCo plan of action that is based around an initial plan of action proposed by the Maintenance Excellence Institute International (TMEII). This team is the decision-making process for the overall project along with top leader approvals for funds.

6. Deliverables:
   a. Team charter
   b. Increased power transmission delivery on existing system and new systems
   c. Improved cost control and accountability to people of Ghana
   d. Right materials/quantities in place at right place
   e. Efficient use of space in the stock room (ease of movement)
   f. Increase levels of productivity in the power transmission work processes
   g. Project resources are in place and used effectively to meet project schedule
   h. Clearly defined roles and responsibilities (defined ownership of processes)
   i. List of standard in-stock parts and all functionalities of a world-class CMMS
   j. Improve levels of communications
   k. Controlled location(s) for parts and parts management via CMMS
   l. Implement a measurement to document project results and validate a ROI (return on investment) acceptable to GRIDCo top leaders

GRIDCo's recommended team members: draft

Team sponsor: Mr. Norbert Anku	Team leader: Mr. Hussaini Adams	Team members:	Team facilitators: Franklin Takyi and Ralph Peters of TMEII
Responsibilities: • Clarity purpose and expectations • Agree with team on metrics to measure results and ROI • Provide support and resources • Resolve issues beyond the team's authority level • Consult with the team leader and/or facilitator as needed • Ensure GRIDCo management team is provided feedback on progress. • Review and approve recommendations • Support and endorse the new shift in the team approach to CMMS and best practice implementation • Liaison with the TMEII support team as required	*Responsibilities:* • Be an active part of the team • Respect the team's decisions and recommendations • Provide honest feedback as it pertains to specific measureable, achievable, realistic, and timely (SMART) goals • Support the team's focused direction with sponsor and GRIDCo management • Make the difficult decisions • Assist the facilitator to keep the team focused and to stay within their scope • Take the initiative on critical project action items • Ensure participation by all • Establish a safe team environment • Bring in the additional resources (SMEs) when required • Liaison with TMEII support team as required	• J.B. Taylor—Area Manager Kumasi • Bennard Gyan—Area Manager Techiman • George Adume—Area Manager Tamale • Mark Aryee—Area Manager Akosombo • George Imprem—Area Manager Takoradi • Jonathan Dateh—Prestea • Storeroom and material manager • IT representative • Maintenance planner • Technician representative • Engineer and control system engineer *Responsibilities:* • Actively participate as a direct part of their job during project period • Share experiences and knowledge of their areas • Keep focused • Communicate to others—provide awareness • Use the model—Assess the situations/Determine the causes/Target solutions/Implement actions/Make it ongoing • Involve the customer as needed • Agree to boundaries (scope) • Be role model for other teams • Provide recommendations with supporting data • Reward each other • Respectful of each other's time • Distribute the work evenly and/or support the assignment of tasks • Support each other • Set SMART goals • Be the pioneers/pilot to others what can be done using the team approach	*Responsibilities:* • Improve the team's group dynamics by focusing on the process • Coach the team leader or team members, transfer knowledge to the team • Facilitate the chartering process • Facilitate the working sessions as well-planned events • Partner with the team leader in the process • Liaison with sponsor

Appendix E

Case Study–Process Mapping for a Refinery

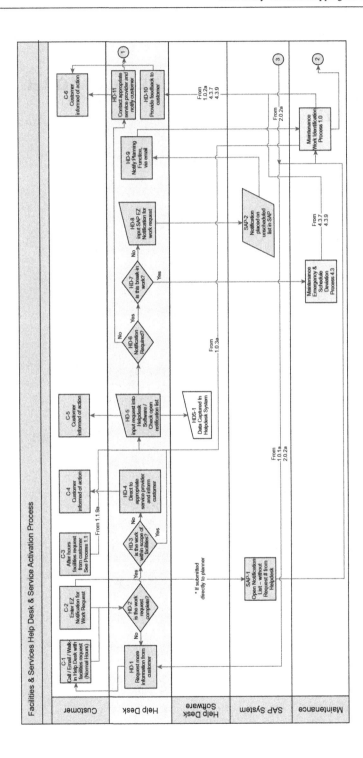

1.0 Facilities & Services Work Identification

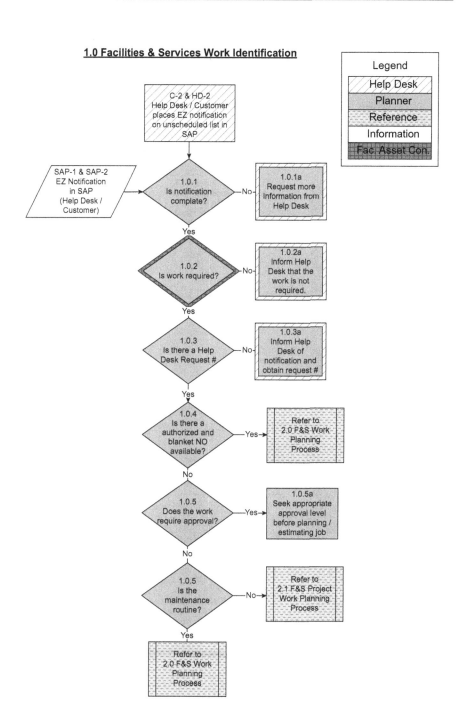

1.1 Facilities & Services Work Identification – After Hours

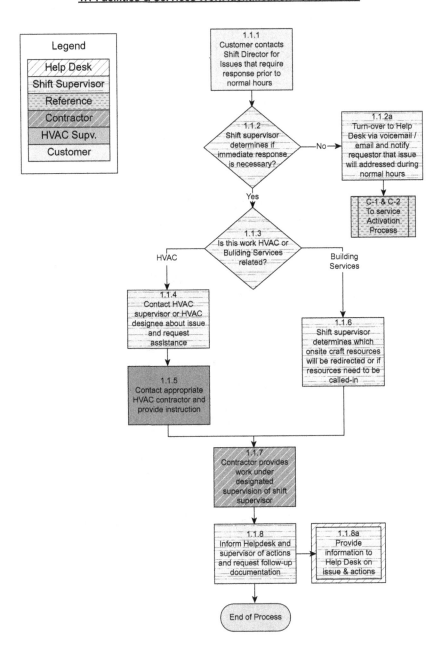

2.0 Facilities & Services Work Planning Process

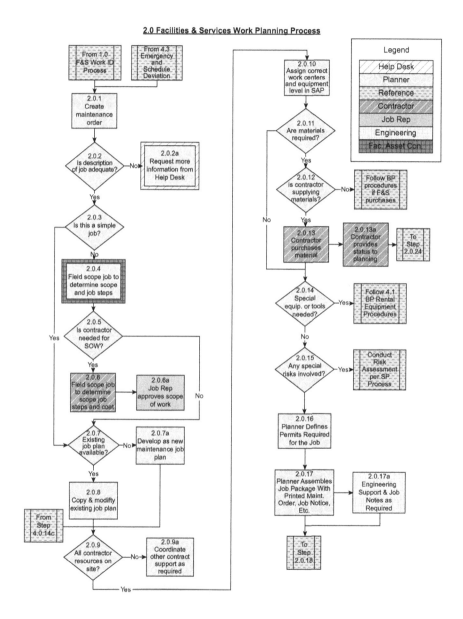

2.0 Facilities & Services Work Planning Process (Continued)

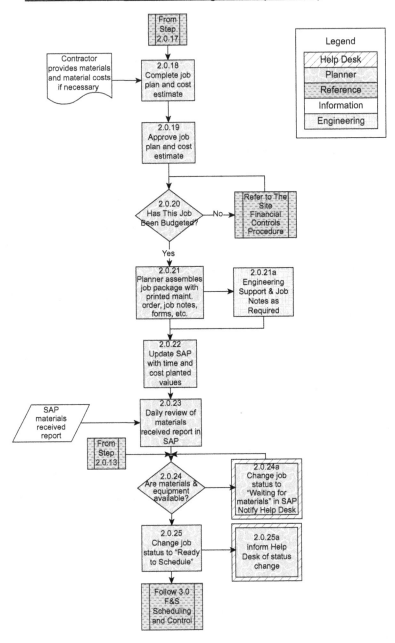

2.1 Facilities & Services Project Work Planning Process

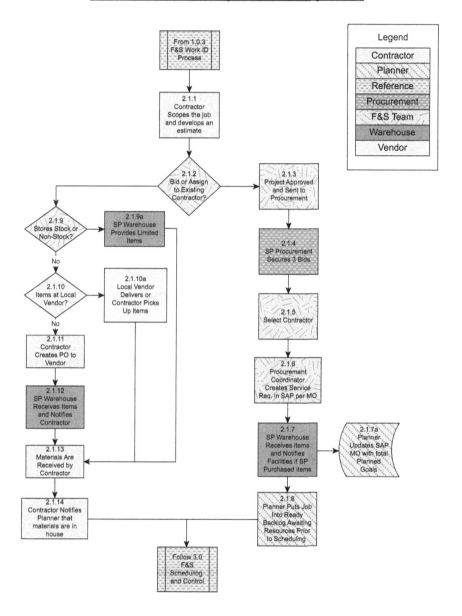

3.0 Facilities & Services Work Scheduling Process

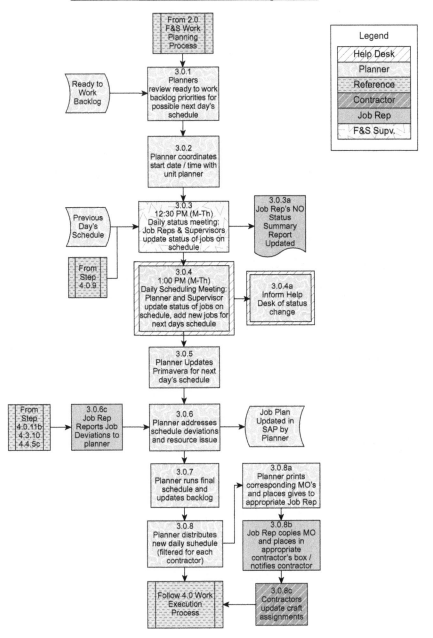

4.0 Facilities & Services Work Execution Process

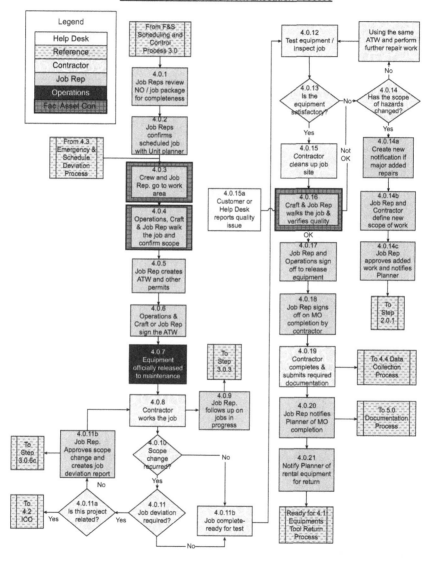

4.1 Facilities & Services Rental Process

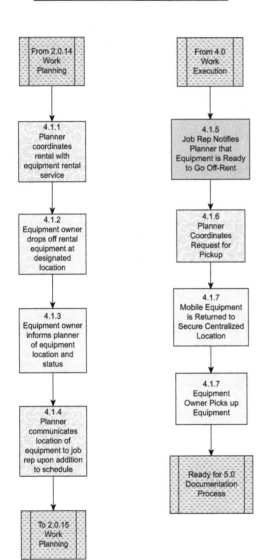

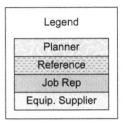

4.2 Facilities & Services Job Change Order Process

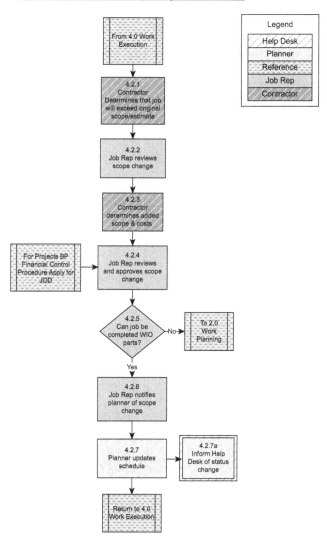

4.3 Facilities & Services Emergency Break-In & Schedule Deviation Process

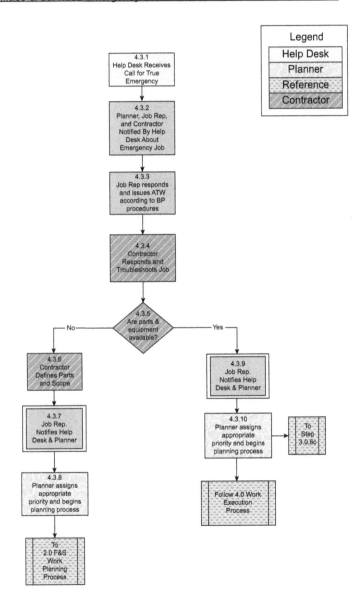

4.4 Facilities & Services Job Site Data Collection Process

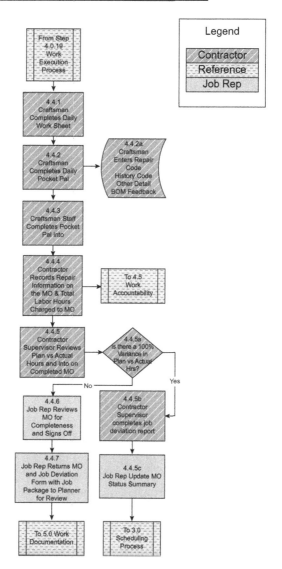

4.5 Facilities & Services Contractor Work Accountability Process

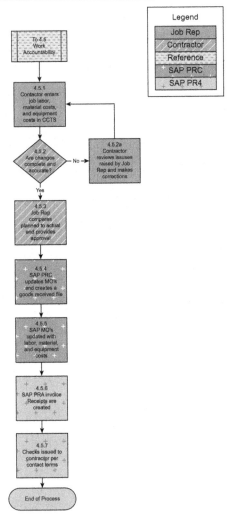

5.0 Facilities & Services Work Documentation Process

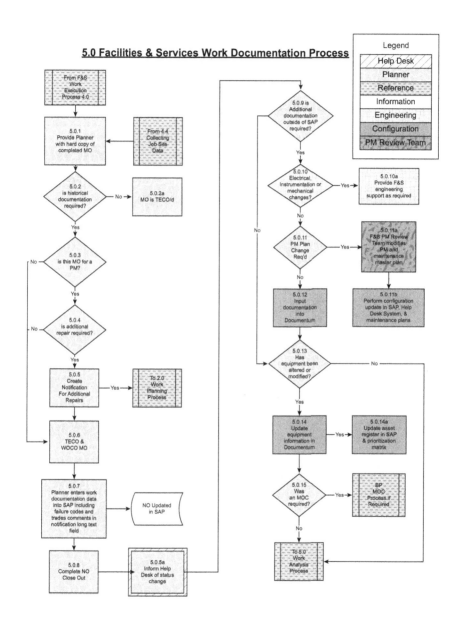

6.0 Facilities & Services Work Analysis Process

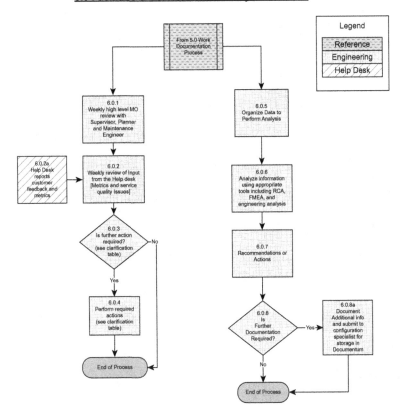

Appendix F

The CMMS Benchmarking System

Introduction

Today's information technology for computerized maintenance management offers the maintenance leader an exceptional tool for asset management and for managing the maintenance process as an internal business and "profit center." However, maintenance surveys and assessments conducted by The Maintenance Excellence Institute International and others validate that poor utilization of existing maintenance software is often a major improvement opportunity.

We will now review each of the nine major categories from the CMMS Benchmarking System and provide key recommendations on each of the 50 benchmark items for getting your computerized maintenance management system (CMMS) implementation started on the right track from day one.

CMMS Data Integrity

1. **Equipment (Asset) History Database**: The equipment database represents one of the essential databases that must be developed or updated as part of implementing a new CMMS. It requires that a complete review of all equipment be made to include all parent/child systems and subsystems that will be tracked for costs, repairs performed, etc. The work to develop or update this database should begin as soon as possible after the data structure of the equipment master file for the new CMMS is known.

   The equipment master information for a piece of equipment (parent/child), manufacturer, serial number, equipment specs, and location will all need to be established. If the installation and removal of components within certain process type operations requires tracking by serial number and compliance to process safety management requirements, then these equipment items will have to be designated in the equipment database.

   If an equipment database exists as part of an old CMMS, then now is the time to review the accuracy of the old equipment database before conversion to the new system. Conversion of the new equipment master database into the new system should be done only after a thorough and complete update of the old database has occurred. Once the new equipment master database has been converted to the new CMMS, a process to maintain it at an accuracy level of 95% or above should be established.

2. **Spare Parts Database**: The spare parts database represents another key database that must be developed or updated as part of implementing a new CMMS. For operations not having a parts inventory system, this will require doing a complete physical inventory of spare parts and materials. All inventory master record data for each item will need to be developed

based on the inventory master record structure for the new CMMS and loaded directly to the inventory module.

Operations that have an existing spare parts database should take the time to do a complete review of it before conversion. Typically, this will allow for purging the database of obsolete parts and doing a complete review of the inventory master record data. This can be a very time-consuming process, but it allows the operation an excellent chance to finally take the time to revise part descriptions, review safety stock levels, reorder points and vendor data, and start the new CMMS with an accurate parts inventory database.

3. **Bill of Materials:** One key functional capability of CMMS is to provide a spare parts listing (bill of materials) within the equipment module. This requires researching where spare parts are used and linking inventory records with equipment master records that are component parts of an equipment asset. This function would also add, change, or delete items from an established spares list or copy a spares list to another equipment master record. In addition, this feature would copy all or part of a spares list to a work order job plan and create a parts requisition or pick list to the storeroom.

The process of establishing a spares list is time-consuming and would involve only major spares that are currently carried in stock. Most CMMS systems have the capability to build the spare parts list as items are issued to or purchased for a piece of equipment. It is recommended that an equipment bill of materials be established, but the conversion of equipment master data can take place without this information being available. Because a bill of materials for spare parts is so beneficial for planning purposes, it is recommended that the process to identify key spares in the equipment master be a priority area.

4. **Preventive Maintenance Tasks/Frequencies**: The preventive and predictive maintenance (PM/PdM) database is another key database necessary for establishing a "Class A" installation. If a current PM/PdM database is present, then it is recommended that the existing procedures be reviewed and updated before conversion to the new system. If the existing PM/PdM database has been updated continuously on the old system, then conversion can probably occur directly from the old to the new PM/PdM database; however, this will depend on the PM/PdM database structure of the new system.

It is recommended that in the very early stages of a new CMMS benchmark/selection process that the status of the current PM/PdM program be evaluated. If a process for the review/update of PM/PdM procedures has not been in place, then it is very important to get something started as soon as possible. This provides an excellent opportunity to establish a team of experienced craftspeople, engineers, and maintenance supervisors to work on PM/PdM procedures to review, to update task descriptions, and frequencies, and to ensure that all equipment is covered by proper procedures.

5. **Maintaining Parts Database**: After a new CMMS is installed, it is highly recommended that one person be assigned direct responsibility for maintaining the parts database. This person would have responsibility for making all additions and deletions to inventory master records, changing stock levels, reorder points and safety stock levels, and changing any data contained in the inventory master records. This person could also be designated responsibility for coordinating the development of the spares list if this information is not available. This person would be responsible for recommending obsolete items on the basis of the monitoring of usage rates or because of equipment being removed from the operation. The practice of having one primary person assigned direct responsibility for the inventory master records can help ensure that parts database accuracy is 95% or greater.

6. **Maintaining Equipment Database**: It is also highly recommended that one person be assigned direct responsibility for maintaining the equipment database. This person would be responsible for making all changes to equipment master records. Information on new equipment would come to this person for setting up parent/child relationships of components in the equipment master records. Information on equipment being removed from the operation would also come to this person to delete equipment master records. Coordination between this person and the person responsible for the parts database would be required to ensure that obsolete parts were identified and/or removed from the inventory system because of removal of equipment.

CMMS Education and Training

7. **Initial CMMS Training**: One of biggest roadblocks to an effective CMMS installation is the lack of initial training on the system. Many organizations never take the time up front to properly train their people on the system. Shop-level people must gain confidence in using the system for reporting work order information and knowing how to look up parts information. The CMMS implementation plan should include an adequate level of actual hands-on training on the system for all maintenance employees before the "go-live" date. It is important to invest the time and expense to "train the trainers" who can, in turn, assist with the training back in the shop. Many organizations set up "conference room pilots" where the CMMS software is set up and training occurs with actual data using CMMS vendor trainers or in-house trainers. It is highly recommended that competency-based training be conducted so that each person trained can demonstrate competency in each function they must perform on the system.

8. **Ongoing CMMS Training**: The CMMS Implementation Plan must consider having an ongoing training program for maintenance and storeroom personnel. After the initial training, there must be someone in the organization with the responsibility for ongoing training. If a good "trainer" has been developed within the organization before the go-live date, then this person can be the key to future internal training on the new system. Ongoing training can include one-on-one support that helps to follow up on the initial training.

9. **Initial CMMS Training for Operations Personnel**: The customers of maintenance must gain a basic understanding of the system and know how to request work, check the status of the work requested, and understand the priority system. During implementation, operations personnel need to get an overview of how the total system will work and the specific things they will need to do to request work. If the organization has a formal planning and scheduling process, then they will also need to know the internal procedures on how this will work.

10. **CMMS Systems Administrator/Backup Trained**: It is important that each site have one person trained and dedicated as the systems administrator with a backup trained whenever possible. This person will typically be from information services and have a complete knowledge of system software, hardware, database structures, and interfaces with other systems as well as report-writing capabilities. The systems administrator will also have responsibility for direct contact with the CMMS vendor for debugging software problems and for coordinating software upgrades.

Work Control

11. **Work Control Function Established**: A well-defined process for requesting work, planning, scheduling, assigning work, and closing work orders should be established. The work control function will depend on the size of the maintenance operation. Work control may involve calls coming directly to a dispatcher who creates the work order entry and forwards the work order to a supervisor for assignment. The work request could also be forwarded directly to an available craftsperson by the dispatcher for execution of true emergency work.

    Work control can also be where work requests are forwarded manually or electronically to a planner who goes through a formal planning process for determining the scope of work, craft requirements, and parts requirements to develop a schedule. PM/PdM work would be generated and integrated into the scheduling process. The status of the work order would be monitored, which might be in progress, awaiting parts, awaiting equipment, awaiting craft assignment, or awaiting engineering support, etc. A work order backlog would also be maintained to provide a clear picture of the work order status.

    Effective work control provides systematic control of all incoming work through to the actual closing of the work order. The work control process should be documented with clearly defined written procedures unique to each maintenance operation.

12. **Online Work Request Based on Priorities**: Requesting work online represents an advanced CMMS functional capability in which the customer enters the work request directly into the system on a local area network or via e-mail. Online work requests would include basic information about the work required, equipment location, the date the work is to be completed by, the name of the requestor, and the priority of the work. This information would go to the work control function where the jobs would be planned, scheduled, and assigned based on the overall workload. The requestor would have the capability to track the status of their jobs online and even give final approval that the work was satisfactorily completed.

13. **Work Order System Accounts for 100% of Craft Hours**: All craft work should be charged to a work order of some type. Accountability of labor resources is an important part of managing maintenance as an internal business. Quick reporting to standing work orders can be established for jobs of short duration within a department or for the reporting of noncraft time such as meetings, delays in getting the equipment to work on, training, and chasing parts.

14. **Backlog Reports**: Maintaining good control of the work to be done is essential to the maintenance process. Having the capability to visually see the backlog helps to effectively plan and schedule craft resources.

    The CMMS reporting system should provide the capability to show the backlog of work in several ways:

• By Type of Work	• By Overdue Work Orders
• By craft	• By parts status
• By department	• By priority

15. **Priority System**: A Class A CMMS installation will have in place a priority system that allows for the most critical repairs to get done first. An effective priority system adds professionalism to the maintenance operation and directly supports effective planning and scheduling. There are two basic systems for establishing priorities:

- Straight numeric priority (1, 2, 3, 4, 5, etc.), in which each priority level is defined by a definition such as Priority 1: a true emergency repair that affects safety, health, or environmental issues.
- RIME system: This system combines the criticality index of the equipment (10 highest to 1 lowest) with the criticality of the work type (10 highest to 1 lowest) to compute the ranking index for maintenance expenditures (RIME) priority number. The RIME priority number equals the equipment criticality index multiplied by the criticality number of the work type.

Budget and Cost Control

16. **Craft Labor, Parts, and Vendor Support Costs**: The equipment history file should provide the source of all costs charged to the asset. Here, it is important to ensure that all labor is charged to the work orders for each asset and that parts are charged to the respective work orders.

17. **Budget Status-Operating Departments**: Operating departments should be held accountable for their respective maintenance budgets. With an effective work order system in place for charging of all maintenance costs, the accounting process should allow for monitoring the status of departmental budgets. One recommended practice is for maintenance to be established as a zero-based budget operation and that all labor and parts be charged back to the internal customer. This practice helps ensure accountability for all craft time, parts, and materials to work orders.

18. **Cost Improvements Due to CMMS**: The effect of a successful CMMS installation should be reduced costs and achieving gained value in terms of greater output from existing resources. The CMMS team should be held accountable for documenting the savings that are achieved from the new CMMS and the maintenance best practices that evolve. The areas that were used to justify the CMMS capital investment (e.g., reduced parts inventory, increased uptime, and increased craft productivity) should all be documented to show that improvements did occur.

19. **Deferred Maintenance Identified**: It is important that maintenance provides management with a clear picture of maintenance requirements that require funding for the annual budget. Deferred maintenance on critical assets can lead to excessive total costs and unexpected failures. Benefits from CMMS will provide improved capability to document deferred maintenance that must be given priority during the budgeting process each year.

20. **Life-Cycle Costing Supported**: A complete equipment repair history provides the basis for making better replacement decisions. Many organizations often fail to have access to accurate equipment repair costs to support effective replacement decisions and continue to operate and maintain equipment beyond its economically useful life. As a result, the capital justification process then lacks the necessary life-cycle costing information to support replacement decisions.

Planning and Scheduling

21. **Planning and Scheduling**: This maintenance best-practice area is essential to better customer service and operations as well as greater utilization of craft resources. For most maintenance operations with 25–30 craftspeople, a full-time planner can be justified.

The CMMS system functionality must support the planning process for control of work orders, backlog reporting, status of work orders, parts status, craft labor availability, etc. The planning and scheduling function supports changing from a "run to failure strategy" to one for proactive, planned maintenance.

22. **Planned Work Increasing**: The bottom-line result for the planning process is to actually increase the level of planned work. The percentage of planned work should be monitored and included as one of the overall maintenance performance metrics. In some organizations with effective PM/PdM programs, the level of planned work can be in the 90% range or more.

23. **Craft Utilization Measured and Improving**: Effective planning and scheduling is essential to increasing the level of actual hands-on wrench time of the craft work force. Improving craft utilization allows more work to get done with current staff by eliminating noncraft activities such as waiting for equipment, searching for parts, and scheduling the right sequence for different crafts on the job.

24. **Work Schedules Available**: One key responsibility of the planning process is to establish realistic work schedules for bringing together the right type craft resources, the parts required, the equipment to be repaired or serviced, and having the time available to complete the job right the first time. The actual schedule may only start with a 1-day schedule and gradually work up to scheduling longer periods of time. Work schedules provide a very important customer service link with operations that helps to improve overall coordination between maintenance and operations.

25. **Spare Parts Status Is Available**: One of the most essential areas to support effective planning is the maintenance storeroom and the accuracy of the parts inventory management system. Jobs should not be put on the schedule without parts being on hand. The planner must have complete visibility of inventory-on-hand balances, parts on order, and the capability to reserve parts for planned work.

26. **Scheduling Coordination with Operations**: As the planning function develops, there will be improved coordination with operations to develop and agree upon work schedules. This may involve coordination meetings near the end of each week to plan week end work or to schedule major jobs for the upcoming week. Direct coordination with operations allows maintenance to review PM/PdM schedules or to review jobs where parts are available to allow the job to be scheduled based on operations scheduling equipment availability.

27. **True Emergency Repairs Tracked**: Many organizations really focus on reducing true emergency repairs, which create uncertainty for operations scheduling and contribute to significantly higher total repair costs than planned work. Improved reporting capabilities of an effective CMMS will allow for better tracking of emergency repairs, document causes for failures, and assist in the elimination of the root causes for failures.

Maintenance Repair Operations Materials Management

The overall area of maintenance repair operations (MRO) parts and materials procurement, storage, inventory management, and issues represents another best-practice area that often needs major work when implementing a CMMS and developing a Class A installation. Many organizations never take the time to set up a well-planned and controlled storeroom operation and often find out that their parts database is a weak link needing major updates before it can truly be effectively used.

28. **Inventory Management Module**: The work order module must be fully integrated with an accurate parts inventory management module to charge parts back to work orders, to check parts availability status for planned work, to reserve parts, and to check the status of direct purchases. A Class A CMMS installation will develop, maintain, and fully utilize the inventory module and ensure that it is fully integrated with the work order module.

29. **Inventory Cycle Counting Established**: Inventory accuracy should be one of the key metrics for MRO materials management, and it can best be accomplished by cycle counting rather than annual physical inventories. Most CMMS systems will allow for your own criteria to be developed, such as doing an ABC analysis of inventory items (based on either usage value or frequency of issue) and then scheduling of periodic counts for each classification of inventory item that you want to cycle count. For example, A items would be counted more frequently than B and C items. The real value of cycle counting is that it is a continuous process that creates a high level of discipline and allows for inventory problems and adjustment to be made throughout the year rather than once after the annual inventory.

30. **Parts Kiting**: This best-practice area is key to the planning process and can evolve over time as the planning process matures to the point of being able to give the storeroom prior notification on the parts required for planned jobs. Controlled staging areas are set for parts that are either pulled from stock or received from direct purchases.

31. **Electronic Parts Requisitioning**: This functional capability can provide paperless work flow for requisitioning of parts directly from maintenance to the storeroom for creation of a pick list for the item or go to purchasing to create a purchase order for a stock item or direct purchase. In some cases, electronic requisitioning might go directly to the vendor using e-commerce capability.

32. **Critical Spares Identified**: Critical spares (or insurance items), which may be a one-of-a-kind, high-cost spare, are often a part of the parts inventory system. It is recommended that these items be classified and identified in the item master record as such. This practice will help to separate these items from the regular inventory management process and identify them as a separate part of the total inventory value that has been fixed. Critical spares should also be identified in the spares list for the equipment they have been purchased for.

33. **Reorder Notification Process**: The capability to determine when and what to reorder on the basis of a review of stock-level reorder points is an important feature for a Class A installation. A recommended reorder report should be generated periodically and reviewed for validity as well as for any future needs that may not be reflected in current on-hand balances. On the basis of a final review of the recommended reorder report, electronic requisitioning then could occur directly to purchasing.

34. **Warranty Information**: Many organizations fail to have a process in place to track warranty information and, in turn, they incur added costs by not being able to get proper credit for items under warranty. Tracking specific high-value parts or components and specific equipment under warranty should be a CMMS functionality of the equipment master or the inventory item master database. The system should provide a quick reference and alert to the fact that the item is still under warranty and that a follow-up claim to the vendor is needed.

Preventive/Predictive Maintenance

35. **PM/PdM Change Process**: This best-practice area simply ensures that PM/PdM procedures are subject to a continuous review process and that all changes to the program are

made in a timely manner. The CMMS system should provide an easy method to update task descriptions and task frequencies and allow for mass updating when the procedure applies to more than one piece of equipment.

36. **PM/PdM Compliance Is Measured**: One key measure of overall maintenance performance should be how well the PM/PdM program is being executed based on the schedule. Measuring PM/PdM compliance ensures accountability from maintenance and from operations. Normally, a scheduling window of 1 week will be established to determine compliance. A goal of 98% or better for PM/PdM compliance should be expected.

37. **Long-Range PM/PdM Scheduling**: As a PM/PdM schedule is loaded to the system, peaks and valleys may occur for the actual scheduling because of frequencies of tasks coming due at the same time period. The CMMS system should provide the capability to level load the actual PM/PdM schedule and to view upcoming PM/PdM workloads to assist in the overall planning process.

38. **Lubrication Services**: Lubrication services, tasks, frequencies, and specifications should ideally be included as part of the PM/PdM module. A continuous change process for this area should also be put in place as well as an audit process established to ensure all lube and PM/PdM tasks are being performed as scheduled.

39. **CMMS Captures Reliability Data**: The elimination of root causes of problems is the goal rather than just more PM/PDM. One important feature of a "Class A" installation is being able to capture failure information that can in turn be used for reliability improvement. This requires that a good coding system for defining causes for failures be developed and that this information be accurately entered as the work order is closed.

40. **Complete PM/PDM Task Descriptions**: PM/PdM task descriptions often provide vague terminology to check, adjust, inspect, etc. and do not provide clear direction for specifically what is to be done. Task descriptions should be reviewed periodically and details added that to the level that a new crafts person would understand exactly what is to be done and be able to adequately perform the stated task.

Maintenance Performance Measurement

41. **Equipment Downtime Reduction**: Another key metric for measuring overall maintenance performance is increased equipment uptime. The improvement in this metric is a combination of many of the previously mentioned best practices all coming together for improved reliability. Downtime due to maintenance should be tracked and positive improvement trends should be occurring within a Class A installation.

42. **Craft Performance**: Two key areas affecting overall craft productivity are craft utilization (wrench time) and craft performance. Measurement of craft performance requires that realistic planning times be established for repair work and PM tasks. A standard job plan database can be developed for defining job scope, sequence of tasks, special tools listing, and estimated times. The goal is measurement of the overall craft work force and not individual performance. Planning times are also an essential part of the planning process for developing a more accurate picture of work load and to support scheduling of overtime and staff additions.

43. **Maintenance Customer Service**: The results of improved maintenance planning must be improved customer service. The overall measurement process should include metrics such as compliance to meeting established schedules and jobs actually completed on schedule.

44. Maintenance Performance Measurement Process: In this area, it is important to have a performance measurement process that includes several key metrics in each of the following major categories:

• Budget and cost	• Planning and scheduling
• Craft productivity	• Customer service
• Equipment uptime	• MRO materials management
• PM/PdM	

The overall maintenance performance process should be established so that it clearly validates the benefits being received from the CMMS and maintenance best-practice implementation.

Other Uses of CMMS

45. Maintenance Managed as a Business: One true indicator for a successful CMMS installation is that it has changed the way that maintenance views it role in the organization. It should progress to the point that maintenance is viewed and managed as an internal business. This view requires greater accountability for labor and parts costs, greater concern for customer service, better planning, greater attention to reliability improvement, and increased concern for the maintenance contribution to the bottom line.

46. Operations Understands Benefits of CMMS: There is direct evidence that operations understands that an improved CMMS is a contributor to improve customer service. The scheduling process is continuously improving through better coordination and cooperation between maintenance and operations within a Class A installation.

47. Engineering Changes: Accurate engineering drawings are essential to maintenance planning and to actually making the repairs. Asset documentation must be kept up to date on the basis of a formal engineering change process. Feedback to engineering must be made on all changes as they occur on the shop floor. Engineering must in turn ensure that master drawings are updated and that current revisions made available to maintenance.

48. Equipment Database Structure: To provide equipment history information in a logical parent/child relationship, the equipment database structure has been developed using an identification of systems and subsystems. Accessing the equipment database should allow for drilldown from a parent level to lower-level child locations that are significant enough for equipment master information to be maintained.

49. Failure and Repair Codes: The reporting capability of the CMMS should provide good failure trending and support analysis of the failure information that is entered from completed work orders. Improving reliability requires good information that helps to pinpoint root causes of failure.

50. Maintenance Standard Task Database: Developing the maintenance standard task database (or standard repair procedures) for recurring jobs is an important part of a planner's job function. This allows for determining the scope of work and special tools and equipment and for estimating repair times. Once a standard repair procedure is established, it can then be used as a template for other similar jobs, resulting in less time for developing additional repair procedures.

Summary

Developing a Class A CMMS installation requires the combination good system functionality and improved maintenance practices. The CMMS team should begin very early during implementation with how it will measure the success of the installation. The recommendations provided here for using the CMMS Benchmarking System can help your organization achieve maximum return on its CMMS investment.

The Maintenance Excellence Institute International is fully committed to ensuring that all clients now using CMMS gain maximum value from their software investment. Maintenance organizations take an important step when they invest in software to help manage the business of maintenance. They often do not achieve maximum return on investment from CMMS software investments. One of the services from The Maintenance Excellence Institute International is to provide Scoreboard for Maintenance Excellence Assessments that includes an independent benchmark evaluation of the current CMMS installation.

Purpose

The CMMS Benchmarking System was developed to support getting maximum value from an investment in CMMS by evaluating how well existing CMMS functionality is being used. The benchmarking system provides a methodology for developing an overall benchmark rating of your CMMS installation as the baseline for determining how well CMMS is supporting best practices within the total maintenance operation. It can also be used as the baseline to measure the success of a future CMMS that is just being installed.

Benchmarking Your CMMS Installation

The CMMS Benchmarking System provides a means to evaluate and classify your current installation as either Class A, B, C, or D. A total of nine major categories are included along with 50 specific evaluation items. Each evaluation item that is rated as being accomplished satisfactorily receives a maximum score of 4 points. If an area is currently being "worked on," a score of 1, 2, or 3 points can be assigned based on the level of progress achieved. For example, if spare parts inventory accuracy is at 92% compared with the target of 95%, a score of 3 points is given. A maximum of 200 points is possible. A benchmark rating of Class A is within the 180- to 200-point range. The complete CMMS Benchmarking Rating Scale is given at the completion of the benchmarking form.

Conducting the CMMS Benchmark Evaluation

The Maintenance Excellence Institute International conducts the CMMS benchmark evaluation to provide an independent and objective evaluation. It is not to make functionality comparisons between different CMMS systems. However, it will identify gaps in functionally, best-practice needs, and support decisions on upgrades. Results from the benchmark evaluation will establish a baseline classification for your current installation. A written report with specific recommendations and a plan of action to improve your CMMS installation to the Class A level will be provided along with guidelines for using the CMMS Benchmarking System in the future as an internal benchmarking tool.

Category number	Category description	Yes (4)	No (0)	Working on It (1, 2, 3)
	CMMS data integrity	**Yes**	**No**	**WOI**
1	Equipment (asset) history data complete and accuracy 95% or better			
2	Spare parts inventory master record accuracy 98% or better			
3	Bill of materials for critical equipment includes critical spare parts denoted in the parts database			
4	Preventive maintenance (PM) tasks/ frequencies data complete for 98% of applicable assets			
5	Direct responsibilities for maintaining parts inventory is assigned			
6	Direct responsibilities for maintaining equipment/asset database is assigned			
	CMMS education and training	**Yes**	**No**	**WOI**
7	Initial CMMS training for all maintenance employees			
8	An ongoing CMMS training program for maintenance and storeroom employees			
9	Initial CMMS orientation training for operations employees			
10	A CMMS systems administrator (and backup) designated and trained			
	Work control	**Yes 4**	**No 0**	**3, 2, or 1 WOI**
11	A work control function is established or a well-defined documented process is being used			
12	Online work request (or manual system) used to request work based on priorities			
13	Work order system used to account for 100% of all craft hours available			

Continued

—cont'd

Category number	Category description	Yes (4)	No (0)	Working on It (1, 2, 3)
14	Backlog reports are prepared by type of work to include estimated hours required			
15	Well-defined priority system is established based on criticality of equipment, safety factors, cost of downtime, etc.			
	Budget and cost control	**Yes 4**	**No 0**	**3, 2, or 1 WOI**
16	Craft labor, parts, and vendor support costs are charged to work order and accounted for in equipment/asset history file			
17	Budget status on maintenance expenditures by operating departments is available			
18	Cost improvements due to CMMS and best-practice implementation have been documented			
19	Deferred maintenance and repairs identified to management during budgeting process			
20	Life-cycle costing is supported by monitoring of repair costs to replacement value			
	Planning and scheduling	**Yes**	**No**	**WOI**
21	A documented process for planning and scheduling has been established			
22	The level of proactive, planned work is monitored and documented improvements have occurred			
23	Craft utilization (true wrench time) is measured and documented improvements have occurred			
24	Daily or weekly work schedules are available for planned work			
25	Status of parts on order is available for support to maintenance planning process			
26	Scheduling coordination between maintenance and operations has increased			
27	Emergency repairs, hours, and costs tracked and analyzed for reduction			
	MRO materials management	**Yes**	**No**	**WOI**
28	Inventory management module fully utilized and integrated with work order module			

—cont'd

Category number	Category description	Yes (4)	No (0)	Working on It (1, 2, 3)
29	Inventory cycle counting based on defined criteria is used and inventory accuracy is 95% or better			
30	Parts kiting is available and used for planned jobs			
31	Electronic requisitioning capability available and used			
32	Critical and/or capital spares are designated in parts inventory master record database			
33	Reorder notification for stock items is generated and used for reorder decisions			
34	Warranty information and status is available			
	PM/PdM	**Yes 4**	**No 0**	**3, 2, or 1 WOI**
35	PM/PdM change process is in place for continuous review/update of tasks/ frequencies			
36	PM/PdM compliance is measured and overall compliance is 98% or better			
37	The long-range PM/PdM schedule is available and level loaded as needed with CMMS			
38	Lube service specifications, tasks, and frequencies included in database			
39	CMMS provides MTBF, MTTR, failure trends and other reliability data			
40	PM/PdM task descriptions contain enough information for new craftsperson to perform task			
	Maintenance performance measurement	**Yes**	**No**	**WOI**
41	Downtime (equipment/asset availability) due to maintenance is measured and documented improvements have occurred			
42	Craft performance against estimated repair times is measured and documented improvements have occurred			
43	Maintenance customer service levels are measured and documented improvements have occurred			

Continued

Category number	Category description	Yes (4)	No (0)	Working on It (1, 2, 3)
44	The maintenance performance process is well established and based on multiple indicators compared with baseline performance values			
	Other uses of CMMS	**Yes**	**No**	**WOI**
45	Maintenance leaders use CMMS to manage maintenance as internal business			
46	Operations staff understands CMMS and uses it for better maintenance service			
47	Engineering changes related to equipment/ asset data, drawings, and specifications are effectively implemented			
48	Hierarchies of systems/subsystems used for equipment/asset numbering in database			
49	Failure and repair codes used to track trends for reliability improvement			
50	Maintenance standard task database available and used for recurring planned jobs			
	Total cmms benchmark rating score			

CMMS, computerized maintenance management system; PM/PdM, preventive and predictive maintenance; MRO, maintenance repair operations; MTTR, mean time to repair; MFBF, mean time between failure.

Summary of CMMS Benchmarking Evaluation

TOTAL CMMS BENCHMARK RATING SCORE: _____
CURRENT CMMS BENCHMARK RATING: CLASS _____
CMMS SOFTWARE VERSION:

DATE OF CMMS INSTALLATION:_____/_____/_____
BENCHMARK RATING PERFORMED BY: _____
LOCATION:_____   DATE: _____/_____/_____

CMMS BENCHMARKING RATING SCALE: INDIVIDUAL ITEM GRADING:
CLASS A = 180 - 200 POINTS (90% +)   YES = 4 points
CLASS B = 140 - 179 POINTS (70% - 89%)   NO = 0 points
CLASS C = 100 - 139 POINTS (50% - 69%)   Working On It = 1, 2 or 3 points
CLASS D =   0 - 99 POINTS (0 - 49%)

EVALUATION COMMENTS:

Appendix G

The ACE Team Benchmarking Process Team Charter Example

CHARTER for The ACE TEAM at_____

Maintenance Strategy Team - Preliminary Review Date   _____

Maintenance Strategy Team - Final Review   _____

ACE Team Review Date   _____

ACE Team Final Acceptance Date   _____

I. **Opportunity:** *What is the reason this team exists?*
The ACE Team (where ACE means "a consensus of experts") exists to provide a well-qualified team of experienced craftspeople, technicians, and supervisors that establish benchmark repairs jobs and work content time for these jobs. The ACE Team is chartered to help develop the ACE System for establishing maintenance performance standards at each work site.

II. **Process:** *What are the steps to be followed, and what are the questions to be answered by this team?*

   a. Orientation, charter review, and charter acceptance or modification.

   b. Ensure that all team members understand team objectives and agree on what needs to be achieved and the criticality of this initiative to the planning and scheduling process.

   c. Understand the current concepts of the ACE System as defined in the XYZ Company's maintenance planning and scheduling standard operating procedure (SOP).

   d. Understand the basics of the new XYZ system for computerized maintenance management system (CMMS)/ enterprise asset management (EAM), the characteristics, functionality, and performance.

   e. Determine the critical repair jobs that should be used as representative benchmark jobs, define key steps and elements for each benchmark job, and define any special tools, safety requirements, and other special requirements for the job.

   f. Determine ways to improve doing the jobs being analyzed as benchmark jobs considering better tools, equipment, skills, and even better predictive/preventative maintenance techniques to avoid this type of failure problem.

   g. Conduct the ACE Team process as outlined in the 10-step approach from the SOP.

   h. Develop a team consensus on work content times for all the benchmark jobs selected.

   i. Continuously improve the ACE Team process within the XYZ organization as an element of our CRI efforts.

III. Evidence of success: *What results are expected, in what periods, for this team to be successful?*

    a. A sufficient number of benchmark jobs will be developed as to individual tasks and steps along with estimated work content times to complete the site's ACE Team spreadsheets.

    b. The actual period to complete the initial spreadsheets will depend on time allocated by the ACE Team at each site.

    c. ACE Teams from one site are expected to share their results with the other sites. Due to the similar nature of equipment, benchmark job write-ups and even work content times that ACE Teams have developed can be shared throughout the operation.

    d. Overall success will be determined by each planner having adequate spreadsheets that cover all construct areas as well as types of crafts work (mechanical, electrical, etc.) so that planning times can eventually be established for 80% or more of the available craft hours.

IV. Resources: *Who are the team members, team leader, and team facilitator? Who will support the team if needed? How much time should be spent both in meetings and outside of meetings?* **The ACE Team should consist of the following representatives:**

    a. One maintenance planner: team leader

    b. One maintenance supervisor

    c. Two or three crafts representatives from Area 1

    d. Two or three crafts representatives from Area 2

    e. Two or three craft representative from Area 3

Note: Crafts representatives should rotate periodically and sufficient numbers designated so as to have at least two to three representatives from each craft area when benchmark jobs from these areas are being reviewed for job steps and estimated for work content time.

    a. The initial ACE Team meeting will be for 1 h. The team should meet initially for at least 3 h each week. This team's activities and success will be considered as part of each team member's job.

V. Constraints: *What authority does the team have? What items are outside the scope of the team? What budget does the team have?*

    a. No changes to organization structure are anticipated.

    b. Benchmark job plans are to be reviewed and approved by the maintenance manager.

    c. Each team must obtain buy-in and overcome concerns from the other crafts on their estimates for benchmark jobs and repair methods recommended for each benchmark job.

    d. The team presents implementation status reports as required and any additional recommendations to the Maintenance Excellence Strategy Team.

    e. The ACE Team has the authority to recommend new and improved repair methods, new tools to help craft productivity and safety, and other ways to improve asset reliability as developed during the ACE Team process.

VI. Expectations: *What are the outputs, when are they expected, and to whom should they be given?*

    a. Spreadsheets for the site that cover all crafts areas and construct types are completed.

    b. Reliable planning times are provided for benchmark jobs so that effective planning, performance measurement, backlog control, and level of PM work can be established with a high level of confidence.

    c. The ACE Team provides a steady source of continuous improvement ideas to make repair jobs safer and easier.

    d. Minutes are to be completed for all team meetings and sent to the maintenance manager and the Maintenance Excellence Strategy Team.

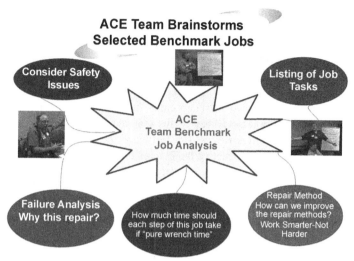

Figure G.1 The ACE team benchmark job analysis.

The Methodology for Applying the ACE Team Benchmarking Process

This section outlines the methodology for applying the ACE Team Benchmarking process. This very easy-to-use procedure allows a planner/scheduler to be the central organizer of the ACE Team within an organization desiring to use this methodology for developing benchmark jobs, the use of slotting, plus allowances to develop reliable planning times for scheduling. A graphical representation of this process is included in Figure G.1.

A New Maintenance Work Measurement Tool from The Maintenance Excellence Institute International

A True Team-Based Approach: Here, we will outline a new and highly recommended methodology for establishing team-based maintenance performance standards, which we call *reliable planning times*. The ACE Team Benchmarking Process (ACE System) was developed by the author in the 1980s. It is a true team-based process that uses skilled craftspeople, technicians, supervisors, planners, and other knowledgeable people to do two things:

1. Improve current repair methods, safety, and quality.
2. Establish work content time for selected "benchmark jobs" for planners and others to use in developing reliable planning times.

Benchmark Jobs: This is a proven process that uses "a consensus of experts," or ACE, who have performed these jobs and can also help to improve them. In turn, a relatively small number of representative benchmark jobs are developed for the major work areas/types within the operation. Benchmark jobs are then arranged into time categories ("time slots") on spreadsheets for the various craft work areas.

Spreadsheets: By using spreadsheets to do what is termed work content comparison or "slotting," a planner is then able to establish planning times for a large number of jobs using a relative small sample of benchmark jobs. This publication also provides the step-by-step process for using the ACE System. Most importantly, it will illustrate how this method supports CRI and quality repair procedures for all types of maintenance repair operations.

Nearly every CMMS allows a user to enter "planned" or "standard" hours on a work order, and then report on actual versus planned hours (the craft performance element of Overall Craft Effectiveness (OCE)) when the job is complete. This holds true for both preventive and corrective maintenance work orders, as well project type work for renovation, major overhauls, and capitalized repairs. Most do not use this for one main reason; they do not have reliable planning times or standard hours available.

Determining the standard hours an average maintenance technician will require to complete a task under standard operating conditions provides everyone involved with a sense of what is expected. The standards provide management with valuable input for backlog determination, manpower planning, scheduling, budgeting, and costing. Labor standards also form the baseline for determining craft productivity and labor savings for improved methods.

The ACE System Supports Continuous Reliability Improvement: Maintenance work, by its very nature, seldom follows an exact pattern for each occurrence of the same job. Therefore, exact methods and exact times for doing most maintenance jobs cannot be established as they can for production-type work. However, the need to have reliable performance measures for maintenance planning becomes increasingly important as the cost of maintenance labor rises and the complexity of production equipment increases.

To work smarter, not harder, maintenance work must be planned, have a reasonable time for completion, use effective and safe methods, be performed with the best personal tools and special equipment possible, and have the right craft skill using the right parts and materials for the job at hand.

Investment for Planners: With an investment in maintenance planners, there must be a method to establish reliable planning times for as many repair jobs as possible. The ACE System provides that method, as well as a team-based process to improve the quality of repair procedures.

Various methods for establishing maintenance performance standards have been used, including reasonable estimates, scientific wild average guess (SWAGs), historical data, and engineered standards such as universal maintenance standards (UMS) using predetermined standard data. These techniques generally require that an outside party establish the standards, which are then imposed upon the maintenance force. This approach often brings about undue concern and conflict between management and the maintenance workforce over the reliability of the standards.

The ACE System: A Team-Based Approach: Rather than progressing forward together in the spirit of continuous improvement, the maintenance workforce in this type of environment often works against the management program for maintenance improvement. The ACE System overcomes this problem with a team-based approach involving craftpeople who will actually do the work that will be planned later as the planning and estimating process matures. As shown later, the ACE System is truly a team-based process that looks first at improving maintenance repair methods, the reliability of those repairs to improve asset uptime, and then secondly to establish a benchmark time the job.

Gaining Acceptance for Performance Standards: To overcome many of the inherent difficulties associated with developing maintenance performance standards, the ACE System is recommended and should be established as the standard process for modern maintenance management. Other methods, such as the use of standard data, can supplement the ACE System. The ACE System methodology relies primarily on the combined experience and estimating ability of a group of skilled craftspeople, planners, and others with technical knowledge of the repairs being made within the operation.

The objective of the ACE Team Benchmarking Process is to determine reliable planning times for a number of selected benchmark jobs and to gain a consensus and overall agreement on the established work content time. This system places a very high emphasis on improving current repair methods, continuous maintenance improvement, and the changing of planning times to reflect improvements in performance and methods as they occur. The ACE System is a very progressive method to develop maintenance performance standards through reliable and well-accepted planning times for maintenance.

The complete 10-step approach to implementing the ACE Team Process within your current planning, estimating, and scheduling can be found at www.Pride-in-Maintenance.com.

The Basic Approach: A Simple Procedure that Works

Generally, the ACE System parallels the concepts of the UMS approach. For both UMS and the ACE System, the range-of-time concept and slotting are used once the work content times for a representative number of benchmark jobs have been established. The ACE System focuses primarily on the development of work content times for representative benchmark jobs that are typical of the craft work performed by the group. An example of an actual UMS benchmark job that has been analyzed with standard data to establish work content time is included as Figure G.1.

For the example illustrated in Figure G.2, we see that through the use of UMS standard data, the eight elements of the job, including oiling of the parts, have been analyzed and assigned time values that total 1.07 h. Because the time value for this benchmark job falls within the time range of 0.9–1.5 h (see Figure G.3), it is assigned a standard work content time of 1.2 h.

What this implies is that the actual work content for this benchmark job will generally be performed within the time range for Work Group G (0.9–1.5 h) with confidence level of 95%. When we refer to "work content," the following applies.

The work content of the benchmark job excludes things such as travel time, securing tools and parts, prints, delays, and personal allowances, etc. The benchmark time that is estimated does not include the typical "make ready" and "put away" activities that are associated with the job. Therefore, a number of allowances must be added to work content time by a planner to get the actual planning time for the job being.

Benchmark Analysis Sheet

Decription: Remove and reinstall 3 oil wiper rings	B.M. No:		
split type of air compressor 1950 cfm at 100 psi	Craft: Mech		
	Dwns: N/A		
	No. of Men: 1	Sh. 1 of 1	
	Analyst: JEB	Date:	

Line	Men	Operation Description	Reference Symbol	Unit Time	Freq.	Total Time
1	1	Remove and reinstall 1 crankcase cover	PWN-10-10	.030	2	.060
2		Remove and reinstall 2 bolts	PWN-10-1	.011	4	.044
3		Slide gland off and on	PWN-10-7	.012	2	.024
4		Unfasten 3 garter springs and refasten	PWN-10-8	.023	6	.138
5		Remove and reinstall 9 wiper ring	PWN-10-9	.012	18	.216
		segments				
6		Fit 9 ring segments to piston rod	PWN-5-2	.040	9	.360
7		Clean 12 springs and rings	PWN-8-1	.016	12	.192
8		Oil 12 parts	PWN-3-9			
		2 squirts per pint				
		.0023 + .0012(N2)		.0023	1	.002
		N2 = number of application		.0012	24	.029

Notes:	Benchmark time	1.07
	Standard work group	E

Figure G.2 An example of a benchmark job analysis using UMS standard data.

Important Note: The estimated time for a benchmark job is for pure work content time and is made under these conditions:

1. The right craft skills and level of competency are available to do the job.
2. An average skilled craftsperson, two-person team, or crew is giving 100% effort to the job—that is, a fair day's work for a fair day's pay.
3. The correct tools are available at the job site or with the craftsperson or crew.
4. The correct parts are available at the job site or with the craftsperson or crew.
5. The machine/process/asset is available and ready to be repaired.
6. The craftsperson or crew is at the job site with all of the above and proceeds to complete the job from start to finish without major interruption.

ACE TEAM Benchmark Job Analysis Sheet						
Benchmark Job Description			Benchmark Job No.: MECH-AC-5			
Remove and reinstall 3 oil wiper rings, split type air compressor 1950 cfm @100psi			Craft: Mechanical			
			Ref. Drawing: AC-9999			
			No. of Crafts: 1			
			Analyst: JEB/ACE Team			
Line No.	No. of Crafts	Operation Description	Ref. Code	Unit Time	Freq	Total Time
1	1	**Remove 2 bolts to remove crankcase cover**				.10
2		Slide gland off and unfasten 3 garter springs				.10
3		Remove 9 wiper ring segments				.10
4		Clean 12 springs and rings and properly oil all 12 parts per Lube Spec #AC-2000	CLN-1			.25
5		Reinstall 9 wiper ring segments and fit to piston rod				.50
6		Slide glide on and fasten 3 garter springs				.10
7		Replace. crankcase cover and fasten with 2 bolts				.10
Notes: CLN-1 = Average benchmark time established to clean and oil this # of small parts per lube specification noted above.			**Benchmark Time for Work Content**			1.25
			Standard Work Group			E

Figure G.3   Example of an ACE team benchmark analysis.

Once a sufficient number of benchmark job times have been established for craft areas and work types, these jobs are categorized onto spreadsheets. They are established on spreadsheets by craft and task area and according to the standard work groups (Figure G.4), which represent various ranges of time. This is exactly the concept behind the UMS approach. Figure G.5 provides an example of a spreadsheet for work groups E, F, G, and H with jobs that have benchmark times of 1.2, 2.0, 3.0, and 4.0 h respectively.

A complete set of ACE Team Spreadsheets for Work Group A (0.1 h) up to Work Group T (30.0 h) is available in the forms section of this appendix.

ACE System Work Groupings and Time Ranges

Figure G.6 includes a listing of the ACE System work groupings, with the respective time ranges for each work group from A to T. Likewise, spreadsheets for Work Groups A, B, C, and D could be developed with benchmark jobs having work content time

CRAFT: _____   CODE:_____

Task Area: Task areas would be mechanical, electrical, hydraulic, etc or by major areas such as fork lift, conveyor systems or building systems types of repairs.

Group E 1.2 Hour	Group F 2.0 Hours	Group G 3.0 Hours	Group H 4.0 Hours
>.9 <1.5	>1.5 <2.5	>2.5...................... <3.5	>3.5......................<4.5
Job Description A	Job Description E	Job Description I	Job Description M
Job Description B	Job Description F	Job Description J	Job Description N
Job Description C	Job Description G	Job Description K	Job Description O
Job Description D	Job Description H	Job Description L	Job Description P

NOTE:

Figure G.4   Example spreadsheets for work groups E, F, G, and H.

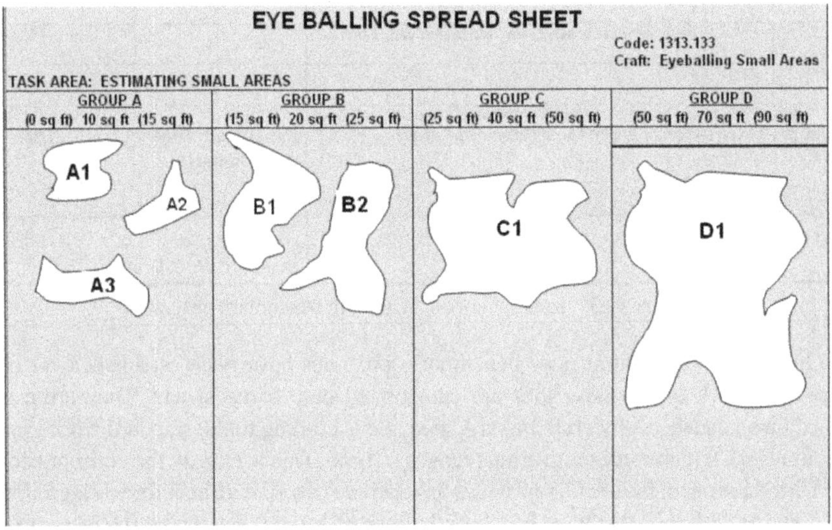

Figure G.5   Example spreadsheets for comparing areas of work groups A, B, C, and D.

below 1.2 h. Spreadsheets for Work Groups I, J, K, and L for benchmark jobs having work content time from 5.0 to 9.0 h, respectively, could also be developed as they were needed. Figure G.7 shows the ACE System standard work groupings and time ranges for Work Group U for 32 h to Work Group CC for 68 h.

The ACE System Time Ranges

	ACE SYSTEM TIME RANGES		
WORK GROUP	FROM	STANDARD TIME (Slot time)	TO
A	0.0	.1	.15
B	.15	.2	.25
C	.25	.4	.5
D	.5	.7	.9
E	.9	1.2	1.5
F	1.5	2.0	2.5
G	2.5	3.0	3.5
H	3.5	4.0	4.5
I	4.5	5.0	5.5
J	5.5	6.0	6.5
K	6.5	7.3	8.0
L	8.0	9.0	10.0
M	10.0	11.0	12.0
N	12.0	13.0	14.0
O	14.0	15.0	16.0
P	16.0	17.0	18.0
Q	18.0	19.0	20.0
R	20.0	22.0	24.0
S	24.0	26.0	28.0
T	28.0	30.0	32.0

Figure G.6 ACE system standard work groupings and time ranges: A to T.

ACE TEAM BENCHMARKING SYSTEM: WORK GROUPS &TIME RANGES Up to 69 Hours			
WORK GROUPS	RANGES FROM (Hrs)	BENCHMARK TIME (Slot Time)	UP TO (Hrs)
U	32.0	34.0	36.0
V	36.0	38.0	40.0
W	40.0+	42.0	44.0
X	44.0	46.0	48.0
Y	48.0	50.0	52.0
Z	52.0	54.0	56.0
AA	56.0	58.0	60.0
BB	60.0	62.0	64.0
CC	64.0+	66.0	68.0

Figure G.7 ACE system standard work groupings and time ranges: 32 to 68 h.

Spreadsheets Provide Means for Work Content Comparison

After sufficient spreadsheets have been prepared based on the representative benchmark jobs from various craft/task areas, a planner/analyst now has the means to establish planning times for many different maintenance jobs using a relatively small number of benchmark jobs as a guide for work content comparison. By using work content comparison (or slotting, as it is called) combined with a good background in craft work and knowledge of the benchmark jobs, a planner now has the tools to establish reliable performance standards consistently, quickly, and with confidence for a large variety of different jobs.

Because the actual times assigned to the benchmark jobs are so critical, it is very important to use a technique that is readily acceptable. The ACE System provides such a technique because it is based on the combined experience of a team of skilled craftspeople and others. It is their consensus agreement on the range of time for the benchmark jobs—a consensus of experts who know the mission-essential maintenance work that is to be done.

The Procedure for Using the ACE System

Select "Benchmark Jobs."

Review past historical data from work orders and select representative jobs that are normally performed by the craft groups. Special attention should be paid to determine the 20% of total jobs (or types of work) that represent 80% of the available craft manpower. Focus on determining repetitive jobs where possible in all craft areas.

Select, Train, and Establish the Team of Experts (ACE Team)

It is important to select craftspeople, supervisors, and planners who as a group have had experience in the wide range of jobs selected as benchmark jobs. All craft areas should be represented in the group. In order to ensure that this group understands the overall objectives of the maintenance planning effort, special training sessions should be conducted to cover the procedures to be used, reasons for establishing performance measures, etc. A total of 6–10 knowledgeable team members is the recommended size for the team.

Develop the ACE Team Charter

At this point, it is highly recommended that a formal ACE Team charter be established as outlined in first part of this appendix. The ACE Team has an important task that will take time to accomplish. However, the task of the ACE Team will be important, and in turn their success as a team can contribute significantly to CRI and increased asset uptime.

Develop Major Elemental Breakdown for Benchmark Jobs

1. For each benchmark job that is selected, a brief element analysis should be made to determine the major elements or steps for completing the total job. Another example is shown in Figure G.3. Here, the elements of the same job that we illustrated in Figure G.2 are used, but they are arranged in a more logical sequence of the actual repair method. In this example,

a standard time allowance for cleaning and oiling a small group of parts had already been established, so reference to CLN-1 task was made. This task referenced back to a standardized lube specification (#AC-2000).

2. This listing of the major steps of the job should provide a clear, concise description of the work content for the job under normal conditions. It is important that the work content for a benchmark job be described and viewed in terms of what is a normal repair and not what may occur as a rare exception. All exceptions, along with make-ready and put-away time, are accounted for by the planner when the actual planned time is completed.

3. An excellent resource to consider for doing the basic element analysis for each benchmark job is the craftspeople (ACEs) who are selected for doing the estimating or even other craftspeople within the operation. Brief training on methods/operations analysis can be included in the initial training for the ACEs. Very significant methods improvements and methods to improve reliability can be discovered and implemented as a result of this important step.

4. The ACE Team process must include and also lead to getting answers to the following questions:
   a. "Are we using the best method, equipment, or tools for the job?"
   b. "Are we using the safest method for doing this job?"
   c. "Are we using the best quality repair parts and materials or is this a part of our problem?"
   d. "What type of preventive task and/or predictive task would help identify or eliminate the root cause of the problem?"
   e. "Where can we work even smarter, not necessarily harder?"

5. Major exceptions to a routine job should be noted if they are significant. Generally, an exception will be analyzed as a separate benchmark job, along with an estimate of time required for such repair.

6. This portion of the ACE Team process ensures that the work content of each benchmark job is clearly defined so that each person/planner doing the estimating has the same understanding about the nature and scope of the job. When the benchmark jobs are finally categorized into spreadsheets, the benchmark job description information developed in this step is then used as key information about the benchmark job on the ACE Team spreadsheets.

Conduct First Independent Evaluation of Benchmark Jobs

1. Each member of the group is now asked to review the work content of the benchmark jobs and to assign each job to one of the UMS time ranges or slots. Each member of the group provides an independent estimate, which represents an unbiased personal estimate of the "pure work content" time for the benchmark job. It is essential that each team member do an independent evaluation of each benchmark job and not be influenced by others on the team with their first evaluation.

2. **Focus on Work Content Time:** It is important here for each member of the team to remember that only the work content of the benchmark job description is to be estimated and not the make-ready and put-away activities associated with the job. This part of the procedure is concerned only with estimating the pure work content, excluding things such as travel time, securing tools and parts, prints, delays, and personal allowances, etc.

3. The estimate should be made for each job under these conditions:
   a. An average skilled craftsman is doing the job giving 100% effort (i.e., a fair day's work for a fair day's pay).
   b. The correct tools are available at the job site or with the craftsperson.
   c. The correct parts are available at the job site or with the craftsperson.

| Make Ready + | Work Time + | Put Away = | Planning Time |

Figure G.8   Planning time elements.

   d. The machine is available and ready to be repaired.

   e. The craftsperson is at the job site with all of the above and proceeds to complete the job from start to finish without major interruption.

4. The work accomplished under these conditions therefore represents the "pure work content" of the job to be performed. Establishing the range-of time estimate for this pure work content is the prime objective of the first evaluation.

5. It is important for each ACE Team member to remember that to develop planning time requires pure work content time plus additional time allowances to cover make-ready and put-away activities associated with each job, as illustrated in Figure G.4. The make-ready time and put-away time will be accounted for as the planner adds time and allowances for these elements as the actual planning time is completed. Make-ready and put-away times are established specifically for each operation and added to the work content time to get the total planned time for the job being estimated (Figure G.8).

Summarize First Independent Evaluation

1. Results of the first evaluation are then summarized to check the agreement among the group as to the time range for each benchmark job. A coefficient of concordance can be computed from the results if required but normally this level of detail is not needed. A coefficient of concordance value of 0.0 denotes no agreement, while a value of 1.0 denotes complete agreement or consensus among the ACEs. Generally, a consensus can be reached by the ACE Team in one, two, or at the most three rounds of evaluations.

2. Define High and/or Low Estimates: Team members who are significantly higher or lower than the rest of the group for a particular benchmark job are then asked to explain their reasons for their respective high or low estimates. They explain their reason for their estimate to the group, discussing the method, condition, or situation for their initial time estimate. This information will then be used during the second evaluation to refine the next round of time estimates from the entire group.

Conduct Second Independent Evaluation of Benchmark Jobs

1. A second evaluation is conducted using the overall results from the first evaluation as a guide for the entire team. Various reasons for high or low estimates from the first evaluation are provided to the group prior to the second evaluation. Normally, this can be done in an open team discussion, with team members making personal notes to use in their second independent evaluation.

2. The second round allows for adjustment to the first estimates if the other ACE Team member's reason for a higher or lower estimated time is considered to be valid. In other words, results from the first evaluation plus reasons for highs and lows will allow each team member to reconsider their first estimate. In many cases, a review of the repair method or scope of work will be more clearly defined, causing a change to the time estimate for the second evaluation.

Summarize Second Evaluation

1. Results of the second evaluation are then summarized to evaluate changes or improvements in the level of agreement. The goal is a consensus among the ACEs as to the time range (Work Group) for each benchmark job. The second round should bring an agreement as to the time range.
2. The second independent evaluation should produce improved agreement among the group. If an extreme variance in time range estimates still exists, further information regarding the work content, scope, and repair method for the job may be needed. Here, those with high/low estimates should again review their reasons for their estimates with the team, describing the scope of work that they see is causing differences from the rest of the team.

Conduct a Third Independent Evaluation if Required

This evaluation is required only if there remains a wide variance in the estimates among the group.

Conduct a Review Session to Establish Final Results

This session serves to finalize the results achieved and to discuss any of the high or low estimates that have not been resolved completely. A final team consensus on all time ranges is the objective of this session.

Develop Spreadsheets

1. The benchmark jobs with good work content descriptions and agreed-upon time ranges can now be categorized onto spreadsheets. From these spreadsheets, which give work content examples for a wide range of typical maintenance jobs, a multitude of individual maintenance performance standards can be established by the planner through the use of work content comparison.
2. The basic foundation for the maintenance planning system is now available for generating consistent planning times that will be readily acceptable by the maintenance workforce that developed them. Figure G.9 provides a graphical illustration of the ACE System.

The ACE Team approach combines the Delphi technique for estimating along with a proven team process plus the inherent and inevitable ability of most people to establish a high level of performance measures for themselves. As used in this application, the objective for the ACE Team process is to obtain the most reliable, reasonable estimate of maintenance-related "work content" time from a group of experienced craftspeople, supervisors and planners. This process provides an excellent means to evaluate repair method, safety practices, and even to do risk analysis on jobs that leads to improved safety practices. The ACE Team process can contribute significantly to continuous reliability.

The ACE Team approach allows for independent estimates by each member of the group, which in turn builds into a consensus of expert opinion for a final estimate. The final results are therefore more readily acceptable because they were developed

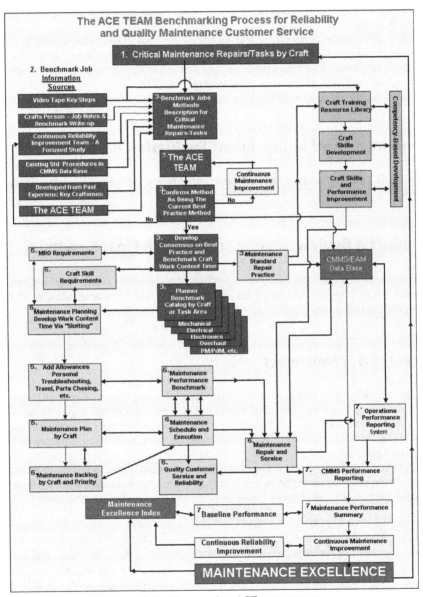

Figure G.9   The complete ACE team process.

by skilled and well-respected craftspeople from within the work unit. Application of the ACE System promotes a commitment to quality repair procedures and provides the foundation for developing reliable planning times for a wide range of maintenance activities (Figure G.9).

FORMS SECTION

ACE TEAM Benchmark Job Analysis				
Benchmark Job Description				
			Benchmark Job No:	
			Craft	
			Ref. Drawing:	
			No. of Crafts:	
			Analyst:	
			Date:	

Line No.	No. of Crafts	Operation Description	Unit Time	Freq	Total Time
Notes:			Benchmark Time for **Work Content**		
			Standard Work Group		

ACE Team Spreadsheet for Work Groups A, B, C and D

CRAFT: _____   CODE:_____

Task Area:

Group A: .1 Hours		Group B: .2 Hours		Group C: .4 Hours		Group D: .7 Hours	
(0.0+)	(.15)	(.15+)	(.25)	(.25+)	(.5)	(.5+)	(.9)

ACE Team Spreadsheet for Work Groups E, F, G and H

CRAFT: _____   CODE:_____

Task Area:

Group E: 1.2 Hours		Group F: 2.0 Hours		Group G: 3.0 Hours		Group H: 4.0 Hours	
(.9+)	(1.5)	(1.5+)	(2.5)	(2.5+)	(3.5)	(3.5+)	(4.5)

ACE Team Spreadsheet for Work Groups I, J, K and L

CRAFT: _____   CODE:_____

Task Area:

Group I: 5.0 Hours		Group J: 6.0 Hours		Group K: 7.3 Hours		Group L: 9.0 Hours	
(4.5+)	(5.5)	(5.5+)	(6.5)	(6.5+)	(8.0)	(8.0+)	(10.0)

ACE Team Spreadsheet for Work Groups M, N, O and P

CRAFT: _____   CODE:_____

Task Area:

Group M: 11.0 Hours		Group N: 13.0 Hours		Group O: 15.0 Hours		Group P: 17.0 Hours	
(10.0+)	(12.0)	(12.0+)	(14.0)	(14.0+)	(16.0)	(16.0+)	(18.0)

ACE Team Spreadsheet for Work Groups Q, R, S and T			
CRAFT: _____　CODE:_____			
Task Area:			
Group Q: 19.0 Hours (18.0+)　　　(20.0)	Group R: 22.0 Hours (20.0+)　　　(24.0)	Group S: 26.0 Hours (24.0+)　　　(28.0)	Group T: 30.0 Hours (28.0+)　　　(32.0)

Appendix H

Shop Load Plan, Master Schedule and Shop Schedules: Example Forms and Steps on How to Use

Shop Load Plan

Sample Form: Shop Load Plan. Figure H.1 is a sample shop load plan form. The computerized maintenance management system (CMMS) used by the center should support computer-aided scheduling, including interactive manpower and other resource scheduling and schedule balancing. The shop load plan should be automated as a standard report in the CMMS. A single database should support all three levels of scheduling (i.e., shop load plan, master schedule, and shop schedule) in a networked system. Although it is possible to examine the shop load plan on a video display terminal, the practical limitations on the number of lines and columns that can be displayed at one time makes a printout on wide paper convenient for use by managers.

Data Elements. The following data elements are shown on the shop load plan. The information either is contained in the CMMS database or is derived from the CMMS database. The information is defined below as an aid to understanding the schedule format. The only item that should require entry in the scheduling process is for the scheduling period. The rest of the information is based on other data entered in the CMMS during the work reception and planning process or is extracted from other databases, such as labor accounting.

1	Period covered	The time period this schedule considers. Normally for the shop load plan, this period is a quarter; however, the immediate next 3 months may be subdivided into a 1-month short-term and a 2-month midterm load plan for additional scheduling control.
2	Shop	The shop or craft group being scheduled, such as carpenters. (The sample form shows 12 shops; this should be adjusted to meet local needs.)
3	Number of employees	The average number of employees available in each shop.
4	Gross work hours available	The total number of work hours available in each shop during the schedule period.
5	Adjustments	The number of work hours that will not be available in each shop for facilities maintenance work due to leave, training, jury duty, and similar nonproduction activities. (This may be presented on more than one line if it is desired to have a line for each type of adjustment.)

Continued

—cont'd

6	Net work hours available	The net work hours available in each shop for facilities maintenance work, computed as item 4 minus item 5.
7	Trouble call (TC) ticket level of effort (LOE)	The number of hours allocated by shop for jobs issued under trouble call tickets. Usually, this is based on past experience.
8	Preventive maintenance (PM) scheduled	The number of hours for scheduled PM work by shop. This is determined from the PM schedule contained in the CMMS database. A typical example is shown as Figure H.1.
9	Predictive testing and inspection (PT&I) scheduled	The number of hours for scheduled PT&I work by shop. This is determined from the PT&I schedule contained in the CMMS database.
10	Scheduled, Recurring	The number of hours by shop for other scheduled recurring work. This may be determined from the CMMS database. It may be presented on more than one line if grouped, such as by type or work order.
11	Total LOE scheduled	Total hours committed to items 7, 8, 9, and 10 above.
12	Carryover from prior period:	Work scheduled or started in the prior period but not completed, and thus carried over to this period. This may be automatically computed by comparing work order labor estimates against labor charges to date.
13	Available to schedule	Net work hours available (item 6) less item total LOE scheduled (item 11), and carryover (item 12). This is the workforce available for scheduling new work orders.
14	Work order number (WO#)	Work order number for each work order listed.
15	Description	An entry giving a short title for each work order. The number of hours estimated for each shop for the work order follows on the same line.
16	Requested start date (RSD)	Requested start date for the work order.
17	Required completion date (RCD)	Required completion date for the work order.
18	Priority (PRI)	Priority of the work order.
19	Material status indicator (MAT)	Normally, this block contains the date on which the required material is expected to be available or that the material is available. This is the overall status of the field "Avail" on a planner's work order material/equipment requirements or checked off as yes on work order form.
20	Work hours	The estimated work hours for the work order for each shop.
21	Total	The total labor hours for the work order for all shops.
22	Labor	Estimated total labor cost for the work order.
23	Mtl	Estimated total material cost for the work order.
24	Other	Estimated total other cost for the work order. This would include items such as equipment rentals and contracted services.
25	Total	The total cost for the work order.

PERIOD COVERED: (1)

SHOP LOAD PLAN

SPECIFIC WORK/ITEM			WORKHOURS													COST EST	REMARKS
		(2) SHOP ->	01	02	03	04	05	06	07	08	09	10	20	30	Total		
WORK FORCE AVAILABILITY	No. of Employees	(3)															
	Gross Workhours Avail	(4)															
	Adjustments	(5)															
	Net Workhours Avail	(6)															

W.O. #	DESCRIPTION	RSD	RCD	PRI	MAT	01	02	03	04	05	06	07	08	09	10	20	30	Total	Labor	Mtl	Other	Total	
COMMITTED WORK FORCE	TC LOE (7)																						
	PM Scheduled (8)																						
	PT&I Scheduled (9)																						
	Scheduled, Recurring (10)																						
	Total LOE, Scheduled, etc. (11)																						
	Carry-over from Prior Period (12)																						
	AVAILABLE TO SCHEDULE (13)																						
(14)	(15)	(16)	(17)	(18)	(19)	(20)												(21)	(22)	(23)	(24)	(25)	
	[list specific work orders]																						
	[as many lines as needed]																						
	Net hours over/under scheduled																						

Figure H.1 Sample form: shop load plan.

Instructions for Use: Shop Load Plan

1. Normally, the shop load plan is prepared covering a quarter. However, shop load plans should be prepared and maintained looking 18 months into the future. The last period should include all work that is in an estimated and approved but unscheduled status. A center also may wish to extract a short-term (next month) and a midterm (following two months) shop load plan for closer work scheduling and management. After final approval of a work order, it is assigned to a shop load plan. Normally, this level of scheduling is done by a senior maintenance planner, not in the shop's organization. This starts the work performance phase and triggers material acquisition to ensure that the required material is available for the assigned start period. Approved work orders remain in the shop load plan until completed or canceled.

2. The primary purpose of the shop load plan is to provide for the orderly scheduling of work in accordance with the center's mission priorities, to assist in resource scheduling and management, and to provide senior managers with information on pending work. It also provides a valuable tool for evaluating the workforce skill mix against workload requirements. If the shop load plan consistently shows a significant amount of overscheduling or unscheduled backlog in a shop coupled with underscheduling in another shop, realignment of workforce assets from the underscheduled to the overscheduled shop may be in order.

Master Schedule

Sample Form: Master Schedule. Figure H.2 is a sample form for a master schedule. The master schedule is based on the shop load plan. However, its focus is on scheduling work performance to a specific week and tracking material status of work orders that are due for master scheduling in the future according to the current and approaching shop load plans. Normally, master schedules are prepared covering 6–10 weeks into the future. Jobs with long lead-time material requirements may be scheduled further in the future. Of special interest is the Work Orders Waiting Material section. This is used to highlight the material status of work orders waiting material that need to start during the master schedule period covered.

    Data Elements. The following data elements are shown on the master schedule form. As with the shop load plan, the information either is contained in the CMMS database or is derived from the CMMS database by manipulation and calculation. The data elements are defined below as an aid to understanding the schedule form. The only data that should require entry in the scheduling process are for the period during which the work order is being scheduled (normally, the specific workweek). The rest is based on other data entered in the CMMS during the work reception and planning process, the material management process, or extracted from other databases such as labor accounting.

1	Period covered	The time period this schedule considers. Normally for the master schedule, this is a workweek.
2	Shop	The shop or craft group being scheduled (e.g., shop 01, carpenters).
3	Number of employees	The average onboard manpower in each shop during the schedule period.

Continued

—cont'd

4	Gross work hours available	The total number of work hours in each shop available during this period.
5	Adjustments	The number of work hours that will not be available for facilities maintenance work due to leave, training, jury duty, and similar nonproduction activities. This may be presented on more than one line if it is desired to have a line for each type of adjustment.
6	Net work hours available	The net available work hours for facilities maintenance work.
7	TC LOE	The level of effort (the number of hours) allocated for jobs, usually issued under trouble call tickets. Usually, this is based on past experience.
8	PM scheduled	The numbers of hours for scheduled PM work.
9	PT&I scheduled	The number of hours for scheduled PT&I work.
10	Scheduled and recurring	The number of hours for other scheduled or recurring work. This may be presented on more than one line if grouped, such as by type or work order.
11	Total LOE scheduled	Total hours committed to items 7, 8, 9, and 10.
12	Available to schedule	Net work hours available (item 6) minus item 11. This is the workforce available for scheduling specific work orders.
13	WO#	Work order number for each specific work order listed.
14	Description	Entry giving a short title for each work order. Also, the numbers of work hours scheduled for the work order by each shop during this schedule period are entered under the shop number on the same line.
15	RSD	Requested start date for the work order.
16	RCD	Requested completion date for the work order.
17	PRI	The work order priority rating.
18	MAT	Material status indicator. Normally, this entry is the date on which material required for the work order is expected to be available, or a code indicating that the material is currently available.
19	Total	Total labor hours for the shops.
20	Labor	Expended labor hours. The total labor hours used or scheduled for the work order prior to this schedule period.
21	Mtl	Cumulative material cost of material used for work order.
22	Other	Cumulative costs of other than labor and material used for the work order. This includes such items as equipment rentals and contracted services.
23	Total	Cumulative total cost.

Instructions for Use: Master Schedule

1. The master schedule is used to direct and coordinate the execution of work in the shops. It provides the coordinating linkage between the shops on jobs involving more than one shop and it highlights the material status of pending work orders. Normally, it is maintained under the direction of the shop supervisor working in close coordination with the shop supervisors and the maintenance planners. Work is scheduled by assigning it to a specific workweek; the

PERIOD COVERED: (1)			MASTER SCHEDULE													
SPECIFIC WORK/ITEM			WORKHOURS												USED TO-DATE	REMARKS
WORK FORCE AVAILABILITY		(2) SHOP ->	01	02	03	04	05	06	07	08	09	10	20	30	Total		
	No. of Employees	(3)															
	Gross Workhours Avail	(4)															
	Adjustments	(5)															
	Net Workhours Avail.	(6)															

W.O. #	DESCRIPTION	RSD	RCD	PRI	MAT	01	02	03	04	05	06	07	08	09	10	20	30	Total	Labor	Mtl.	Other	Total	
COMMITTED WORK FORCE	TC LOE (7)																						
	PM Schedule (8)																						
	PT&I Schedule (9)																						
	Scheduled and Recurring (10)																						
	Total LOE, Scheduled, etc. (11)																						
	AVAILABLE TO SCHEDULE (12)																						
(13)	(14)	(15)	(16)	(17)	(18)													(19)	(20)	(21)	(22)	(23)	
	(list specific work orders)																						
	(as many lines as needed)																						
	Net hours over/under scheduled																						

WORK ORDERS WAITING MATERIAL																							
	(as many lines as needed)																						
	Total waiting material																						

Figure H.2 Sample form: master schedule.

automation program used should perform all necessary calculations, including computing estimated carryover work and resources expended (or projected to be expended) up to the period under consideration.

2. It is essential for the master schedule to give close attention to balancing the work to each shop to ensure that all forces are productively employed. To this end, the master scheduler will assign labor hours to each scheduled job within available manpower and job phasing requirements.

3. Although it is possible to examine the master schedule on a video display terminal, the practical limitation on the number of lines and columns that can be displayed at one time makes it difficult to see all work that is subject to scheduling. Accordingly, printouts on wide paper and wall-mounted scheduling boards normally are used to display job status.

Shop Schedule

Sample Form: Shop Schedule. A form for a shop schedule is provided as Figure H.3. The shop schedule provides the day-to-day scheduling/assignment of workers and equipment to work orders. It is used by the shop supervisor as an aid in scheduling personnel and shared equipment assets.

Data Elements. The following data elements are shown on the shop schedule form. The information either is contained in the CMMS database or is derived from the CMMS database by manipulation and calculation. The data elements are defined below as an aid to understanding the schedule form. The only data elements that should be entered during the scheduling process are the assigned hours for each work order and employee being scheduled. The remaining data elements should be provided by the computer based on other data entered in the CMMS during the work reception and planning process, the material management process, or extracted from other databases. For example, hours for PM on the schedule shown be the top priority based upon level loading PM across the entire year as shown in Figure H.4.

1	Period covered	The time period this schedule considers. Normally for the shop schedule, this is a specific day.
2	Shop	The shop or craft group being scheduled (e.g., shop 01, carpenters).
3	Employee	The name or other identifier of the worker being scheduled.
4	Gross work hours available	The total number of work hours available for each worker during this period, normally 8 h.
5	Adjustments	The number of work hours that will not be available for facilities maintenance work due to leave, training, jury duty, and similar nonproduction activities. This may be presented on more than one line if it is desired to have a line for each type of adjustment.
6	Net work hours available	The net available work hours for each employee for facilities maintenance work.
7	WO#	Work order number for each work order listed.
8	Description	An entry giving a short title for each work order. The hours assigned to each employee for each work order number follows on the same line under the employee's identification.

Continued

—cont'd

9	RSD	Requested start date for the work order.
10	RCD	Requested completion date for the work order.
11	PRI	The work order priority rating.
12	MAT	Material status indicator. Normally, this is a code or symbol indicating that the material is currently available.
13	Total	Total hours for all employees.
14	Labor	Cumulative labor hours for the work order prior to this schedule period. This information is provided as part of the labor distribution/timekeeping process.
15	Mtl	Cumulative cost of material used for the work order.
16	Other	Cumulative cost of other than labor and material used for the work order. This includes such items as equipment rentals and contracted services.
23.	Total	Cumulative total cost.

Instructions for Use: Shop Schedule

The shop supervisor uses the shop schedule for scheduling and managing craft personnel. It is typically prepared on a weekly basis for each day of the following week, based on jobs scheduled in the master schedule. The shop supervisor enters the hours each employee is scheduled to work on each assigned job for each day. The workforce availability is determined from leave, training, and related activities that are also scheduled through the shop supervisor. A planner/supervisor's projection of resources over a 4-week period is shown in Figure H.5. Also, a weekly schedule example with 90% of the availability is shown in Figure H.6.

SHOP SCHEDULE

PERIOD COVERED: (1)   For Shop   (2)

SPECIFIC WORK/ITEM

		EMPLOYEE -> (3)
Gross Workhours Avail.	(4)	
Adjustments	(5)	
Net Workhours Avail	(6)	

W.O. #	DESCRIPTION	RSD	RCD	PRI	MAT		WORKHOURS									Total	USED				REMARKS
(7)	(8)	(9)	(10)	(11)	(12)											(13)	Total (13)	Labor (14)	Mtl (15)	Other (16)	Total (17)
	(assigned work orders)																				
	(as many lines as needed)																				
	Net hours over/under scheduled																				

Figure H.3   Sample shop schedule.

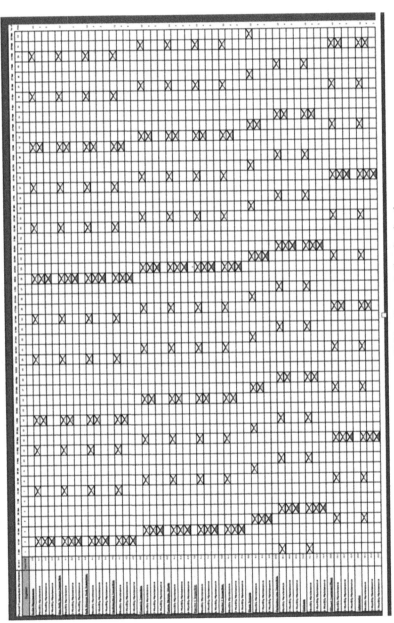

Figure H.4 Sample PM leveling and planning by shop area.

Operating unit	Craft	04-Jul-05 week 1	11-Jul-05 week 2	18-Jul-05 week 3	25-Jul-05 week 4
Week commencing					
Area 1	Mech	8 M	7 M	6 M	5.5 M
	Elec	4.5 E	5 E	3 E	4 E
	Inst	1.5 –	2 –	0	3 –
Area 2	Mech	10 M	9 M	12 M	7 M
	Elec	5.5 E	3.5 E	7 E	4 E
	Inst	3 –	3 –	3 –	2.5 –
Area 3	Mech	0	1.5 M	0	1 M
	Elec	0	0.5 E	0	1 E
	Inst	0	0.5 –	0	0
Area 4	Mech	7 M	8.5 M	7 M	10 M
	Elec	4.5 E	5 E	5 E	5 E
	Inst	2.5 –	2 –	2 –	1.5 –
etc					
Weekly totals	Mech	25.0 M	26.0 M	25.0 M	23.5 M
	Elec	14.5 E	14.0 E	15.0 E	14.0 E
	Inst	7.0 –	7.5 –	5.0 –	7.0 –

Monthly totals		
Mech	99.5	Mechanical
Elec	57.5	Electrical
Inst	28.5	Instrumentation

Figure H.5   Sample 4-week shop look-ahead of resources by trade type (mechanical. Electrical, and instrumentation).

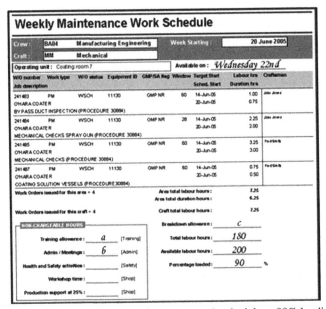

Figure H.6   Sample weekly maintenance work schedule at 90% loading.

Appendix L

Routine Planner Training Checklist

The purpose of this document is to ensure new planners are trained on all aspects of planning/scheduling a maintenance order in SAP/P3.

Personal Details

Name: Position: Level: Team Leader Name:	Employee Number: Team/Work Group: Performance/Functional Unit:

Section 1 (Basic)

Activity	Responsibility	Comments	Trainer initials/Date
Work identification			
• Retrieve approved notifications from notification backlog			_____
• Ability to review/edit notification for needed corrections			_____
• Understand the RIME system and how to edit to assign (provide employee with the RIME document)			_____
Planning/Estimating			
• Create maintenance order from approved notification			_____
• Explain the job note process			_____
• Review reusable job plan process (creation, use, retrieval, etc.)			_____
• Review tasking process in SAP			_____
• Review PM activity field in SAP			_____
• Review revision coding system in SAP for EBB purposes			_____
• Provide copy of revision codes used for EBB			_____

Continued

—cont'd

Activity	Responsibility	Comments	Trainer initials/Date
• Review estimating process (supply employee with the fluor estimating template)			_____
• Generation of WCC associated with the ISSOW			_____
Approval process			
• Explain cost approval process EBB			_____
• Provide employee with the different levels of approval per work process flow process			_____
• Explain approval process for opportunity contractors			_____
• Provide CWR to employee for approval to use tier 2 and Tier 3 contractors, if required			
Initiating approved work			
• Forward cost estimate to cost analyst			
• Inform employee which cost analyst is assigned to his area of responsibility			
• Order needed materials and rental equipment per job requirements			
• Know how to search SAP identified materials			
• Know how to search material by use of BOM			
• Know how to use the VAM system			
• Know how to create a free text requisition			
• Know how to use the total control system			
• Have employee set up in the total control system			
• Inform employee of their assigned procurement coordinator and procurement associate			
Scheduling			
• Review backlog			
• Review the different planner groups			
• Facilitation of weekly/daily scheduling meetings. Provide employee with meeting agenda document			
• Assignment of scheduled work			
Pre for Execution			
• Generation of CWR for opportunity contractors once they have been approved			

Activity	Responsibility	Comments	Trainer initials/Date
• Minimum requirements for job packages. Provide employee with job package requirement document			
Scope Change			
• Review JCO process and supply employee with JCO form			
Post Execution			
• Returning of rental equipment, materials, and repair items • Entry of history in SAP • Close maintenance orders to maintenance • Revision of reusable job plan if required • Update of BOM if required			

BOM, bill of materials; VAM, vendor agreement manager; EBB, events based Budgeting; RIME, ranking index of maintenance expenditures; PM, preventive maintenance; ISSOW, integrated risk management and safety system; WCC, work control center; JCO, job control order; CWR, corrective work order.

Section 2 (SAP)

Activity	Responsibility	Comments	Trainer initials/Date
SAP user menu			
• How to access maintenance systems (logistics, plant Maintenance, Maintenance processing, etc.)			_____
• Adding and deleting favorites			_____
• Renaming and organizing favorites			_____
Creating notification			
• Know the difference between Maintenance request and malfunction report			_____
• Entering short and long text			
• Assigning functional location			_____
• Assigning main work centers			_____
• Assigning the priority. Explain how RIME is assigned and give copy of RIME document			_____
• Approving the notification. Knowing what level of approval is required based on the estimated cost (ex. 10K approved @ unit level, above 10K approved at Asset Supt. Level, etc.)			_____

Continued

—cont'd

Activity	Responsibility	Comments	Trainer initials/Date
Closing Out/Documenting work history			
• Know how to capture and input history of work performed and condition of repair			———
• Know how to close a notification in conjunction with closing out an MO			———
Creating maintenance order			
• Know how to create a maintenance order from a notification			———
• Understand the different order types and assignment			———
• Understand the different PM Activity type codes and assigning the one that best fits the description of the MO			———
• Transferring long text from the notification to the MO			———
• Understanding and assigning revision codes for EBB. Furnish employee with EBB list			———
• Using the radial buttons in the control tab to identify emergency/break-in or planned jobs.			
• Selecting approver in the partner field of the MO			
Tasking operations			
• Operation activity number			———
• Work centers			———
• Short and long text			———
• Craft number (number of craft per operation)			———
• Duration hours of operation			———
Material procurement			
• Search for SAP-identified materials in the component field			———
• Search material through the use of the VAM system. Explain what the VAM is, how it is used, and what the benefits are			———
• Create requisition from beginning to end (ex. assigning purchasing group, vendor, offset material date, etc.)			———
• Search for material via BOM			———
• How to request a BOM to be added in SAP			———

Activity	Responsibility	Comments	Trainer initials/Date
Scheduling			
• Know the different planner group codes used for status of work orders and scheduling (ex. 124=daily, 123=weekly, etc.)			_____
Reusable job plans			
• Have a basic understanding of how to create, edit, and assign reusable job plans (job templates)			_____
Closing of MO			
• See notification closeout and document work history			_____
Notifications/Maintenance orders/Order operations			
• Understand how to create, modify, and execute variants			_____
• Know how to use "current function" for viewing layouts			_____
• Know how to transport reports form SAP to excel			_____

BOM, bill of materials; VAM, vendor agreement manager; EBB, events based Budgeting; MO, maintenance order.

Sign Off

Date	Employee Signature	Team Leader Signature

Index

Note: Page numbers followed by, "f" and "t" indicate, figures and tables respectively.

A

A Strategy for Developing a Corporate-Wide Scoreboard, 43
ACE (a consensus of experts)
  system
    basic approach, 246–251
    benchmark jobs, 251. *See also* ACE benchmark jobs
    performance standards, gaining acceptance for, 246
    review session, 254
    spreadsheets, developing, 254
    team charter, 252
    team-based approach, 245–246
    team of experts, 251–252
    time ranges, 249, 250f–251f
    work content comparison, 249–251
    work content time, 253, 253f
    work groupings, 249, 250f–251f
  Team Benchmarking Process™, 19, 69, 91, 208–211, 242, 255f
    benchmark jobs, 245
    constraints, 243
    continuous reliability improvement, 245
    evidence of success, 242
    expectations, 243
    investment for planners, 245
    maintenance work measurement tool, 244–246
    methodology for applying, 244
    note, 243
    resources, 243
    spreadsheets, 245
    team-based approach, 244
ACE benchmark jobs
  benchmark jobs, 251
    elemental breakdown for, 252
    first independent evaluation of, 253
    second independent evaluation of, 254
    third independent evaluation of, 254

ACE system work groupings and time ranges, 459–460
  spreadsheets, 460f
    work content comparison, means for, 461–462
    for work groups, 468f–470f
    standard work groupings, 461f
    10-step procedure, 462
    planning time elements, 464f
ACE team benchmarking process
  analysis sheet, 458f
  complete team process, 466f
  example process, 459f
  job analysis, 455f, 467f–468f
  maintenance work measurement tool, 455–457
  methodology for, 455–457
  planning time elements, 464f
  10-step procedure, 457–459
Adjusted averages, 241
Adjustment/alignment, 387
Allowances
  application of, 260f
  for determining total planned time, 258f
  miscellaneous, 259f
  personal, fatigue and delay, 259f
American Petroleum Institute (API), 372
Analysis paralysis scheme, 73
Anticline, 372
Applications parts list, 372
Apprentice, 372
Aquifer, 372
Area maintenance, 372
Asset care, 372
Asset list, 372
Asset management, 372
Asset number, 372
Asset register, 372

Asset utilization, 372
Assets, 372
   criticality, 237t–240t
   history, 103f, 103t, 120
   management, definition of, 11
   numbering structure, 140t
   tracking, 279
Availability, 372
Available hours, 372
Average life, 372

B

Backlog, 92, 92, 372
   calculating weeks of, 164f
   definition of, 158
   integrity, maintaining, 162f–163f
   job status codes, 164f
   maintenance resources with, balancing,
      163f
   management, 159f
   monitoring, 162f
   ready, 92, 96, 158
   reports, 118
   total, 91, 96, 158
   trends, monitoring, 163f
Bar code, 373
Barrel (BBL), 373
Barrel of oil equivalent (BOE), 373
Baseline information, preassessment check-
      list for, 60–64
   craft labor rates/overtime history, 61
   craft skills development, 61
   maintenance budget and cost accounting,
      61
   maintenance storeroom, 62
   MRO purchasing operations, 62
   organization charts/job descriptions, 61
   predictive maintenance, 39–40, 47–48,
      57, 61–62
   preventive maintenance, 39–40, 47–48,
      57, 61–62
   reliability centered maintenance, 61–62
   total productive maintenance, 61–62
   work control, 62
   work orders, 62
Basin, 373
Bathtub curve, 154f
Being stationary, effects of, 280–281, 280f

Benchmark(ing), 40–45, 40f, 373
   analysis sheet, 248f
   baseline value assignment, 54–55
   definition of, 40–41
   evaluation criteria, format of, 55, 57f
   external, 44
   functional, 42
   internal, 43
   international, 44
   jobs, 245–246, 251
      analysis, 247f
      elemental breakdown for, 252
      first independent evaluation of, 253
      independent evaluation of, 253
      second independent evaluation of, 254
      third independent evaluation of, 254
   levels of, 285f
   maintenance, 40f
   performance, 41–42
   process, 42
   results, 54
   strategic, 41
Bill of materials (BOM), 109–110, 373
Bitumen, 373
Blanket work orders, 92
Bohai Bay, China, 381
Borehole, 373
Bowers, Terry, 36
Breakdown, 373
Breakdown maintenance, 298, 373
British thermal unit (BTU), 373
Budget, 120
   status-operating departments, 121
Built-in test equipment (BITE), 373
Business vacation, 373

C

Calibrate, 373
Call-out, 373
CAPEX, 373
Capital, 373
Carbon capture and storage (CCS), 373
Carbon dioxide equivalents (CO_2e), 373
Carbon intensity, 373
Carbon sequestration, 373
Carbon sink, 374
Carlin, Joel, 36
Carrying costs, 374

Casing, 374
Cause of failure, 201
Central maintenance, 374
Centum cubic feet (CFF), 374
Change out, 374
Checkout, 374
Chief maintenance officer (CMO), 3, 374
    as maintenance leader, 3–4
    pride in ownership, 5
City gate stations, 374
Clean, 374
Closed-loop-type organization, 184f
CMMS benchmarking system, 374
    budget and cost control, 443
    data integrity, 439–441
        bill of materials, 440
        equipment (asset) history database, 439
        maintaining equipment database, 441
        maintaining parts database, 440
        preventive maintenance tasks/ frequencies, 440
        spare parts database, 439–440
    education and training. See CMMS education and training
    evaluation, 448–452, 449t–452t
    installation, 448
    maintenance performance measurement, 446–447
    maintenance repair operations materials management, 444–445
    planning and scheduling, 443–444
    preventive/predictive maintenance, 445–446
    purpose, 448
    uses of, 447
    work control
        backlog reports, 442
        function established, 442
        online work request based on priorities, 442
        priority systems, 442–443
        work order system, 442
CMMS education and training
    initial training, 441
        for operations personnel, 441
    ongoing training, 441
    systems administrator/backup trained, 441

Coal bed methane (CBM), 374
Code, 374
Commodity code, 374
Communication, and leadership, 22
Competitive benchmarking, 41–42
Completion, 374
Component, 375
Component number, 375
Compound annual growth rate (CAGR), 375
Compressed natural gas (CNG), 375
Compression, 405
Compressor, 405
Computerized maintenance management system (CMMS), 14–15, 42–43, 71, 79–80, 82, 87, 92, 241, 285, 305, 307, 375, 439, 471
    backlog reports, 118
    benchmark evaluation, conducting, 101–120
    bill of materials, 109–110
    captures reliability data, 134
    cost improvements due to, 121
    craft hours, work order system accounts for, 118
    data integrity, 103t
    for developing maintenance operation as profit center, 143
    education and training, 112t
    equipment database, maintaining, 111–112
    equipment history database, accuracy of, 103–104
    initial training, 112
    ongoing training, 112–113
    online work request based on priorities, 118
    operations personnel, initial training for, 113
    operations understands benefits of, 141
    parts database, maintaining, 110–111
    preventive maintenance tasks/frequencies, 110
    priority system, 120
    spare parts database, accuracy of, 107–108
    systems administrator/backup trained, 113
    uses of, 140t
    work control, 114, 114t, 115f–117f
Condensate, 375
Condition monitoring, 375

Conditional probability of failure, 375
Condition-based maintenance (CBM), 375,
    389
Conditioning, 387
Confidence, 375
Configuration, 375
Configuration data, 278
Consequence of failure, 224
Construction trade estimates, 241
Consumables, 375
Contingency, 375
Continuous reliability improvement (CRI),
    73, 74f, 76, 82–83, 245, 375
Contract acceptance sheet, 376
Contract maintenance, profit and
    customer-centered, 2–3
Contractor management, 181, 199f, 299
Convenience work, 160f
Conventional resources, 376
Coordination, 167, 168f, 170f, 376
Core competencies, for maintenance, 8
Core requirement, for maintenance, 8
Corrective maintenance (CM), 376
Cost
    avoidance, 71
    center, 1–2
    control, 120
    improvements due to CMMS, 121
    life cycle, 122
Craft availability, 376
Craft labor, 121
    resources, 69–70, 71f, 76
Craft leaders, 376
    PRIDE in Excellence within, 29–30
    PRIDE in Maintenance and Construction
        within, 36–37
    PRIDE in Maintenance within, 31–35,
        32f, 34f
Craft performance (CP), 71–72, 77, 79–85,
    87–88, 139, 241
Craft productivity, 69–71, 85–87, 447t
    improved by planning and scheduling, 276f
Craft service quality (CSQ), 26, 72, 77,
    82–84
Craft utilization (CU), 68–70, 70f, 76–79, 376
    measuring and improving, 128, 128f
Craftsperson, 377
Crescent-Xcelite Plant
    leadership buy-in, obtaining, 59–60
    operating goals, 57–60

Critical, 377
Critical equipment, 377
Critical path method (CPM), 377
Critical spare, 377
    identifying, 132, 133f
Criticality, 377
Crossdocking, 377
Customer-centered benefits, of planning and
    scheduling, 1–10
Cycle count, 377
Cycle counting, 131, 131f

D

Data integrity, 103t
Data migration, 99
Davis, Rodney, 36
Dead stocks, 378
Defect, 378
Deferred maintenance, 5–6, 12–13,
    120–121, 378
DELPHI technique, 254
Demand, 378
Detailed planning, 207–222
Deterioration rate, 378
Developed acreage, 378
Developed reserves, 378
Development well, 378
Direct cost savings, 71
Direct costs, 378
Directional drilling, 378
Discard task, 378
Disposal, 378
Distribution line, 378
Distribution system, 378
Do it now (DIN) work, 160f
Down, 378
Downtime (DT), 278, 282, 378
Drilling rig, 378
Dry gas, 378
Dry hole, 378

E

Economic life, 378–379
Economic order quantity (EOQ), 379
Economic repair, 379
Economically producible, 379
Educated guesses, 241
Effective planning, benefits of, 9
Effectiveness factor, 68, 68f, 72

Efficiency, 12
Electronic parts requisitioning, 132, 132f
Elke system, 298
Emergency maintenance, 379
Emergency maintenance task, 379
Emergency work, 161–162
Emotional intelligence, 26
Empowerment, 23
Engineering change, 379
Engineering change notice (ECN), 379
Engineering support, 187f
Engineering work order, 379
Enhanced oil recovery (EOR), 379, 383
Environmental assessment, 379
Environmental consequences, 379
Equipment configuration, 379
Equipment
    database structure, 141, 141f
    downtime reduction, 139
    failure patterns, 154f
    history, 208
    history database, accuracy of, 103–104,
        110–111
    records, 207–208
Equipment maintenance strategies, 379
Equipment repair history, 379
Equipment use, 379
Estimated plant replacement value, 379
Estimated ultimate recovery (EUR), 379
Estimating, 11–14, 19
Estimating index, 379
Everhart, Robbie, 36
Examination, 379
Execution, 27
Expediting, 379
Expense, 380
Expensed inventory, 380
Expert system, 380
Exploratory well, 380
External benchmarking, 44

F

Facilities Management Excellence™, 45
Failsafe, 380
Failure analysis, 380. *See also* Failure
    modes, effects, and critical analysis
    (FMECA)
Failure cause, 380. *See also* Failure mode
Failure coding, 380

Failure consequences, 380
Failure effect, 380
Failure finding interval, 380
Failure finding task, 380
Failure mode, 380
Failure modes and effects analysis (FMEA),
    151f, 343t–345t, 380
    of reliability-centered maintenance,
        145–156
Failure modes, effects, and critical analysis
    (FMECA), 380
Failure pattern, 381
Failure rate, 381
Failure reporting analysis and corrective
    action system (FRACAS), 343t–345t
Failure, 380
Farm-in, 381
Farm-out, 381
Fault tree analysis (FTA), 381
Feedback, accepting, 23
Field, 381
Fill rate, 381
First in still here (FISH), 381
First in–first out (FIFO), 381, 385
Flaring, 381
Floating production, storage, and offloading
    (FPSO), 381
Forecast, 381
Formation, 382
Forward workload, 382
Fossil fuel, 382
Frequency, 382
Front-end engineering and design (FEED),
    382
Fugitive emissions, 382
Function, 382
Functional benchmarking, 42
Functional failure, 382
Functional levels, 382
Functional maintenance structure, 382
Functional test, 387

G

Gained value, 71
Gantt chart, 382
Gas cost recovery (GCR), 382
Gathering lines, 393
Gathering system, 382
General support equipment (GSE), 382

Generic benchmarking, 42
Global warming potential (GWP), 382
Golden rule, 30, 30
Go-line, 382
Greenhouse gas, 382
GRIDCo Ghana, 420f–421f
    CMMS Selection and Best Practices
        Implementation Team-Draft, 417
    leadership driven-self-managed team at,
        419

H

Hardware, 382
Health, safety, security, and environmental
    (HSSE) challenges, 39, 223
Healthcare Facilities Scoreboard for
    Maintenance Excellence, 43
Heavy oil, 383
Hidden failure, 383
High quality maintenance contractors,
    in-house maintenance plus, 5–6
Higher throughput, 297
Historical averages, 241
Hold for disposition stock, 383
Horizontal drilling, 383
Hot work, 383
Human capital, planning and scheduling of,
    295f
Hydraulic fracturing, 383
Hydraulic fracturing fluids, 383
Hydrocarbons, 383

I

Identification, 383
Improved oil recovery (IOR), 383
Indirect costs, 383
Ineffective planning, 171f
    symptoms of, 8–9
Infant mortality, 383
Infill wells, 383
Information resources, 53
Inherent reliability, 384
In-house maintenance plus high quality
    maintenance contractors, 5–6
Injection, 403
In-situ recovery, 384
Inspection plan, creating, 237t–240t
Inspection, 384, 387

Insurance items, 384
Interchangeable, 384
Interface, 384
Internal benchmarking, 43
International benchmarking, 44
Interval-based, 384
Inventory, 384
    cycle counting, 131, 131f
    management module, 131
Inventory control, 384
Inventory turnover, 384
ISO 55000, 39–41, 45
Issues, 384
Item of supply, 384
Item, 384

J

Jack-up rig, 384
Job close out, 270f
Job follow up, 270f
Job packages, 90–91, 95, 211
    items, 221f
    planned, 211
Job plan, 95, 201, 206, 304–307
    database, 201
    labor needs, 305
    parts, 305
    permits, 305
    procedures, 305
    special tools and equipment, 305
    specifications, 305
Job scope, 304
    inspection sheet, 306f–307f
Job status codes, 164f
Joint operating agreement (JOA), 384
Just-in-time (JIT), 384

K

KAHADA (Keep a half a day ahead), 94
Keep full, 384. *See also* Shop stock
Key performance indicators (KPIs), 384
Knowing your total maintenance
    requirements, 386
Knuckle buster, 385

L

Labor estimating, 208–210
Labor libraries, 208

Last in–first out (LIFO), 385
Lead by example, 23–24
Lead time, 385
Leaders vs managers, 21, 22f
Leadership, 21–28
  buy-in, obtaining, 59–60
  communication, 22
  definition of, 21
  empowerment, 23
  execution of, 27
  feedback, accepting, 23
  lead by example, 23–24
  motivation, 23
  opportunity, 26
  patience and, 25
  perseverance and, 25
  personal goals, 26
  place in history, believe in, 27
  power, 24–25
  power of optimism, 22
  professional goals, 21
  publicity, 22
  quality, 22, 26–27
  quit from, 27
  respect and, 27
  responsibilities, 24
  self-confidence, 28
  style, 25–26
  success of, 27–28
  vision, 21
Lease, 385
Ledger inventory, 385
Level of repair (LOR), 385
Level of services (stores), 385
Library of planning aids, 208
Life, 385
Life cycle, 297, 385
  costing, 122
Life cycle analysis (LCA), 385
Life cycle cost (LCC), 385
Liquefied natural gas (LNG), 385
Local distribution company (LDC), 385, 405
Logistics engineering, 385
Logistics support analysis (LSA), 385
Lubrication services, 134

M

Main, 385
Maintainability, 385–386

Maintainability engineering, 386
Maintenance, 386
  challenges, 173, 175f–176f
  contract maintenance, profit and
    customer-centered, 2–3
  core requirement versus core
    competencies for, 8
  costs, 168f
  customer service, 139
  deferred, 5–6, 12–13, 121
  failure of, 313–314
  indiscriminate cutting of, 85–86
  managed as business, 141
  organization, benefits of, 177f–178f
  performance measurement process, 139t,
    140, 140f
  predictive, 14, 39–40, 47–48, 57, 61–62,
    71, 73, 89, 92, 94–95, 95f, 128,
    133t, 134
  preventative, 12–14, 39–40, 47–48, 57,
    61–62, 67–69, 71, 73, 89, 92–95,
    95f, 110, 128, 133t, 134
  profit in, 6–8, 6f
  reliability centered, 61–62, 138f
  standard task database, 142
  storeroom, 62
  strategies, development of, 224f
  technical library, 207, 210f
  total productive, 61–62
  as traditional cost center, 1–2
  winning, 388, 404
Maintenance business process improvement,
  386
Maintenance business system, planner
    review of, 99–144
  computerized maintenance management
    system
    backlog reports, 118
    benchmark evaluation, conducting,
      101–120
    bill of materials, 109–110
    craft hours , work order system
      accounts for, 118
    data integrity, 103t
    education and training, 112t
    equipment database, maintaining,
      111–112
    equipment history database, accuracy
      of, 103–104

Maintenance business system, planner
    review of (*Continued*)
  initial training, 112
  ongoing training, 112–113
  online work request based on priorities,
    118
  operations personnel, initial training
    for, 113
  parts database, maintaining, 110–111
  preventive maintenance tasks/
    frequencies, 110
  priority system, 120
  spare parts database, accuracy of,
    107–108
  systems administrator/backup trained,
    113
  work control, 114, 114t, 115f–117f
Ranking Index of Maintenance
    Expenditures (RIME) system,
    120–143, 120f
  budget and cost control, 120
  budget status–operating departments, 121
  CMMS captures reliability data, 134
  CMMS, operations understands benefits
    of, 141
  complete PM/PDM task descriptions,
    134
  cost improvements due to CMMS, 121
  craft labor, 121
  craft performance, 139
  craft utilization, measuring and
    improving, 128, 128f
  critical spares, identifying, 132, 133f
  deferred maintenance, 121, 121f
  electronic parts requisitioning, 132,
    132f
  engineering changes, 141
  equipment database structure, 141, 141f
  equipment downtime reduction, 139
  inventory cycle counting, 131, 131f
  inventory management module, 131
  life cycle costing, 122, 122f
  long range PM/PdM scheduling, 134
  lubrication services, 134
  maintenance customer service, 139
  maintenance managed as business, 141
  maintenance performance measurement
    process, 139t, 140, 140f
    maintenance standard task database,
      142
  MRO materials management, 130t–131t
  parts kitting and staging, 131, 132f
  parts of, 121
  planned work, increasing, 128, 128f
  planning and scheduling, 122,
    123f–127f
  PM/PdM change process, 134
  PM/PdM compliance, 134
  reliability centered maintenance, 138f
  reorder notification process, 133
  scheduling coordination with
    operations, 130
  spare parts status, 130
  true emergency repairs, tracking, 130
  vendor support costs, 121
  warranty information, 133
  work schedules, 128, 129f
Maintenance coordinator, 308–311
Maintenance coordinator
  duties and responsibilities, 409
  job evaluation form, 409–413
  primary purpose, 409
  task list, 414f–415f, 415, 416
Maintenance engineering, 386
Maintenance excellence, scoreboard for
  computerized maintenance management
    systems (CMMS) and business
    systems, 354t
  continuous reliability improvement,
    355t–356t
  contractor management, 332t–334t
  craft skills development and technical
    skills, 321t–322t
  critical asset facilitation and overall
    equipment effectiveness (OEE),
    356t–357t
  energy management and control, 350t
  health, safety, security, and environmental
    (HSSE) compliance, 351t–352t
  maintenance and quality control, 352t
  maintenance business operations, budget
    and cost control, 324t–325t
  maintenance engineering and reliability
    engineering support, 351t
  maintenance leadership, management and
    supervision, 323t

maintenance organization, administration and human resources, 319t–321t
maintenance performance measurement, 353t
maintenance strategy, policy and total cost of ownership, 316t–318t
management of change (MOC), 362t–363t
manufacturing facilities planning and property management, 334t–335t
MRO materials management and procurement, 336t–337t
operator-based maintenance and pride in ownership, 322t
organizational climate and culture, 318t–319t
overall craft effectiveness (OCE), 357t–358t
predictive maintenance and condition monitoring technology applications, 338t–339t
preventive maintenance and lubrication, 337t–338t
pride in maintenance, 366t–367t
process control and instrumentation systems technology, 349t–350t
process safety management (PSM), 362t–363t
production asset and facility condition evaluation program, 335t
reliability analysis tools, 343t–345t
reliability-centered maintenance (RCM), 339t–343t
risk mitigation, 363t–366t
risk-based inspections (RBI), 363t–366t
risk-based maintenance (RBM), 345t–349t
shop facilities, equipment, and tools, 354t–355t
shop-level reliable planning, estimating and scheduling (M/R), 328t–331t
STO and major planning/scheduling with project management, 331t–332t
storeroom operations and internal MRO customer service, 335t–336t
sustainability, 358t–359t
top leaders' support to management, 315

traceability, 359t–361t
work management and control: maintenance and repair (M/R), 325t–326t
work management and control, shutdown, turnarounds, and outages (STO), 326t–328t
Maintenance Excellence Institute International, 32, 73, 244–246
Maintenance excellence strategy team, 45–47
Maintenance leaders, 3–5, 8, 12–14, 185f, 386
distinguished from maintenance managers, 21, 22f
gambling, 1
Maintenance Leadership Team, 35
Maintenance managers, 4
distinguished from maintenance leaders, 21, 22f
Maintenance planner/scheduler
duties and responsibilities, 409
job evaluation form, 409–413
primary purpose, 409
task list, 414f–415f, 415, 416
Maintenance policy, 386
Maintenance repair operations (MRO), 115f–116f, 386, 444
materials management, 444–445. See also Maintenance repair operations materials management
parts and material resources, 53
purchasing operations, 62
Maintenance repair operations materials management, 277f, 130t–131t, 277–284
data, 278–279
spare parts ownership, 281–282
spares maintenance program, establishing, 282
spares storage, 279–281
being stationary, effects of, 280–281, 280f
exposure to environment, 279–280
storeroom function, 281
Maintenance requirements, 386
Maintenance schedule, 387

Maintenance shutdown, 387
Maintenance strategy, 387
Maintenance supervisor, 173
  role of, 174, 185f–186f
Maintenance task routing file, 387
Management of change (MOC), 388
Master schedule, 474–477, 474t–475t
  instructions for use, 475–477
Material library, 210
Maximum corrective time (MCT), 397
Maximum preventive time (MPT), 397
McDonald, Larry, 30
Mean active maintenance time (MAMT), 397
Mean downtime (MDT), 397
Mean time between failures (MTBF), 372, 388, 397
Mean time between maintenance (MTBM), 397
Mean time to repair (MTTR), 388, 397
Measure contractors, 91
Mercaptan, 388
Miscellaneous allowances, 259f
Mitigation measures, 237t–240t
Model work order, 388, 402. *See also* Standard job
Modification, 388
Modularization, 388
Most valuable people (MVP), 388
Most valuable player (MVP), 32f
Motivation, 23
MP2, 298
MRO materials management, 388

N

Natural gas, 388
Natural gas liquids (NGLs), 388
Natural gas vehicle (NVG), 388
Net acres, 388
No scheduled maintenance, 388–389. *See also* Run-to-failure
Nondestructive testing (NDT), 388
Nonoperational consequences, 388
Nonrepairable, 388
Nonroutine maintenance, 388

O

Obsolescence, 389
Obsolete, 389

Odorant, 389
Oil analysis. *See* Teratology
Oil sands, 389
On-condition maintenance. *See* Condition-based maintenance (CBM)
One million barrels (MMBBL), 389
One million cubic feet (MMCF), 389
One thousand cubic feet (MCF), 389
Online work orders, 118
Operating context, 228f, 389
Operating hours, 389
Operational consequences, 389
Operational efficiency, 389
Operations leaders, defining results to, 11–20
Operator, 389
Operator-based maintenance (OBM), 389
Opportunity, 26
Optimization, 67
Order point, 389
Order quantity, 389
Original equipment manufacturer (OEM), 406
Outage, 389
Overall craft effectiveness (OCE), 12, 15–16, 72, 72f, 241, 389
  calculation of, 74f
  comparison with overall equipment effectiveness, 73f
  data collection, technology for, 87–88
  elements of, 83
  gained value, 85
Overall equipment effectiveness (OEE), 68, 390, 404
  comparison with overall craft effectiveness, 73f
Overhaul, 390

P

Pareto's principle, 390
Part numbers, 390
Parts and material resources, 53
Parts inventory master, 103f, 103t, 130t–131t
Parts kitting, 131, 132f
Parts maintenance, 167–168, 169f
PASS 55: 2008, 39–40
Patience, 25
Pay-me-later syndrome, 5–6

People resources, 51
Percent planned work, 390
Performance benchmarking, 41–42
Performance factor, 68, 68f, 72
Periodic maintenance, 390
Permeability, 391
Perseverance, 25
Personal, fatigue and delay allowance
    (PF&D), 259f
Peters, Pete, 36
P–F interval, 1, 2f, 153f, 391
Physical asset and equipment resources,
    53
Physical asset management, 1–3, 11
Pick list, 391
Planned job packages, 211
Planned maintenance, 391
Planned work, increasing, 128, 128f
Planner skills, 307
Planner worksheet, 222f
Planners/schedulers (P/S), 5, 14, 18–19, 31,
    34–35
  factors influencing, 191f
  hiring, 89–98
    position goals and relationships, 96
    roles and responsibilities, 89–96
    selection criteria, 96–97
  needs of, 188f–189f
  qualities of, 192f
  ratio of crafts to, worksheet for,
    190f–191f
  responsibilities of, 187f–188f
  right things, 179f
Planning, 11–19, 302–303
  aids, library of, 208
  detailed, 202, 207–222
  failure of, 313–314
  of human capital, 295f
  ineffective, 171f
  job, 304–307
  profit and customer-centered benefits of,
    1–10
  promoted by storerooms, 182f
  skills, 303
  maturing process, signs of, 209f
  RIME system, 122, 123f–127f
  storerooms support to, 182f
  times, 79–82, 84
Plant engineering, 90, 391
Plant operating goals, 57–60

Play, 391
Porosity, 391
Possible reserves, 391
Potential failure, 391
Pounds per square inch (Psi), 391
Power, 24–25
Power of optimism, 22
Predictability, 313
Predictive maintenance (PDM/PdM), 14,
    39–40, 47–48, 57, 61–62, 71, 73, 89,
    92, 94–95, 95f, 128, 133t, 392
  change process, 134
  complete task descriptions, 134,
    135f–138f
  compliance, 134
  long range scheduling, 134
Preplanned job packages, 305
Preventive maintenance (PM), 12–14,
    39–40, 47–48, 57, 61–62, 67–69, 71,
    73, 89, 92–95, 95f, 128, 133t
  change process, 134, 392. See also
    Scheduled maintenance (SM)
  complete task descriptions, 134
  compliance, 134
  long range scheduling, 134
  tasks/frequencies, 110
PRIDE in Excellence, 29–30
PRIDE in Maintenance, 31–33, 32f, 73–74,
    75f, 76, 392
  gaining support from craft workforce,
    33–35
  positive and proven approach, 33, 34f
PRIDE in Maintenance and Construction,
    36–37
PRIDE in Work, 30
Priority, 392
Priority systems, 120, 159f
Proactive maintenance, 392
Proactive planned work, 180f
Probabilistic risk assessment, 392
Probabilistic safety assessment, 392
Probability of failure (POF), 223
Probable reserves, 392
Process benchmarking, 42
Process mapping, 42, 168
Process safety management (PSM), 392
Processing, 392
Processing natural gas, 393
Procurement, 393
Produced water, 393

Production, 393
Production sharing contract (PSC), 393
Productive well, 393
Productivity, 12
    categories of, 67
    definition of, 67–68
Profit
    ability, 3
    in maintenance, 6–8, 6f
    optimization, 3–4
Profit and customer-centered maintenance
        (PCCM), 393–394
Profit-centered benefits, of planning and
        scheduling, 1–10
Profit-centered maintenance (PCM), 393
Project evaluation and review technique
        (PERT) chart, 394
Proppant, 394
Protective device, 394
Proved developed reserves, 394
Proved reserves, 394
Provisioning, 394
Publicity, 22
Purchase order, 394
Purchase requisition, 394
Purchasing benefits, 178f
Purchasing/stores, 210

Q

Quality factor, 68, 68f, 72
Quality of leadership, 22, 26–27
Quality rate, 395
Questa, Carole, 36

R

Range of time concept, 246
Ranking Index for Maintenance
        Expenditures (RIME) system,
        120–143, 120f, 159f–160f, 161, 397
    benefits of, 161f
    budget and cost control, 120
    budget status-operating departments, 121
    CMMS
        operations understands benefits of, 141
        reliability data, 134
    complete PM/PDM task descriptions, 134
    cost improvements due to CMMS, 121
    craft labor, 121

craft performance, 139
craft utilization, measuring and
        improving, 128, 128f
critical spares, identifying, 132, 133f
deferred maintenance, 121, 121f
electronic parts requisitioning, 132, 132f
engineering changes, 141
equipment
    database structure, 141, 141f
    downtime reduction, 139
inventory cycle counting, 131, 131f
inventory management module, 131
life cycle costing, 122, 122f
long range PM/PdM scheduling, 134
lubrication services, 134
maintenance customer service, 139
maintenance managed as business, 141
maintenance performance measurement
        process, 139t, 140, 140f
maintenance standard task database, 142
MRO materials management, 130t–131t
parts kitting and staging, 131, 132f
parts of, 121
planned work, increasing, 128, 128f
planning and scheduling, 122, 123f–127f
PM/PdM
    change process, 134
    compliance, 134
reliability centered maintenance, 138f
reorder notification process, 133
scheduling coordination with operations,
        130
spare parts status, 130
true emergency repairs, tracking, 130
vendor support costs, 121
warranty information, 133
work schedules, 128, 129f
RCM++, 229, 229f–232f
Reaction time/response time, 395
Reactive planned work, 180f
Ready backlog, 92, 96, 158
Ready line, 395
Reasonable certainty, 395
Rebuild, 395
Rebuild/recondition, 395
Recompletion, 395
Recordable cases, 395
Redesign, 395
Redundancy, 395

Refinery process flow, 173, 174f–175f
Refinery work flow, 211, 212f–220f
Refinery, process mapping for
  facilities and services work analysis
    process, 431f
  facilities and services work
    documentation process, 437f
  facilities and services work execution
    process, 431f
    contractor work accountability process,
     436f
    emergency deviance and schedule
     deviation process, 434f
    job change order process, 433f
    job site data collection process,
     435f
    rental process, 432f
  facilities and services work identification,
    425f
    after hours, 426f
  facilities and services work planning
    process, 427f–428f
    project work, 429f
  facilities and services work scheduling
    process, 430f
Refurbish, 395
Reliability, 395
Reliability analysis, 395
Reliability centered maintenance (RCM),
    61–62, 73, 88, 138f, 145–156,
    395–396
  bathtub curve, 154f
  decision diagram, 152f
  default actions, 155f
  elements of, 146f
  equipment failure patterns, 154f
  failure consequences of, 152f
  failure effects of, 150f
  failure modes and effects analysis of, 151f
  failure modes of, 149f
  functional failure of, 148f–149f
  functions and performance standards of,
    148f
  implementation of, 147f
  operating context, 146
  P–F interval, 153f
  principles of, 146f
  proactive maintenance of, 153f
Reliability engineering, 396

Reliability improvement, 29, 33
Reliability performance indicators (RPI),
    397
Reliable Maintenance Excellence Index
    (RMEI), 17–19, 42–44, 43f, 95,
    285–296, 286f–290f
  customer-centered, 295f
  development of, 287f, 290f–291f
  human capital, planning and scheduling
    of, 295f
  performance data collection, 295f
  performance metrics and purpose,
    291f–295f
  profit-centered, 295f
Reliable planning times, 244
ReliaSoft, 229
Remove, 387
Remove and reinstall, 387
Remove and replace, 387
Reorder notification process, 133
Reorder point (ROP), 396. *See also* Order
    point
Repair, 396
Repair parts, 396
Repairable, 396
Repetitive jobs, 202
Replaceable item, 396
Reserves, 397
Reservoir, 397
Resources, 397
Respect, 27
Responsibilities of leadership, 24
Restoration, 397
Return-on-assets, 397
Return on capital employed (ROCE), 397
Risk, 397
  definition of, 223
  matrix, 226f
  ranking, 237t–240t
Risk-based inspection (RBI), 223–240,
    397–398
  equipment, 229f–230f
  software, 229–232
Risk-based maintenance (RBM), 223–240,
    225f–226f, 395
  case study example, 232–236, 232f–236f
  category 22 for, 237t–240t
  factors for implementation, 227f
  operating context, 228f

Risk-based maintenance (RBM)
    (*Continued*)
    outline of, 225f
    overview of, 227f
    quantitative analysis, 228f
    results, 236–240, 236f
    risk matrix, 226f
Robinson, Ricky, 36
Root cause analysis (RCA), 343t–345t
Root cause corrective action (RCCA),
    343t–345t
Rotable, 398
Routine maintenance task, 398
Routine maintenance work
    acceptance process, 219f–220f
    execution process, 219f
    planning process, 215f–216f
    scheduling process, 217f–218f
Routine planner training checklist
    personal details, 483
    basic, 483–485
    SAP, 485–487
    sign off, 487
Running maintenance, 398
Run-to-failure, 398
Run-to-failure strategy, 122

S

Safety consequences, 398
Safety stock, 398
Salvage, 398
Schedule compliance, 398
Scheduled discard task, 398
Scheduled maintenance (SM), 392, 398
Scheduled operating time, 398
Scheduled restoration task, 398
Scheduled work order, 399
Scheduling, 11–19, 257–276, 308–313, 310f
    allowances, application of, 260f
    compliance, 270f–271f, 273f–274f
    coordination meeting, preparation for,
        261f–264f
    coordination required by planners for,
        260f–261f
    coordination with operations, 130
    function, measures for, 274f–276f
    guidelines for completing, 264f–270f
    of human capital, 295f

job close out and follow up, 270f
key procedures for, 263f
long range PM/PdM, 134
miscellaneous allowances, 259f
non-compliance, reasons for, 271f–273f
performance, 273f–275f
personal, fatigue and delay allowance,
    259f
profit and customer-centered benefits of,
    1–10
RIME system, 122, 123f–127f
total planned time, allowances for deter-
    mining, 258f
travel time, 258f–276f
Scope of work, 90–92, 201, 207–222
Scoping, 399
Scoreboard for Excellence. *See* Scoreboard
    for Facilities Management
    Excellence; Scoreboard for
    Maintenance Excellence
Scoreboard for Facilities Management
    Excellence, 43, 399
Scoreboard for Fleet Management
    Excellence, 43
Scoreboard for Maintenance Excellence,
    13–17, 31–33, 95
    baseline information, preassessment
        checklist for, 60–64
    craft labor rates/overtime history, 61
    craft skills development, 61
    maintenance budget and cost
        accounting, 61
    maintenance storeroom, 62
    MRO purchasing operations, 62
    organization charts/job descriptions, 61
    predictive maintenance, 39–40, 47–48,
        57, 61–62
    preventive maintenance, 39–40, 47–48,
        57, 61–62
    reliability centered maintenance, 61–62
    total productive maintenance, 61–62
    work control, 62
    work orders, 62
    benchmark evaluation criteria, format of,
        57f
    continuous reliability improvement, 48f,
        59f
    current position of, determining, 47–48,
        49f–50f, 52f

definition of, 41
evolution of, 45–47, 46f
external benchmarking, 44
functional benchmarking, 42
internal benchmarking, 43
maintenance benchmarking, 40f
maintenance resources, 51f
  hidden resources, 53–57
  information resources, 53
  parts and material resources, 53
  people resources, 51
  physical asset and equipment resources,
    53
  technical skill resources, 53
1993 version of, 47f
performance benchmarking, 41–42
physical asset management strategy with,
  39–66
plant operating goals, 57–60
process benchmarking, 42
rating summary comments, 56f
self-assessment, 60f
strategic benchmarking, 41
Scoreboard for Maintenance Excellence,
  399–400
Secondary damage, 401
Secondary failures, 401
Secondary function, 401–402
Self-assessment, 60f
Self-confidence, 28
Selling benefits, 172f, 173, 177f
Serial number, 402
Service contract, 402
Service level, 402
Service line, 402
Serviceability, 402
Servicing, 402
Shale, 402
Shelf life, 402
Shop load plan, 471–474, 471t–472t
  instructions for use, 474
Shop schedule, 477–478
  sample, 479f
    4-week shop look-ahead of resources
      by trade type, 481f
    PM leveling and planning by shop area,
      480f
    weekly maintenance work schedule at
      90% loading, 482f

Shop stock, 384, 402
Short, Eddie, 36
Shutdown, 402
Shutdown maintenance, 402
Shutdown, turnaround, or outage (STO),
  116f–117f, 126f–127f, 174–181,
  198f, 402
Slotting, 242, 246
Snea, Keith, 36
Social life cycle analysis (S-LCA), 402
Society for Maintenance and Reliability
  Professional (SMRP) Metrics and
  Definitions, 314
Solomon auditing, 42
Sour gas, 402
Spacing, 402
Spare parts
  database, accuracy of, 107–108,
    111–112
  ownership, 281–282
  status, 130
Spares maintenance program, establishing,
  282
Spares storage, 279–281
  being stationary, effects of, 280–281,
    280f
  exposure to environment, 279–280
Spares tracking, 279
Special skills, 93
Specifications, 402
Spreadsheets, 245, 249–252, 249f–250f,
  257f
  developing, 254
Staffing, 168f
Standard item, 402
Standard job, 402. See also Model Work
  Order
Standard operating procedure (SOP), 211,
  242
Standardization, 402
Standby, 402
Standing work orders, 92, 402–403
Steam-assisted gravity drainage (SAGD),
  403
Stock keeping unit (SKU), 403
Stock number, 403
Stock out, 403
Storage, 403
Storage underground, 403

Storeroom
  benefits, 178f
  function, 281
  planning process promoted by, 182f
  reporting
    to finance or purchasing, 183f
    to maintenance, 183f
  success, 184f
  support to planning process, 182f
Stores issue, 403
Stores requisition, 403
Strategic benchmarking, 41
Supply, 403
Support equipment, 404
Sustainability, 16–17, 404
Sweet gas, 403
Systems Performance Team, 36

T

Teamwork works, 404
Technical data and information, 404
Technical skill resources, 53
Teratology, 389, 405
Terotechnology, 404
Test and support equipment, 404
The Maintenance Excellence Institute
    (TMEI), 386
The Maintenance Excellence Institute
    International (TMEII), 12, 39–41,
    43, 48–51, 59
Thermography, 404
Throwaway maintenance, 404
Thurston, Julie, 36
Tight gas, 405
Time standards for maintenance work
  adjusted averages, 241
  construction trade estimates, 241
  educated guesses, 241
  historical averages, 241
  slotting, 242
  universal maintenance standards, 242, 246
Top leaders, 404
  defining results to, 11–20
Top management support, 297–298, 300
Total asset management, 404
Total backlog, 91, 96, 158
Total cost of ownership, 206
Total maintenance requirements (TMR),
    12–13, 19, 91, 157–158, 162–165

Total operations success, 12
Total planned time, allowances for
    determining, 258f
Total productive maintenance (TPM),
    61–62, 73, 404
Total recordable rate (TRR), 405
Total shareholder return (TSR), 405
Total system support (ToSS), 405
Traceability, 404–405
Tradesperson, 405
Transmission, 405
Transmission lines, 405
Transportation, 405
Travel time, 258f–276f
Troubleshooting, 92, 201, 387, 405
True emergency repairs, tracking, 130
TrueWorkShops™, 32
Turnaround time, 405
Turnover, 405

U

Unconventional reservoirs, 405–406
Undeveloped acreage, 406
Unit, 406
Universal maintenance standards (UMS), 242
Unplanned downtime, 297, 299
Unplanned maintenance, 406
Unscheduled maintenance (UM), 406
Up, 406
Uptime, 406
Usage, 406
Useful life, 406
Utilization, 406

V

Value engineering, 406
Value-adding investments, 4
Variance analysis, 406
Vendor support costs, 121
Vibration analysis, 406
Vision of leadership, 21

W

Wall, Joe, 36
Warranty, 406
  claims, 406
  information, 133
Waterflood, 406

Wear out, 406
Wellhead, 393
Wet gas, 406
Willis, Ed, 36
Withdrawal, 403
Work content, 246–247
    comparison, 249–251
Work control, 62, 114t, 180,
    193f–197f
Work identification, 212f–214f
Work order (WO), 62, 201–206, 202f–205f,
    278, 407
  blanket, 92
  cause of failure, 201
  criteria for, 206f
  as prime source for reliability
    information, 206
  processing, 213f–214f
  standing, 92

Work requests, 89–90, 92, 407
Work schedules, RIME system,
    128, 129f
Working interest, 407
Workload, 407
World-class wrench time, 302
Wrench time, 68–71, 76–84, 83f, 87–88,
    257f, 258, 307–308, 314
  definition of, 302
  world-class, 302

X

X-maintenance, 407

Z

Zero-based budgeting, 407